Reactive Intermediates

Edited by
Leonardo S. Santos

Related Titles

Downard, K. (ed.)

Mass Spectrometry of Protein Interactions

2007
ISBN: 978-0-471-79373-1

Platz, M. S., Moss, R. A., Jones, M. (eds.)

Reviews of Reactive Intermediate Chemistry

2007
ISBN: 978-0-471-73166-5

Hillenkamp, F., Peter-Katalinic, J. (eds.)

MALDI MS

A Practical Guide to Instrumentation, Methods and Applications

2007
ISBN: 978-3-527-31440-9

Niemantsverdriet, J. W.

Spectroscopy in Catalysis

An Introduction

Third, Completely Revised and Enlarged Edition
2007
ISBN: 978-3-527-31651-9

de Hoffmann, E., Stroobant, V.

Mass Spectrometry

Principles and Applications

Third Edition
2007
ISBN: 978-0-471-48565-0

Reactive Intermediates

MS Investigations in Solution

Edited by
Leonardo S. Santos

WILEY-VCH Verlag GmbH & Co. KGaA

The Editor

Dr. Leonardo S. Santos
Talca University
Chemistry Institute of Natural Resources
Avenida Lircay s/n
Talca
Chile

■ All books published by Wiley-VCH are carefully produced. Nevertheless, authors, editors, and publisher do not warrant the information contained in these books, including this book, to be free of errors. Readers are advised to keep in mind that statements, data, illustrations, procedural details or other items may inadvertently be inaccurate.

Library of Congress Card No.: applied for

British Library Cataloguing-in-Publication Data
A catalogue record for this book is available from the British Library.

Bibliographic information published by the Deutsche Nationalbibliothek
Die Deutsche Nationalbibliothek lists this publication in the Deutsche Nationalbibliografie; detailed bibliographic data are available on the Internet at http://dnb.d-nb.de.

© 2010 WILEY-VCH Verlag GmbH & Co. KGaA, Weinheim

All rights reserved (including those of translation into other languages). No part of this book may be reproduced in any form – by photoprinting, microfilm, or any other means – nor transmitted or translated into a machine language without written permission from the publishers. Registered names, trademarks, etc. used in this book, even when not specifically marked as such, are not to be considered unprotected by law.

Cover Grafik-Design Schulz, Fußgönnheim
Typesetting Thomson Digital, Noida, India
Printing and Binding T.J. International Ltd, Padstow, Cornwall

Printed in the Great Britain
Printed on acid-free paper

ISBN: 978-3-527-32351-7

To Gisa, Guilherme and Larissa
whose patience and love enable them
to hold out the hours of homework
and the moments of my absence

Contents

Foreword *XIII*
Preface *XXI*
List of Contributors *XXIII*

1 A Brief Overview of the Mechanisms Involved in Electrospray Mass Spectrometry *1*
Paul Kebarle and Udo H. Verkerk
1.1 Introduction *1*
1.1.1 Origins of Electrospray Mass Spectrometry *1*
1.1.2 Aims of this Chapter *2*
1.2 Production of Gas-Phase Ions by Electrospray and Electrospray Ionization Mass Spectrometry *3*
1.2.1 Overview *3*
1.2.2 Production of Charged Droplets at the Capillary Tip *5*
1.2.3 Electrospray as an Electrolytic Cell *7*
1.2.4 Required Electrical Potentials for ES. Electrical Gas Discharges *8*
1.2.5 Current, Charge and Radius of Droplets Produced at the Capillary Tip *10*
1.2.6 Solvent Evaporation from Charged Droplets Causes Coulomb Fissions of Droplets *10*
1.2.7 Evaporation of Droplets Leading to Coulomb Fissions Producing Progeny Droplets that Ultimately Lead to Ions in the Gas-Phase; Effects of the Concurrent Large Concentration Increase *11*
1.2.8 Mechanism for the Formation of Gas-Phase Ions from Very Small and Highly Charged Droplets. The Ion Evaporation Model (IEM) *15*
1.2.9 Observed Relative Ion Intensity of Small Analytes. Dependence on the Nature of the Analyte, its Concentration and Presence of Other Electrolytes in the Solution. High Sensitivities of Surface-Active Analytes *17*
1.2.10 Large Analyte Ions such as Dendrimers and Proteins are Most Probably Produced by the Charged Residue Model (CRM) *22*

1.2.11	Nanospray and Insights into Fundamentals of Electro and Nanospray	26
1.2.12	Consequences of the Increase in Concentration Caused by Extensive Evaporation of Solvent in ESI Process. Promotion of Bimolecular Reactions Involving Analyte Ions	28
1.2.12.1	Positive-Negative Ion-Pairing Reactions Involving Impurities such as Na^+	28
1.2.12.2	Determination of Equilibrium Constants in Solution via ESI-MS	29
	References	31
2	**Historical Perspectives in the Study of Ion Chemistry by Mass Spectrometry: From the Gas Phase to Solution**	**37**
	Hao Chen	
2.1	A Brief History and Recent Advances in Mass Spectrometry	38
2.1.1	Early Developments	38
2.1.2	Recent Advances	40
2.2	Overview of the Study of Ion/Molecule Reactions in the Gas Phase by Mass Spectrometry	41
2.2.1	Brief History	41
2.2.2	Basic Types of Ion/Molecule Reactions	42
2.2.3	Relationship to Reaction Analogies in Solution	43
2.2.3.1	Mechanism Elucidation of Classical Organic Reactions	43
2.2.3.2	Mechanism Elucidation of Organometallic Reactions	44
2.2.3.3	Catalyst Screening	49
2.2.3.4	Synthesis of Elusive Ionic Species	49
2.2.3.5	Probing Reactivity of Microsolvated Cluster Ions	50
2.2.4	Experimental Methods for the Study of Ion/Molecule Reactions	50
2.2.4.1	Low-Pressure Ion/Molecule Reactions	50
2.2.4.2	High-Pressure Ion/Molecule Reactions	51
2.3	Future Perspectives	52
	References	53
3	**Organic Reaction Studies by ESI-MS**	**63**
	Fabiane M. Nachtigall and Marcos N. Eberlin	
3.1	Introduction	63
3.2	Reaction Mechanisms	65
3.2.1	Morita-Baylis-Hillman Reaction	65
3.2.2	Morita-Baylis-Hillman Reaction Co-catalyzed by Ionic Liquids	67
3.2.3	α-Methylenation of Ketoesters	71
3.2.4	Unexpected Synthesis of Conformationally Restricted Analogs of γ-Amino Butyric Acid (GABA) via a Ring Contraction Reaction	73
3.2.5	The Heck Reaction	75
3.2.6	Suzuki Reaction	81
3.2.7	Stille Reaction	81
3.2.8	Three-Component Pd(0)-Catalyzed Tandem Double Addition-Cyclization Reaction	83

3.2.9	Alkynilation of Tellurides Mediated by Pd(II)	84
3.2.10	TeCl$_4$ Addition to Propargyl Alcohols	88
3.2.11	S$_N$2 Reactions	89
3.2.12	Allylic Substitution Reaction	91
3.2.13	Heterogeneous Fenton Reaction	93
3.2.14	Mimicking the Atmospheric Oxidation of Isoprene	93
3.2.15	Advanced Oxidation Processes of Environmental Importance	95
3.2.16	Tröger's Bases	96
3.2.17	The Three-Component Biginelli Reaction	98
3.2.18	Modeling the Ribonuclease Mechanism	103
3.2.19	Oxidative Cleavage of Terminal C=C bonds	105
3.3	General Remarks	108
	References	108

4 Studies of Reaction Mechanism Intermediates by ESI-MS 113
Rong Qian, Jing Zhou, Shengjun Yao, Haoyang Wang, and Yinlong Guo

4.1	Introduction	113
4.2	Studies on the Intermediates and Mechanisms of Pd-Catalyzed Reactions	113
4.3	Studies on Some Reactive Intermediates and Mechanisms of Radical Reactions	115
4.4	Studies on the Intermediates and Mechanism of Organocatalysis Reactions	121
4.5	Studies on the Intermediates and Mechanism of Transition Metal-Catalyzed Polymerization Reactions	123
	References	129

5 On-line Monitoring Reactions by Electrospray Ionization Mass Spectrometry 133
Leonardo S. Santos

5.1	Introduction	133
5.2	Preservation of the Charge in the Transit of Ions from Solution to the Gas Phase Using the ESI Technique	134
5.3	Developing Methods to Study Reaction Mechanisms	135
5.3.1	Monitoring Methods	135
5.3.1.1	Off-Line Monitoring	135
5.3.1.2	On-Line Monitoring	136
5.3.2	Microreactors	136
5.3.2.1	PEEK Mixing Tee as Microreactor	136
5.3.2.2	Capillary Mixer Adjustable Reaction Chamber	137
5.3.2.3	Photolysis Cell	138
5.3.2.4	Photochemical Reactor	139
5.3.2.5	Nanospray Photochemical Apparatus	140
5.3.2.6	Electrochemical Cell	141
5.4	Probing Reactivity of Intermediates	142
5.4.1	Reaction Mechanism Studies	143

5.4.1.1	Radical Fenton Reaction 143
5.4.1.2	Heterogeneous Fenton System 144
5.4.1.3	Radical Cation Chain Reactions 145
5.4.1.4	[2 + 2]-Cycloaddition of Trans-Anethole 145
5.4.1.5	Electron Transfer Initiated Diels–Alder Reactions 148
5.4.1.6	Radical Chain Reactions 149
5.4.1.7	Photochemical Reactions 151
5.4.1.8	Photochemical Switching Reaction 151
5.4.1.9	Photoinitiated Polymerization Reaction 153
5.4.2	Electrochemical Reactions 154
5.4.3	Heck Reaction 154
5.4.4	Suzuki Reaction 156
5.4.5	Pd-Catalyzed Enantioselective Allylation Reaction 156
5.4.6	Stille Reaction 157
5.4.7	Alkynilation of Tellurides Mediated by Pd(II) 158
5.4.8	Lewis Acid-Catalyzed Additions 162
5.4.9	C–H Activation and Hydrogenations 162
5.4.10	Oxidation Reactions 163
5.4.11	Epoxidation 164
5.4.12	The Baylis-Hillman Reaction 166
5.4.13	The Baylis-Hillman Reaction Co-Catalyzed by Ionic Liquids 167
5.4.14	Ring Contraction Reaction 169
5.4.15	Nucleophilic Substitution Reactions – The Meisenheimer Complex 169
5.4.16	Oxidative Degradation of Caffeine 171
5.4.17	Mimicking Atmospheric Oxidation of Isoprene 173
5.4.18	α-Methylenation of Ketoesters 175
5.4.19	Transient Intermediates of Petasis and Tebbe Reagent 176
5.4.20	On-Line Screening of the Ziegler–Natta Polymerization Reaction 178
5.4.21	On-Line Screening of the Brookhart Polymerization Reaction 181
5.4.22	$TeCl_4$ Addition to Propargyl Alcohols 181
5.4.23	Mechanism of Tröger's Base Formation 186
5.5	Conclusion 187
	References 188

6 Gas Phase Ligand Fragmentation to Unmask Reactive Metallic Species 199
Richard A. J. O'Hair

6.1	Introduction and Scope of the Review 199
6.2	Unmasking Reactive Metallic Intermediates via Collision-Induced Dissociation 201
6.2.1	Formation and Reactivity of Organometallics 201
6.2.1.1	Formation and Reactions of Organolithium Ions 202
6.2.1.2	Formation and Reactions of Alkaline Earth Organometalates 203

6.2.1.3	Formation and Reactions of Organocuprates and Organoargentates *205*	
6.2.1.4	Formation of Metal Carbenes *208*	
6.2.2	Formation and Reactivity of Metal Hydrides *210*	
6.2.2.1	Mononuclear Metal Hydrides *210*	
6.2.2.2	Multinuclear Metal Hydrides *212*	
6.2.3	Formation and Reactivity of Metal Oxides *215*	
6.2.3.1	Bond Heterolysis *215*	
6.2.3.2	Bond Homolysis of Metal Nitrites and Nitrates *216*	
6.2.4	Formation and Reactivity of Metal Nitrides and Related Species *220*	
6.3	Conclusions *224*	
	References *224*	

7   Palladium Intermediates in Solution *229*
Anna Roglans and Anna Pla-Quintana

7.1	Introduction *229*
7.2	ESI-MS Studies in Suzuki-Miyaura Cross-Coupling and Related Reactions *231*
7.3	ESI-MS Studies in the Identification of Oxidative Addition Intermediates *237*
7.4	ESI-MS Studies in Mizoroki-Heck and Related Reactions *240*
7.5	ESI-MS Studies in Stille Cross-Coupling Reactions *251*
7.6	ESI-MS Studies in Palladium-Catalyzed Reactions Involving Allenes *254*
7.7	ESI-MS Studies in Palladium-Catalyzed Alkynylation Reactions *258*
7.8	ESI-MS Studies in Palladium-Catalyzed Allylic Substitution Reactions *260*
7.9	ESI-MS Studies in Palladium-Catalyzed Oxidation of 2-Allylphenols *268*
7.10	ESI-MS Studies in Palladium-Catalyzed Polymerization Reactions *269*
7.11	Conclusions *272*
	References *273*

8   Practical Investigation of Molecular and Biomolecular Noncovalent Recognition Processes in Solution by ESI-MS *277*
Kevin A. Schug

8.1	Introduction *277*
8.2	Methods and Applications *280*
8.3	Practical Aspects of Titration Analysis *290*
8.4	Summary and Outlook *298*
	References *298*

Index *307*

Foreword

Mass spectrometry could be described, without implying any criticism, as an example of work in progress. Each time it appears to be approaching 'maturity,' another breakthrough occurs to expand its usefulness in new areas of science. As this volume clearly demonstrates, that process is still going on. In the early twentieth century, mass spectrometry was principally a tool for physicists to study particles and petroleum chemists to characterize petroleum mixtures. Wider use by chemists began with the ability to obtain structural information from the spectra of pure organic molecules. The analytical application of mass spectrometry truly came of age with its use as a detector for gas chromatography. Indeed, up to the present, advances in chromatography and mass spectrometry have leapfrogged each other, combining to create analytical tools that have steadily advanced in selectivity and detection limit for over two decades. The power of these tools is such that they have found critical applications in virtually every area of science, engineering, and medicine.

From the standpoint of mass spectrometric instrumentation, the story of this evolution has taken place on four fronts:

(1) Methods of separating ions of different atomic or molecular masses
(2) Methods of obtaining more chemical information by tandem mass spectrometry
(3) Methods of ionizing analyte molecules
(4) Methods that improve sensitivity and throughput.

Spectacular advances in all four of these areas have facilitated the remarkable expansion of mass spectrometry in diverse areas, including the subject of this volume. Sometimes new applications are introduced by mass spectrometrists recognizing an area of opportunity and sometimes by researchers in that area who have the temerity and opportunity to try a new technique. In any case, mass spectrometry has become a central tool in scientific investigation, and facilities for its use have become a critical part of virtually all scientific research organizations.

Reactive Intermediates: MS Investigations in Solution. Edited by Leonardo S. Santos.
Copyright © 2010 WILEY-VCH Verlag GmbH & Co. KGaA, Weinheim
ISBN: 978-3-527-32351-7

Mass Analyzers

Most analytical methods use a bulk property to distinguish, separate, or identify an analyte. Properties such as chemical reactivity, chromatographic retention time, and optical absorbance or emission reveal information about an aggregate of analyte molecules, giving a collective response value. A remarkable thing about a mass spectrum, and one of the unique attributes of mass spectrometry, is that the analyte molecules or atoms are separated by mass, and the detector records the mass of each individual analyte molecule. The isotopic composition of the analyte is revealed as are mass shifts due to modifications of very large molecules, even when only a fraction of them have been modified.

To perform a separation of sample molecules or atoms according to their individual masses, all mass spectrometers rely on the fact that the trajectory of a charged particle in the presence of electric and magnetic fields is mass-dependent, or, more exactly, dependent on the mass-to-charge ratio (m/z) of the particle. To avoid distortion of the differentiating trajectory by collisions with molecules, this separation must be carried out in a vacuum, though, as we shall see, sometimes such collisions can be used to advantage.

Chapter 1 in this volume reviews the historical development of mass spectrometers in some detail. I will here introduce the general concepts of mass separation in the context of some seminal developments in the instrumentation. Ions accelerated to a nearly constant kinetic energy ($1/2\ mv^2$) will, in a region with uniform magnetic field, have a curved trajectory dependent on the ion momentum. Magnetic sector mass spectrometers, based on this principle, held a dominant position for many years. The addition of an electric sector greatly improved the mass resolution, and these 'double-focusing' mass spectrometers were the gold standard into the 1990s. Obtaining mass resolutions in the tens of thousands enabled the development of 'exact mass' determination, whereby the amount of the mass defect in the elements could be used to determine the chemical formula of an analyte based on the measurement of the ion m/z to within some few parts per million. The champion, however, for mass resolution has been the Fourier transform ion cyclotron resonance (FTICR) mass spectrometer introduced by Marshall and Comisarow in 1974. Ions in a very high magnetic field move in a circle on a plane orthogonal to the magnetic field flux. The frequency of their rotation is a function of their m/z value. A batch of ions is excited to rotation and the resulting signal is analyzed by Fourier Transform to obtain the frequencies and thus the m/z's of the ions in the batch. Mass accuracies in small fractions of parts per million have been achieved.

Though mass analysis based on ion flight time was an early innovation, the lack of good high-speed electronics and its poor mass resolution prevented its wide adoption. What did bring mass spectrometry into the main stream of chemical analysis was the development of the quadrupole mass spectrometer. This simple device provided unit resolution mass spectra in a relatively compact, low-cost format. The partnership of the quadrupole mass analyzer with gas chromatography resulted in a powerful analytical tool that continues to see wide use in a variety of areas. Wolfgang Paul, who invented the familiar quadrupole with four rods and an RF generator, also developed a circular variation called the ion trap, composed of a ring electrode and two end cap electrodes The conversion of the ion trap into a mass spectrometer involved stafford's invention of the means of scanning the stored ions out of the trap in order of their m/z values after first reducing their kinetic energies by the introduction of helium gas at low pressure to act as a cooling collision partner.

Meanwhile, three developments brought time-of-flight mass spectrometry back into contention. The first was the high-speed signal analysis available with solid-state electronics, and the second was the ion mirror introduced by Mamyrin, which improved the mass resolution by focusing ions that had been spatially disperse at the time of ion acceleration along the flight tube. In 1989, Dawson and Gilhaus reduced the initial ion kinetic energy dispersion by forming a low-energy beam of ions from the source so that their main kinetic energy was along the axis of the beam. A section of the beam was then accelerated orthogonally to the beam axis for the measurement of flight time. Time-of-flight instruments that combine the ion mirror with orthogonal acceleration have achieved resolutions in the tens of thousands – sufficient for exact mass measurements.

An important distinction among methods of mass analysis is whether they operate in continuous or batch mode. In the continuous mode, ions are continuously being generated, sorted, and detected. Magnetic sector and most linear quadrupole mass analyzers are in the continuous category. If a single detector is used with a continuous method, the mass analyzer is acting as a mass filter, passing only a narrow range of m/z values at each time. To obtain a mass spectrum, the mass filter is scanned across the range of m/z values of interest. At any time, ions outside the immediate range are lost. Batch instruments, on the other hand, perform their mass analysis on sets or batches of ions. Ions from continuous ion sources must therefore be 'bunched' for batch analysis. On the other hand, discontinuous methods of ion generation are well matched to batch instruments. Time-of-flight, FTICR, and ion trap instruments are of the batch type. A mass spectrum is generated from each batch analysis, so that all ions over a wide m/z range in each batch are detected. Among batch instruments, the time required to analyze each batch sets the maximum spectral generation rate. Time-of-flight instruments can generate thousands of spectra per second, while ion trap instruments begin to lose resolution at spectral generation rates above a few per second. Higher spectral generation rates are, of course, useful when following the output of a fast chromatographic separation or the rate of a rapid reaction.

Progress in mass analyzer instrumentation continues on two fronts, evolutionary and revolutionary. Instruments of the classic types discussed above are improving in efficiency, simplicity, portability, and user friendliness with every 'new model' cycle. Revolutionary changes have included Makarov's orbitrap mass analyzer which achieves near FTICR resolutions using only RF fields in a unique design. The linear quadrupole has now been implemented as a linear ion trap by Welling where ions can be bunched, cooled, fragmented, or reacted with background gas, prior to or between stages of mass analysis.

Mixture Composition and Molecular Structure by Mass Spectrometry

Arguably one of the most significant advances in adding analytical power to mass spectrometry was the development of tandem mass spectrometry, or the application of two or more sequential stages of mass analysis. Obviously, no new information is obtained by repeating a mass analysis unless some reaction which changes the analyte ion's mass or charge occurs between the stages of mass analysis. Knowing the m/z of the ion prior to the reaction (the precursor ion) and then knowing the m/z's of the product ions can give important clues to the structure of the precursor ions as well as the process by which the product ions are formed. This possibility was first noticed by the observation of metastable ions that fragmented between the electric and magnetic sectors of a double-focusing mass spectrometer. Such fragmentations caused spurious peaks in the mass spectrum. Once their origin was explained, the phenomenon was used to study the ion fragmentation process. However, the intentional production and mass analysis of fragment ions for analytical applications began in 1976 with Cooks, who used a magnetic sector to select the precursor ion m/z, intentional collision with neutral gas target molecules to induce ion fragmentation, and an electric sector to sort out the product ions. In these sector-based tandem instruments, the mass resolution of the electric sector was low. Very high collision energies were required to achieve even relatively poor fragmentation efficiencies. Therefore, it was a surprise to many when, in 1978, Yost, Morrison, and I discovered high ion fragmentation efficiencies at low collision energies in a linear, non-mass-selective quadrupole. Our goal had been to obtain fragmentation between two quadrupole mass spectrometers for analysis by separation and identification. The triple-quadrupole, as it became known, brought tandem mass spectrometry (now called MS/MS) into the mainstream of analytical instruments because of its simplicity, high fragmentation efficiency, and unit mass resolution for both precursor selection and product analysis. In current nomenclature, this is now called a QqQ instrument, where the lower case q represents the linear quadrupole collision cell.

In the wake of the introduction of the triple-quadruple instrument, both mixture analysis and structure determination applications of MS/MS were pursued simultaneously. For mixture analysis, it was desirable to use 'soft' ionization so that each analyte in the mixture represented only one m/z value. When this value was selected with the first mass analyzer and fragmented, its product m/z pattern could be mined for identification information much as a primary mass spectrum had been previously. Clearly this could be done for each component of the mixture in turn. At first, this appeared to allow the first mass analyzer to take the place of prior chromatographic separation, but it quickly became clear that MS/MS was an even more powerful chromatographic detector than MS, providing greater selectivity and more information for identification. Investigators were intrigued by this new concept. If MS/MS was good, would MS/MS/MS be better? Pentaquadrupole (QqQqQ), dual double-focusing (BEqBE), and many other configurations were constructed during this phase of tandem MS exploration. Another aspect explored was the scan modes other than the one that produces the product spectrum. Scanning the first mass analyzer produces a spectrum of all the precursor m/zs that can produce the product m/z to which the second mass analyzer is set. This is called a precursor scan and is useful for the identification of all species in the sample which produce a particular ionic fragment. Scanning both mass analyzers with the second lagging behind the first by a constant m/z gives a spectrum of all the precursor ions that produce a neutral fragment of the set mass difference. The precursor and neutral loss scans are powerful analytical tools as was well demonstrated at the time. As we shall see, later tandem instrument designs produce product ion scans more efficiently than the QqQ instruments. Precursor and neutral loss scan information is available, but only with additional experiment time and extra data processing. As a result, these alternative scans are now too often overlooked. They could be especially useful in the study of gas phase reactions occurring in the collision chambers of tandem instruments.

If chromatographic sample introduction is used with soft ionization and MS/MS, the precursor mass spectrometer must be tuned to pass the precursor ion for the eluting analyte. If the peak contains just one component, the chromatography and the precursor mass analyzer are redundant. This can be alleviated by using collisional fragmentation before the first mass analyzer to obtain structural information for each eluant or by choosing a characteristic fragment ion to increase selectivity. If the chromatogram is relatively crowded with peaks, it is very likely that the peak is not just a single component, but rather contains a number of minor components. These can be discovered and analyzed using the first mass analyzer to exclude the major component.

A normal mass spectrum using a hard ionization source already contains many fragment ion m/z values. In fact, it may not contain the molecular ion at all. MS/MS can be used to further fragment each of the primary fragments to further characterize the molecule, fragment by fragment and also to determine which fragments arise from which larger fragments. If the product analyzer has exact mass capabilities, this method of structural analysis is all the more powerful.

Methods of Ion Formation

We know that ionization methods are either hard, producing fragments of the ionized molecules, or soft, producing principally positively charged ions by the addition of a proton or other cation, or negatively charged ions by the abstraction of a proton or the addition of an anion to the analyte molecule. They are also categorized by being vacuum or atmospheric. Vacuum techniques work well with volatile components, as with the electron impact ionization sources used with gas chromatographic or membrane inlet systems. Ionization can be from the electron impact directly (EI, a hard ionization technique) or from reaction with a reagent ion such as CH_5^+ that was created by electron impact with a reagent gas molecule (chemical ionization, a soft ionization technique). A nonvolatile sample subjected to heating generally decomposes rather than evaporates unless the heating is performed very rapidly. In general, the more rapid the heating, the less decomposition occurs. Thus a pulsed laser focused on a solid sample matrix can cause evaporation and ionization of nonvolatile molecules, even very large ones. Using an appropriate matrix to facilitate laser absorption and the necessary proton exchange helps greatly, and thus we have matrix-assisted laser desorption ionization (MALDI), first effectively developed by Hillenkamp and Karas in 1985.

Atmospheric pressure ionization sources have become more widely used since efficient methods of getting the ions from the atmosphere into the vacuum of the first mass analyzer have been developed. In 1984, John Fenn discovered how to use electrospray as a means of ionization for mass analysis. This development would be the impetus for many developments in atmospheric sources. Its ability to ionize very large molecules from a liquid matrix has proven to be an immensely powerful tool. It is one of those discoveries that became widely used before it was very well understood. Fundamental work is still ongoing regarding the nature of the electrospray ionization (ESI) process. My work showing how surface activity of the analyte is related to the response factor for the ion evaporation mechanism and how increasing the excess charge density reduces the potential for inter-analyte interference explained why reducing the emitter diameter (microspray and nanospray) leads to improved mass sensitivity and why metal ions, not being surface active, have such poor response factors. I believe, however, that there is still a potential for inorganic ESI through the development of complexing agents that would complex the metal ions to form surface-active ions. Such complexation would also desirably shift the ion m/z to higher values. One of the more remarkable ESI developments was the discovery of Graham Cooks' group that solvent ions produced by electrospray could produce ions from analyte on solid surfaces on which they impinged and that these ions could be collected for mass analysis. This technique, called DESI for desorption ESI, is remarkably simple to apply, has a wide range of interesting applications, and has spawned several spin-off variations. Other atmospheric pressure ionization sources include atmospheric pressure chemical ionization (APCI), and atmospheric pressure matrix-assisted laser desorption/ionization (MALDI).

Each ionization technique has its own characteristic strengths and limitations. I do not believe there will ever be a perfect, universal ionization method. The chemistry of ionization is different for each class of compounds. ESI works for compounds that are ionized in solution or can undergo proton (or other cation) addition or abstraction more readily than the solvent vapor molecules. ESI produces ions of large biomolecules with a variety of charge states, whereas MALDI produces principally singly charged ions of even very large molecules. EI will work only with compounds that are volatile, and so on. This means that mass spectrometrists dealing with a variety of analyte types need to understand the chemistry of ionization in order to select the optimum method of achieving it.

Improving Sensitivity and Throughput

When I first started in mass spectrometry in 1976, after previous work in electrochemistry and early computer-controlled instrumentation, it was with the goal of making an intelligent analytical instrument that would be not only computer controlled, but would devise the experimental strategy based on the analytical goals and then automate the analysis with program branching based on the results of previous tests. To a limited extent, this has been accomplished in the automated tuning routines of modern instruments and the ability to instruct an MS/MS instrument to automatically collect product spectra from the principal peaks it finds in the set of precursor m/z values. I never envisioned that the process of data collection (e.g., peak finding) and analysis (e.g., library matching) would be done by hidden, proprietary algorithms the user cannot know. My goal to provide a more powerful, faster analytical instrument now too often leads to an opaque barrier between the scientist and the raw data. On the other hand, computer optimization has now become very sophisticated, so that an instrument's operating parameters can change during a scan in order to provide optimum response at every point in the scan. Data collection and presentation are greatly facilitated by computer, so that much more can be accomplished in a given time.

Both computerized optimization and improved instrument design have contributed to an increase in the number of ions detected per mole of analyte. Programs such as SIMION for ion optics simulation enable accurate prediction of ion efficiency of hypothetical designs. They also reveal the characteristics of various design approaches. The increasing use of RF-only ion guides has improved ion transmission between sections of instruments by increasing the acceptance angle and reducing the deflections due to fringing fields of linear quadrupoles. Among the most dramatic improvements in transmission efficiency has been that from atmospheric pressure ionization sources to the first mass analyzer. The initial ESI sources used very fast pumps to carry the gas away as soon as possible and electrostatic fields to attract the ions to each succeeding element. Efficiencies of transmission were less than 1 part in 10^4. In 1992 Douglas and French discovered that when using an ion guide in the first evacuated stage, a higher pressure in the guide and a lower kinetic energy ion introduction from the higher pressure region actually increased the ion transmission substantially. This was counter-intuitive, as those of us working with collision chambers had long known that many collisions within a quadrupole led to a loss of axial kinetic energy and the ions could get 'stuck' in the collision chamber. At higher pressures, the cooling effect of the collisions brought the ions to the central axis of the quadrupole where they pushed each other along in response to the fields at the entrance and exit of the quadrupole. This set the stage for high-efficiency collection of ions from atmospheric sources, so that modern instruments can now detect ions equal to several percent of the analyte molecules introduced. It also paved the way for the development of the linear ion trap, which is useful as an ion-molecule reaction chamber, a mass analyzer, and an interstage bunching device. When one realizes that many types of mass spectrometers detect single ions, and that even a femtomole (10^{-15} moles) of analyte has 6 billion molecules, one understands that sensitivity (response factor) is no longer the limiting factor for detectability. Rather it is selectivity that determines limits of detection, and gains in selectivity are obtained by prior separations, chromatographic introduction, chemical selectivity by prior reaction, higher m/z resolution, use of MS/MS, and so forth. In all of these areas, there is still room to improve.

Sample throughput has become increasingly important as the number of analytes increases in survey and screening applications. Reducing the time required to detect a statistically significant number of ions for each analyte requires not only efficient ionization and transmission and the elimination of m/z scanning, but also the detector's ability to handle a relatively high ion flux. Ion counting and arrival time detectors offer great sensitivity by counting each ion detected, but are subject to pile-up losses if the ion flux is even a few percent of the maximum count frequency. Analog detectors, on the other hand, have limited dynamic range and lack the sensitivity of counting detectors, but may be the best for high-throughput applications at this time.

The Future

The pace of evolution in mass spectrometric instrumentation and methodology has been greatly increased by the widespread adoption of mass spectrometry as a principal tool for biomedical research. It has multiplied the market for mass spectrometers many fold and continues to bring waves of new researchers into the field of mass spectrometry. Incremental improvements in instrumentation will be in the areas of simplicity, portability, reduced detection limits, and compatibility with separation and selectivity methodology. Improvements in efficiency of ionization, transmission and detection will be important for increasing throughput and improving quantitation. In particular, I believe that mass spectrometers which are compatible with array detectors will assume a greater importance because of their simplicity and potential for high throughput. Currently only magnetic sector mass analyzers are used with array detectors, but others, including my current interest, development of distance-of-flight mass spectrometry, are on the horizon. New methods of ionization will open mass spectrometry to new classes and matrices of analytes. New forms of chemical reactivity may be combined with ionization and intermediate m/z changes to give us needed improvements in selectivity. Over the last two decades we have watched mass spectrometry improve the limits of detection by an order of magnitude every few years. It is our own form of Moore's Law. Given the motivation and the creativity of the people working in this field, I see no reason why this rate of progress will not continue for many years to come.

Albuquerque, New Mexico, U.S.A., 2009   *Chris Enke*

Preface

If a history of science in the 20th century is ever written, one of the highlights may well be a chapter about mass spectrometry. Over the decades, the development of mass spectrometry instrumentation closely followed other innovations in science. Although mass spectrometry (MS) played a crucial role in many of the most exciting and important breakthroughs of the last century (proteomics, metabolomics, polymers, etc), the last decade has witnessed such an unprecedented flourishing in the field of probing reaction mechanisms in solution. Organic mass spectrometrists enthusiastically embraced the new ideas of John Fenn and suddenly incorporated them to the MS problem-solving ability to probe and intercept several types of intermediates and transient species directly from solution. In that way, several mechanisms have been corroborated and discarded by using ESI-MS reaction monitoring. The main goal of this book is to add to our reading pleasure and to further inspire and encourage new generations of chemists to dare the impossible. Who could imagine that Woodward's, Corey's, Curran's, Stork's, Brookhart's, Evans', and others' mechanistic proposals for reactions are being elucidated through trapping the empiric postulated species directly from solution? There is too much to be learned and discovered in organic chemistry. Today, for new reactions, a mechanistic proposal can be thought not only with both sides of them (reactants and products), but also determining the intermediates that are being formed and consumed during the processes. We will be enriched beyond measure if the 21st century is a period of profound development of mass spectrometry together with chemistry – the life science.

Throughout each chapter, clear fundamental, principles, and mechanisms accompany the text. Special emphasis is given by Paul Kebarle and Udo H. Verkerk on introducing the ESI-MS technique (mechanism and process). Then, Hao Chen and R. Graham Cooks show an overview of the study of ion/molecule reactions from gas phase to solution. Finally, we describe the concepts giving ideas and tools about *"what can I expect and what shall I look out for when doing my experiments"* using ESI-MS to probe organic reactions mechanisms. Therefore, one may ask how can meaningful information be obtained in an area as large as organic chemistry with a technique that sounds too good to be true from an operator or technician's point of view. We have two answers for that. This book testifies to the fact that ESI indeed transfers intact molecules, including non-covalently bound and transient species, to a very powerful, but straightforward mass spectrometry system. A wealth of data and information can be obtained, even with transient or unstable species, which quite often cannot be realized by other analytical systems. These characteristics make for an ideal enterprise in traveling around the mechanism of reactions directly from solutions.

Needless to say, of course, that bringing *Reactive Intermediates: MS Investigations in Solution* to completion required the assistance of many talented individuals. It has been a great pleasure and distinction for us to assemble a diverse group of distinguished international authors. What you are currently holding in your hands is the first treatise that is not an update of reaction mechanisms proposed by ESI-MS, but rather an invitation to see what can be done with ESI-MS. In little words, this makes for very useful reading by any account.

As mass spectrometry is a live technique, we also apologize in advance for the inevitable errors, statements and assumptions that a book like this may contain (or not), and new chemists are dare to comment and correct such errors in future. In closing, we wish to dedicate these chapters to our mentors who embedded within us the passion for chemistry and particularly for mass spectrometry. If this book can educate somehow and inspire more young researchers in the same way that they did, then we will regard this book as a great success and our personal satisfaction.

Talca, Chile *Leonardo Silva Santos*
December 2009

List of Contributors

Hao Chen
Ohio University
Department of Chemistry and
Biochemistry
Center for Intelligent Chemical
Instrumentation
Athens, Ohio
USA

Marcos N. Eberlin
State University of Campinas
Thomson Mass Spectrometry
Laboratory
Institute of Chemistry
Campinas, 13085-850
Brazil

Yinlong Guo
Chinese Academy of Science
Shanghai Mass Spectrometry Center
Shanghai Institute of Organic
Chemistry
Fenglin Road 354
Shanghai, 200032
People's Republic of China

Richard A. J. O'Hair
The University of Melbourne
School of Chemistry
Victoria 3010
Australia

and

The University of Melbourne
Bio21 Institute of Molecular Science and
Biotechnology
Victoria 3010
Australia

and

The University of Melbourne
ARC Centre of Excellence in Free
Radical Chemistry and Biotechnology
Victoria 3010
Australia

Paul Kebarle
University of Alberta
Department of Chemistry
Edmonton
Alberta, T6G 2G2
Canada

Reactive Intermediates: MS Investigations in Solution. Edited by Leonardo S. Santos.
Copyright © 2010 WILEY-VCH Verlag GmbH & Co. KGaA, Weinheim
ISBN: 978-3-527-32351-7

Fabiane M. Nachtigall
State University of Campinas
Thomson Mass Spectrometry
Laboratory
Institute of Chemistry
Campinas, 13085-850
Brazil

Anna Pla-Quintana
University of Girona
Department of Chemistry
Campus de Montilivi, s/n
17071 Girona
Spain

Rong Qian
Chinese Academy of Science
Shanghai Mass Spectrometry Center
Shanghai Institute of Organic
Chemistry
Fenglin Road 354
Shanghai, 200032
People's Republic of China

Anna Roglans
University of Girona
Department of Chemistry
Campus de Montilivi, s/n
17071 Girona
Spain

Leonardo S. Santos
Universidad de Talca
Chemistry Institute of Natural
Resources
Laboratory of Asymmetric Synthesis
Avenida Lircay s/n
Talca
Chile

Kevin A. Schug
The University of Texas at Arlington
Department of Chemistry &
Biochemistry
Arlington
TX, 76019
USA

Udo H. Verkerk
York University
Centre for Research in Mass
Spectrometry
CB 220, Chemistry Building
4700 Keele Street
Toronto
Ontario, M3J 1P3
Canada

Haoyang Wang
Chinese Academy of Science
Shanghai Mass Spectrometry Center
Shanghai Institute of Organic
Chemistry
Fenglin Road 354
Shanghai, 200032
People's Republic of China

Shengjun Yao
Chinese Academy of Science
Shanghai Mass Spectrometry Center
Shanghai Institute of Organic
Chemistry
Fenglin Road 354
Shanghai, 200032
People's Republic of China

Jing Zhou
Chinese Academy of Science
Shanghai Mass Spectrometry Center,
Shanghai Institute of Organic
Chemistry
Fenglin Road 354
Shanghai, 200032
People's Republic of China

1
A Brief Overview of the Mechanisms Involved in Electrospray Mass Spectrometry

Paul Kebarle and Udo H. Verkerk

1.1
Introduction

1.1.1
Origins of Electrospray Mass Spectrometry

Electrospray Ionization (ESI) is a method by which solutes present in a solution can be transferred into the gas phase as ions. The gas-phase ions can then be detected by mass spectrometric means (ESI-MS). Remarkably, ESI can handle a vast variety of analytes from small inorganic or organic species to polymers, nucleic acids, and proteins of very high molecular mass. The analytes present in the solution may be positive or negative ions, or compounds that are not ionized in the solution that is sprayed. In that case the analyte is charged by association with one or more of the ions present in the solution. This charging process is part of the electrospray mechanism. ESI-MS is an excellent method for detection of analytes generated by high-pressure liquid chromatography or capillary electrophoresis. As a result, scientists in biochemical, biomedical, and pharmaceutical research were early users of the new technology. A more recent and rapidly developing area is the study of homogeneous catalysis in solution via ESI-MS, and this involves the detection of ionic catalytic intermediates [1]. This is the central area of the present book.

The significance of ESI-MS was recognized by the award of a Nobel Prize in 2002 to John Fenn [2], who was the major developer of the method. The initial development of the method is due to Malcolm Dole. In the nineteen sixties Malcolm Dole was interested in the determination of the molecular mass of synthetic polymers. But how could one get large polymers into the gas-phase without decomposing them? Dole reasoned that if one used a very dilute solution of the analyte and nebulized such a solution into extremely small droplets one could obtain many droplets that contain only a single analyte molecule. Evaporation of such droplets would then lead to a transfer of the analyte molecules to the gas phase. If the analyte was

Reactive Intermediates: MS Investigations in Solution. Edited by Leonardo S. Santos.
Copyright © 2010 WILEY-VCH Verlag GmbH & Co. KGaA, Weinheim
ISBN: 978-3-527-32351-7

not charged, as was the case for some polymers, the presence of an electrolyte such as Na^+ could lead to addition of Na^+ to the polymer molecule on evaporation of droplets that happen to contain a single polymer molecule and one Na^+ ion. Such statistical charging was known to occur [3] but was a rather inefficient source of ionized analytes. Something better was required. While working as a consultant of a paint company, Dole witnessed the use of electrostatic spraying to apply paint to car bodies [4a]. In this spray process, the spray nozzle was kept at a high voltage, and this led to the production of very small charged paint droplets which were attracted to the car body kept at ground potential. Dole and coworkers were able to produce small charged droplets by applying electrospray to polystyrene solutions. The evaporation of the very small droplets led to polystyrene ions that could be detected using ion mobility or kinetic energy analysis of the produced ions [4b]. While Dole's methods and results had some flaws they clearly indicated that electrospray is a very promising soft ionization method for macromolecules [4b].

Following Dole's [4] work, John Fenn introduced some decisive improvements that allowed a mass spectrometer to be interfaced to an electrospray source [5, 6] and clearly demonstrated that ESI-MS could be used very effectively for the analysis of small ions and molecules [5] as well as peptides and proteins with a molecular mass extending into the megadalton range [6]. This work had a big impact and started the ESI-MS revolution that is continuing to this day.

1.1.2
Aims of this Chapter

This chapter is written for users of ESI-MS. It presents an account of 'how it all works.' Such understanding is desirable because the observed mass spectra depend on a large number of parameters. These start with a choice of solvent and concentrations of the analyte, choice of additives to the solution that may be beneficial, choice of the flow rates of the solution through the spray capillary, the electrical potentials applied to the spray capillary (also called 'needle') and the potentials of ion optical elements that are part of the mass analyzer. Proper choice of these parameters requires not only some understanding of conventional mass spectrometry but also of the electrospray mechanism. In early work on ESI-MS many of these parameters were established by trial and error, but now that a better understanding of the mechanism is at hand more rational choices are possible. The present chapter provides an up to date account of Electrospray. For a broader coverage, which is somewhat dated but still relevant, the review by Smith and coworkers is recommended [7].

As mentioned already, electrospray existed long before its application to mass spectrometry. It is a method of considerable importance for the electrostatic dispersion of liquids and creation of aerosols. Much of the theory concerning the mechanism of the charged droplet formation was developed by researchers in aerosol science. A compilation of articles devoted to electrospray can be found in a recent special issue of the Journal of Aerosol Science [8].

1.2 Production of Gas-Phase Ions by Electrospray and Electrospray Ionization Mass Spectrometry

1.2.1 Overview

There are three major steps in the production of gas-phase ions from electrolyte ions in solution: (a) production of charged droplets at the ES capillary tip; (b) shrinkage of the charged droplets due to solvent evaporation and repeated charge-induced droplet disintegrations that ultimately lead to small highly charged droplets capable of producing gas-phase ions, and (c) the actual mechanism by which gas-phase ions are produced from these droplets. All stages occur in the atmospheric pressure region of the apparatus, see Figure 1.1.

A small fraction of the ions resulting from the preceding stages enter the vacuum region of the interface leading to the mass spectrometer through a small orifice or capillary. Two types of apparatus using a capillary are shown in Figure 1.2. The created gas-phase ions may be clustered with solvent molecules and other additives that

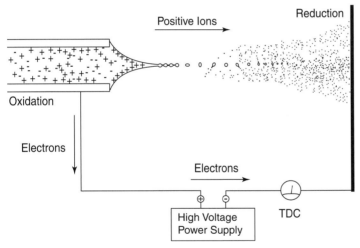

Figure 1.1 Schematic of major processes occurring in the atmospheric pressure region of electrospray. TDC stands for total droplet current (I). The figure illustrates major processes occurring in the atmospheric pressure region of an ESI run in the positive ion mode. Penetration of the imposed electric field into the liquid leads to formation of an electric double layer at the meniscus. The double layer is due to the polarizabilty and dipole moments of the solvent molecules and an enrichment near the meniscus of positive ions present in the solution. These cause a destabilization of the meniscus and formation of a cone and a jet charged by an excess of positive ions. The jet splits into droplets charged with an excess of positive ions. Evaporation of the charged droplets brings the charges closer together. The increasing Coulombic repulsion destabilizes the droplets, which emit a jet of smaller charged progeny droplets. Evaporation of progeny droplets leads to destabilization and emission of a second generation of progeny droplets, and so on until free gas-phase ions form at some point.

(a)

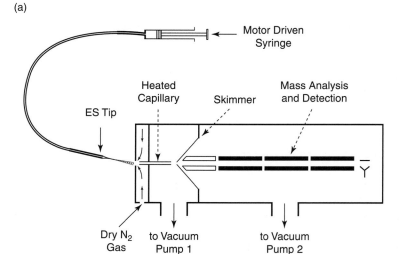

(b)

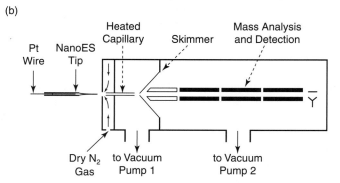

Figure 1.2 (a) Schematic of Electrospray (ES) and interface to mass spectrometer. Solution containing analyte is supplied to the ES spray tip by a motor-driven syringe via flexible glass capillary tubing. A positive potential is applied to the spray tip (positive ion mode). The spray of positively charged droplets emerges from the spray capillary tip (see Figure 1.1). Solvent evaporation of the charged droplets leads to gas-phase ions. A mixture of ions, small charged droplets, and solvent vapor in the ambient gas enters the orifice leading to the nitrogen counter-current chamber. The weak nitrogen counter-current removes the solvent vapor, but the ions, driven by an electric potential and pressure difference, enter the heated capillary pathway into the low pressure chamber. An electric field between this capillary and the skimmer cone accelerates the ions for a further collision-activated 'clean-up' of the ions. The potential difference over the cone orifice and downstream ion optical elements transports the ions into the high vacuum region of the mass analysis chamber. (b) Same as Figure 1.2a but showing Nanoelectrospray. Large diameter end of NanoES tip capillary is 'loaded' with μL amounts of solution. The electrical potential is supplied to the nano tip either by a Pt wire or by a metal film coating the outside of the capillary. A spray of charged nano droplets results from the pull of the electric field on the polarized meniscus of the solution at the capillary tip.

would broaden m/z peaks excessively. The gas-phase ions are therefore subjected to a *thermal* declustering or 'clean-up' stage by heating the capillary. Often a countercurrent flow of an inert gas is used to minimize entrance of solvent vapor into the vacuum region. A second clean-up stage is obtained through *collisional* activation by applying an electric potential difference between the capillary exit and the skimmer. The chamber past the capillary is at a vacuum of a few torr, so that acceleration of ions by the applied field results in multiple collisions with neutral gas molecules. The accumulating internal energy of the ions leads to ion desolvation, and at higher applied fields to ion fragmentation. The mass-selecting ion optics of the mass spectrometer are placed beyond the skimmer because they require high vacuum conditions.

1.2.2
Production of Charged Droplets at the Capillary Tip

As shown in Figure 1.1, a voltage (V_c) of 2–3 kV is applied to the spray capillary. The counter electrode in ESMS may be a plate with an orifice leading to the mass spectrometric sampling system or a sampling capillary, mounted on the plate, which leads to the MS, as shown in Figure 1.2a. Because the spray capillary tip has a very small diameter, the electric field (E_c) at the capillary tip is very high ($E_c \approx 10^6$ V/m). The value of the field at the capillary tip opposite a large and planar counter electrode can be evaluated with the approximate relationship [9];

$$E_c = 2\, V_c / [r_c \ln(4d/r_c)] \tag{1.1}$$

where V_c is the applied potential, r_c is the capillary outer radius, and d is the distance from the capillary tip to the counter electrode. For example, the combination of $V_c = 2000$ V, $r_c = 5 \times 10^{-4}$ m and $d = 0.02$ m leads to: $E_c \approx 1.6 \times 10^6$ V/m. As indicated by Eq. (1.1) the field E_c is proportional to V_c, with the most important geometry parameter being r_c. E_c is essentially inversely proportional to r_c, while it decreases slowly with the electrode separation d.

A typical solution supplied to the capillary is a polar solvent in which the analyte is soluble. Because ESI-MS is a very sensitive method, low concentrations, $10^{-7} - 10^{-3}$ mol L^{-1} (M), of analyte need to be used. Methanol or methanol/water, acetonitrile or acetonitrile/water are often used as the solvent. However, apolar and nonprotic solvents like toluene, nitromethane, dichloromethane and formamide can be used as well although ionic additives may be required in order to obtain stable spray conditions. For an overview of some of the solvents used in electrospray see Ref. [10]. For simplicity in the subsequent discussion, we will assume that the analyte is ionic, and only the positive ion mode will be considered.

The field E_c when turned on, will penetrate the solution near the spray capillary tip. This will cause a polarization of the solvent near the meniscus of the liquid. In the presence of even traces of an electrolyte, the solution will be sufficiently conducting and the positive and negative electrolyte ions in the solution will move under the influence of the field. This will lead to an enrichment of positive ions on or near the surface of the meniscus and enrichment of negative ions away from the meniscus. The forces due to the polarization cause a distortion of the meniscus into a cone pointing downfield (see Figure 1.1). The increase of surface due to the cone formation is resisted by the surface tension of the liquid. The cone formed is called a Taylor cone (see Taylor [11] and Fernandez de la Mora [12]). If the applied field is sufficiently high,

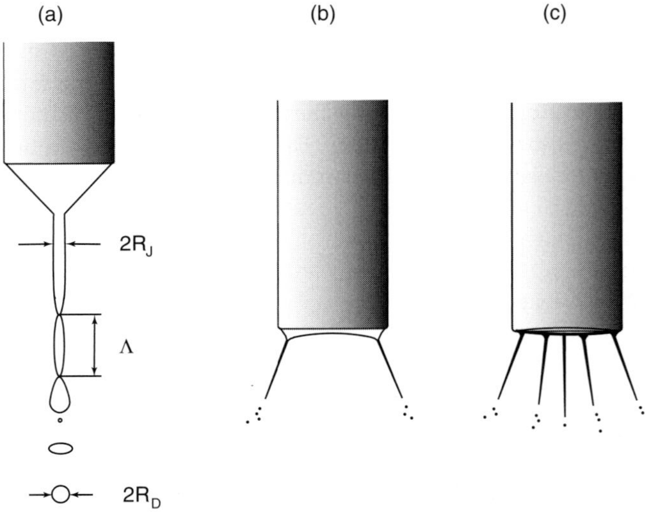

Figure 1.3 Different forms of Electrospray at the tip of the spray capillary. (a) cone jet mode. Relationship between radius of droplets and radius of jet: $R_D/R_J \approx 1.9$. (b) and (c). Multijet modes result as the spray voltage is increased, and the flow rate imposed by the syringe is high. (After Cloupeau, Ref. [13].)

a fine jet emerges from the cone tip, whose surface is charged by an excess of positive ions. The jet breaks up into small charged droplets (see Figure 1.1 and Figure 1.3a, due to Cloupeau [13a]).

It is apparent from Figure 1.3a that the size of the droplets formed from the cone jet is dependent on the jet diameter $2R_J$. The droplets initially produced therefore could be expected to be approximately of the same size. This was proposed by Cloupeau [13] and confirmed by studies of Tang and Gomez [14]. The formed droplets are positively charged because of an excess of positive electrolyte ions at the surface of the cone and the cone jet. Thus, if the major electrolyte present in the solution is ammonium acetate, the excess positive ions at the surface will be NH_4^+ ions. This mode of charging, which depends on the positive and negative ions drifting in opposite directions under the influence of the electric field, has been called the electrophoretic mechanism [13b,c].

The charged droplets drift downfield through the air toward the opposing electrode. Solvent evaporation at constant charge leads to droplet shrinkage and an increase in the repulsion between the charges. At a given radius, the increasing repulsion overcomes the surface tension at the droplet surface. This causes a coulomb fission (also called a coulomb explosion) of the droplet. The droplet fission occurs via formation of a cone and a cone jet that splits into a number of small progeny droplets. This process bears a close resemblance to the cone jet formation at the capillary tip (see de la Mora [12] and references therein). Further evaporation of the parent droplet leads to repeated fissions. The progeny droplets also evaporate and break up. Very small charged droplets result that ultimately lead to gas-phase ions by processes which will be described in detail in subsequent sections.

The cone-jet mode at the spray capillary tip described and illustrated in Figures 1.1 and 1.3a is only one of the many possible ES modes. For a qualitative description of this and other modes, see Cloupeau [13a–c]. More recent studies by Vertes and coworkers [15] using fast time-lapse imaging of the Taylor cone provide details on the evolution of the Taylor cone into a cone jet and pulsations of the jet. These pulsations lead to spray current oscillations. The current oscillations are easy to determine with conventional equipment and can be used as a guide for finding conditions that stabilize the jet and improve signal-to-noise ratios of the mass spectra. The cone-jet mode is the most used and best characterized mode in the electrospray literature [12, 13].

The magnitude of currents obtained with ES (see Figure 1.1) in the cone jet mode at a typical flow rate of 6 μL/min are around 0.1 μA. Only a fraction of this current will enter the first chamber after passage through the heated capillary and skimmer (see Figures 1.1 and 1.2). A charge loss by a factor of 100 would reduce the ion current to 10^{-9} A; fortunately, modern current detection technology employing ion-electron multipliers allows the detection of much lower currents typically in the range of $10^{-12} - 10^{-16}$ A.

1.2.3
Electrospray as an Electrolytic Cell

At a steady operation of the electrospray in the positive ion mode (see Figure 1.1), the positive droplet emission continuously carries off positive charge. The requirement for charge balance together with the fact that only electrons can flow through the metal wire that supplies the electric potential to the electrodes (Figure 1.1) leads to the conclusion that the ES process must include an electrochemical conversion of ions to electrons. In other words, the ES device can be viewed as a special type of electrolytic cell [16]. It is special because the ion transport does not occur through uninterrupted solution, as is normally the case in electrolysis, but through the gas phase. Thus, in the positive ion mode where the charge carriers are positively charged droplets (and subsequently gas-phase positive ions) a conventional electrochemical oxidation reaction should be occurring at the positive electrode, that is at the liquid/metal interface of the spray capillary (Figure 1.1). This reaction replenishes positive ions to the solution and prevents the build-up of a charge imbalance. Concurrently, gas-phase ions are reduced when they hit a downstream metal surface. The nature of these ions depends on the experimental conditions. If the spray capillary is made of metal (M), the metal can become oxidized and enter the solution as cations, while releasing electrons to the metal electrode, see Eq. (1.2).

$$M(s) \rightarrow M^{2+}(aq) + 2e \text{ (on metal surface)} \tag{1.2}$$

$$4OH^-(aq) \rightarrow O_2(g) + 2H_2O + 4e \text{ (on metal surface)} \tag{1.3}$$

The other alternatives for restoring the charge balance are the removal of negative ions present in the solution by an oxidation reaction as illustrated in Eq. (1.3) (for aqueous solutions) or oxidation of sacrificial additives such as I^- or hydroquinone [17, 18].

One expects that the reaction with the lowest oxidation potential will dominate, and that the oxidation reaction will be dependent on the material present in the metal electrode, the solutes/ions present in the solution, and the nature of the solvent. Proof of the occurrence of an electrochemical oxidation at the metal capillary was provided by Blades et al. [16]. When a Zn spray capillary tip was used, release of Zn^{2+} to the solution could be detected. Furthermore, the amount of Zn^{2+} release to the solution per unit time when converted to coulomb charge per second was found to be equal to the measured electrospray current (I) in amperes (coulomb/s, Figure 1.1). Similar results were observed with stainless steel capillaries [16]. These were found to release Fe^{2+} to the solution. These quantitative results provided the strongest evidence for the electrolysis mechanism. These oxidation reactions introduce ions which were not previously present in the solution (see Eq. (1.2)). However, they also provide an opportunity to generate reactive intermediates that can be studied by mass spectrometry.

Van Berkel and coworkers have examined the consequences of the electrochemical processes to ESI-MS in a series of publications [17]. They were able to demonstrate that ions produced by the electrolysis process can in some cases have unintended and undesired effects on the mass spectra obtained with pH- or oxidation-sensitive analytes [17].

Ions introduced into the solution by inadvertent or deliberate electrolysis amount to very low concentrations. Taking the Zn capillary tip as example, a solution of 10^{-5} M NaCl in methanol at a flow rate $V_f = 20\,\mu L\,min^{-1}$ was found to lead to an electrospray current of 1.6×10^{-7} A. The Zn^{2+} concentration produced by the Zn-tipped capillary evaluated from the current was 2.2×10^{-6} M. Assuming that the Na^+ ion was the analyte ion, the concentration of the ions produced by the oxidation at the electrode is only $\sim 1/5$ of that of the analyte. It will be shown later that the electrospray current increases very slowly with the total electrolyte concentration. Therefore, ions produced by oxidation at the electrode may not be noticed in the mass spectrum at higher analyte or additive concentrations.

1.2.4
Required Electrical Potentials for ES. Electrical Gas Discharges

D.P.H. Smith [19] was able to derive a useful approximation for the electric field at the capillary tip (E_{on}) required for the *onset* of instability of a static Taylor cone, see Eq. (1.4). Instability of the Taylor cone is required for the formation of a jet at the apex of the cone. The equation for the onset field, when combined with Eq. (1.1), leads to an equation (for the potential, V_{on}, required for the start of electrospray;

$$E_{on} \approx \left(\frac{2\gamma\cos\theta}{\varepsilon_o r_c}\right)^{1/2} \tag{1.4}$$

$$V_{on} \approx \left(\frac{r_c \gamma \cos\theta}{2\varepsilon_o}\right)^{1/2} \ln(4d/r_c) \tag{1.5}$$

Table 1.1 Onset voltages[a], V_{on} for ESI of solvents with different surface tension γ.

Solvent	CH$_3$OH	CH$_3$CN	(CH$_3$)$_2$SO	H$_2$O
γ (N m^{-1})	0.0226	0.030	0.043	0.073
V_{on} (Volt)	2200	2500	3000	4000

[a]Calculated with Eq. (1.6).

where γ is the surface tension of the solvent, ε_o is the permittivity of vacuum, r_c is the radius of the capillary, and θ is the half angle for the Taylor cone. Substituting the values $\varepsilon_o = 8.8 \times 10^{-12}$ J^{-1} C^2 and $\theta = 49.3$ (see Taylor [11]), one obtains

$$V_{on} = 2 \times 10^5 (\gamma r_c)^{1/2} \ln(4d/r_c) \quad (1.6)$$

where γ must be substituted in N/m and r_c in m to obtain V_{on} in volts. Shown in Table 1.1 are the surface tension values for four solvents and the calculated electrospray onset potentials for $r_c = 0.1$ mm and $d = 40$ mm. The surface of the solvent with the highest surface tension (H$_2$O) is the most difficult to stretch into a cone and jet, and this leads to the highest value for the onset potential V_{on}. As a result, use of neat water as solvent can lead to the initiation of an electric discharge from the spray capillary tip.

Experimental verification of Eqs. (1.5) and (1.6) has been provided by Smith [19], Ikonomou et al. [20] and Wampler et al. [21] For *stable* ES operation one needs to go a few hundred volts higher than the calculated V_{on}. The electrospray onset potential is the same for both the positive and negative ion modes, but the electric discharge onset is lower when the capillary electrode is negative [19, 20] and metallic. This is probably due to emission of electrons from the negative capillary, which initiate the discharge. Use of glass capillaries reduces the risk of electric discharge. In this case the electric potential is applied via an internal metal wire or external metal coating in contact with the solution (see Figure 1.2b). Neat water as solvent can be used with this 'nanospray' arrangement without the occurrence of electric discharges.

The occurrence of electric discharge leads to a sudden increase in the ion current together with a visible glow around the spray capillary [22]. Currents above 10^{-6} A are generally due to the presence of an electric discharge. A more specific test is provided by the appearance of discharge-characteristic ions in the mass spectrum. Thus, in the positive ion mode the appearance of protonated solvent clusters such as H$_3$O$^+$(H$_2$O)$_n$ from water or CH$_3$OH$_2^+$(CH$_3$OH)$_n$ from methanol solvent indicates the presence of a discharge [20]. In the absence of an electrical discharge, protonated solvent ions are only produced at high abundance when the solvent has been acidified, that is when H$_3$O$^+$ or CH$_3$OH$_2^+$ are present in the solution. The presence of an electrical discharge severely degrades the performance of ESMS. The electrospray ions are observed at much lower intensities than was the case prior to the discharge, while discharge-generated ions appear with very high intensities [20, 21].

Air at atmospheric pressure is not only a convenient but also a very suitable ambient gas for ES, particularly when solvents with high surface tension are to be electrosprayed. Initiation of gas discharges occurs when free electrons are accelerated by the high electric field near the capillary to velocities where they can ionize the gas molecules. At near atmospheric pressures, the collision frequency of the electrons with the gas molecules is very high, limiting the electron acceleration process and minimizing the initiation of an electric discharge. In addition to the pressure effect, oxygen molecules in air have electron affinity and readily capture free electrons. Trace gases such as SF_6, with higher electron affinity and electron capture cross sections than O_2, can be added to assist the electron capture and suppress the electrical breakdown at higher electrospray voltages, when the atmospheric oxygen effect is insufficient [21].

1.2.5
Current, Charge and Radius of Droplets Produced at the Capillary Tip

Fernandez de la Mora and Locertales [23] have proposed the following approximate relationships that correlate the current, droplet size and charge on the generated droplets. Assuming a flow rate below 1 µL min^{-1} and operation of electrospray in the *cone jet mode*,

$$I = f\left(\frac{\varepsilon}{\varepsilon_o}\right)\left(\gamma K V_f \frac{\varepsilon}{\varepsilon_o}\right)^{1/2} \qquad (1.7)$$

$$R \approx (V_f \varepsilon / K)^{1/3} \qquad (1.8)$$

$$q \approx 0.7[8\pi(\varepsilon_o \gamma R^3)^{1/2}] \qquad (1.9)$$

where γ is the surface tension of solvent; ε the permittivity of solvent; ε_o the permittivity of vacuum (free space); the $\varepsilon/\varepsilon_o$ ratio the dielectric constant of solvent; K the conductivity of solution; E the applied electric field at capillary tip (see Eq. (1.2)); R the radius of droplets produced at capillary tip; q the charge of droplets and V_f the flow rate (volume/time); $f(\varepsilon/\varepsilon_o)$ is a numerical function tabulated by the authors [23]. For liquids with a dielectric constant $(\varepsilon/\varepsilon_0) \geq 40$ (water, water/methanol mixtures, acetonitrile, formamide) the value of $f(\varepsilon/\varepsilon_o)$ is approximately 18. The described relationships were obtained for solutions with a conductivity larger than 10^{-4} S m^{-1}. Assuming electrolytes that dissociate completely, this requirement corresponds to solutions with concentrations higher than $\sim 10^{-5}$ mol L^{-1}, that is a concentration range commonly present in ESMS. A more recent theoretical treatment by Cherney [24] has confirmed the deductions of Fernandez de la Mora and Locertales and has provided a more detailed description of the conditions existing in the cone jet. The derived relationships are also in agreement with experimental data by Chen and Pui [25].

1.2.6
Solvent Evaporation from Charged Droplets Causes Coulomb Fissions of Droplets

The charged droplets produced by the spray needle shrink due to solvent evaporation while the charge remains constant. As the droplet gets smaller, the repulsion between the charges at the surface increases and at a certain droplet radius the repulsion of the charges overcomes the cohesive force of the surface tension. This instability results in fission of the droplet, at which point the droplet typically releases a jet of small, highly charged, progeny droplets. The condition for the instability is given by the Rayleigh [26] equation;

$$Q_{Ry} = 8\pi(\varepsilon_0 \gamma R^3)^{1/2} \tag{1.10}$$

where Q_{Ry} is the charge on the droplet; γ is the surface tension of the solvent; R is the radius of the droplet and ε_o is the electrical permittivity. The fission at or near the Rayleigh limit, with release of a jet of small monodisperse charged progeny droplets, has been confirmed by a number of experiments.

Most experiments used Phase Doppler Interferometry (PDI), a method well suited for volatile solvents as used in ESI-MS [27]. A series of PDI measurements using various solvents are given in Table 1.2. One can deduce from this table that the dependence on the type of solvent is relatively small. Thus, droplets from all solvents experience Coulomb fissions close to, or at, the Rayleigh limit. The loss of mass on fission is between 2 and 5% of the parent droplet mass but the loss of charge is much larger, that is, some 15–25% of the charge of the parent droplet.

Beauchamp and coworkers [28] provide information on the charge of the parent droplet immediately after the droplet fission. An example of such data is given in Figure 1.4, where the charge of the droplets before and after the fission is given as a percentage of the Rayleigh condition, Eq. (1.10). These, and results for the other solvents studied [28], show that the evaporating charged droplets oscillate at all times between fairly narrow limits of the Rayleigh condition. This finding has a bearing on the discussion of the mechanism by which large molecules enter the gas phase (see the Charged Residue Mechanism in Section 1.2.10). Notable also, see Figure 1.4a, is the observation that the diameter of the charged parent droplet undergoing evaporation *and* Coulomb fissions remains very close to the diameter of an uncharged droplet that loses mass solely because of evaporation. This result supports the observations of a series of authors (see Table 1.2 and Refs. [26, 29–31]).

When the sprayed solution contains a solute, such as a salt, the continuous evaporation of the droplets will lead to very high concentrations of the salt and finally to charged solid particles – 'skeletons' of the charged droplets that can reveal some aspects of the droplet evolution. Fernandez de la Mora and coworkers [32] have used this approach to study charged droplet evolution. This work is of special relevance to the ion evaporation model and is discussed in Section 1.2.8.

1.2.7
Evaporation of Droplets Leading to Coulomb Fissions Producing Progeny Droplets that Ultimately Lead to Ions in the Gas-Phase; Effects of the Concurrent Large Concentration Increase

It is clear that the process of repeated droplet fissions of both parent and progeny droplets ultimately will lead to very small charged droplets that are the precursors of the gas-phase ions. The mechanisms by which the gas-phase ions are produced from

Table 1.2 Experimental observations of Rayleigh fissions of charged droplets.

Reference	Solvent	Droplet diameter range (μm)[a]	Onset of Instability (% of Rayleigh limit)	% of mass lost in breakup[a]	% of charge lost in breakup
(Smith et al. Ref. [28a])	Water	10–40	90	nd	20–40
	Methanol	10–40	110	nd	15–20
	Acetonitrile	10–40	100	nd	15–20
(Grimm and Beauchamp Ref. [28b])	n-Heptane	35–45	100	nd	19
	n-Octane		87	nd	17
	p-Xylene		89	nd	17
(Gomez and Tang Ref. [27])	Heptane	20–100	70	nd	nd
(Taflin et al. Ref. [29])	Low vapor pressure oils	4–20	75–85	2	10–15
(Richardson et al. Ref. [30])	Dioctyl phthalate	nr	102–84	2.3	15–50
(Schweitzer et al. ref. [31])	n-octanol	15–40	96–104	5	23

[a]nr: not reported, nd: not determined.

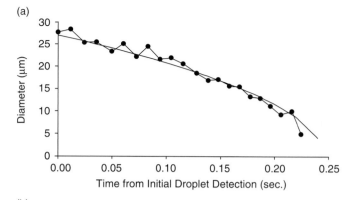

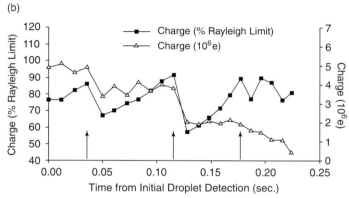

Figure 1.4 Evaporation and discharge of a positively charged water droplet in nitrogen gas at ambient pressure and 317 K and a weak (51 V cm^{-1}) electric field. (a) Variation of droplet diameter with time. Also plotted (smooth curve) is the predicted change in diameter due to evaporation of a neutral water droplet in a vapor-free N_2 gas at 317 K. (b) Variation of droplet charge with time, represented as number of elementary charges and as percent of the Rayleigh limit. Arrows indicate discharge events. Note that water droplets undergo a Coulomb fission at approximately 90% of the Rayleigh limit and are at approximately 65% of the limit after the Coulomb fission. (Reprinted from Ref. [28a] with permission from the American Chemical Society.)

the very small 'final' droplets is considered in Sections 1.2.8 and 1.2.10. Here we examine some of the details of the evolution of the initial droplets formed at the spray capillary. The whole process is driven by the decrease in droplet volume by solvent evaporation. The continuous evaporation is possible because the thermal energy required for the evaporation is provided by the ambient gas at near atmospheric pressure. As will be shown below, a large loss of solvent by evaporation occurs before the final droplets are formed that lead to ions. It is instructive and for certain applications desirable to estimate the increase in solute concentration due to the volume lost by evaporation.

A droplet evolution scheme is shown in Figure 1.5. It deals with droplets produced by nanoelectrospray. Nanospray (see Section 1.2.4) is a technique that considerably

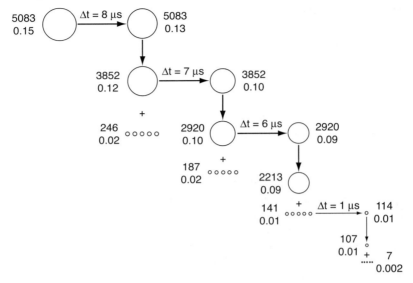

Figure 1.5 Droplet history of charged water droplets produced by nanospray. First droplet is one of the droplets produced at spray needle. This parent droplet is followed over three evaporation and fission events. The first generation progeny droplets are shown as well as the fission of one of the progeny droplets that leads to second-generation progeny droplets. R is the radius of the droplets and Z gives the number of charges on the droplet. Z corresponds to the number of excess singly charged ions near the surface of the droplet. The parent's charge is $Z = 0.9\ Z_R$ just before the fission and $Z = 0.7\ Z_R$ just after the fission (as observed in Figure 1.4), while the progeny droplets have $Z = 0.7\ Z_R$ just after the fission of the parent. (Based on Figure 1.1 in Peschke, Verkerk and Kebarle [33].) It should be noted that the droplets' history presented in Figure 1.6 is only a qualitative model. Thus, the assumption that only five progeny droplets could be formed at each fission is quite uncertain. The actual number could be much larger and the progeny droplets much smaller.

reduces the droplet diameter by performing electrospray using a glass capillary. Because the charged droplets are very much smaller than the droplets produced by electrospray, these droplets reach the final droplet stage much sooner.

In deriving Figure 1.5, the used stability limits for droplet fission (at droplet charge $Z = 0.9\ Z_R$) and droplet charge just after the droplet fission ($Z = 0.7\ Z_R$) were due to Beauchamp and coworkers [28a]. Further assumptions with which Figure 1.5 was obtained are described in the section 'Calculations and Experimental' of Peschke et al. [33]. Using the droplet radii, one can calculate that approximately 40% of the volume is lost *between* each fission, while only 2–5% of the parent droplet mass is lost *in* each fission event. This means that after 10 successive fissions of the parent droplet a 29-fold volume decrease will have taken place corresponding to a 29-fold solute concentration increase in the parent droplet. Such very large increases in concentration will promote the occurrence of undesired bimolecular reactions, like ion pairing, involving analyte ions and impurities present in the solution, see Section 1.2.12.

1.2.8
Mechanism for the Formation of Gas-Phase Ions from Very Small and Highly Charged Droplets. The Ion Evaporation Model (IEM)

Two mechanisms have been proposed to account for the formation of gas-phase ions from the final droplets. The first mechanism was proposed by Dole [4], who was interested in analytes of very high molecular mass (see introduction). For such macromolecules, droplet evolution as described in the preceding section would lead to some droplets containing one analyte molecule in addition to the ionic charges on the surface of the droplet. Solvent evaporation from such a droplet will lead to a gas-phase analyte ion whose charge originates from the charge at the surface of the vanished droplet. This assumption is now known as the Charged Residue Model (CRM) and is discussed in detail in Section 1.2.10.

Iribarne and Thomson [34], who worked with small ionic analytes such as Na^+ and Cl^-, proposed a second mechanism, the Ion Evaporation Model (IEM). This model predicts that, after the radii of the droplets have decreased to a very small size (≤ 10 nm), direct ion emission from the droplets will occur. At this point, the *ion evaporation* process replaces Coulomb fission. Iribarne and Thomson supported their model by experimental [34a] and theoretical [34b] results. The experimental results involved measurements of the relative abundance of the ions produced by ESI of solutions containing NaCl as the only solute. The authors found that there was a large number of ion aggregates of the type $[(NaCl)_n(Na)_m]^{m+}$, including $[(NaCl)_n(Na)]^+$, whose abundance decreased rapidly as n increased. However, the ion with lowest mass in that series, Na^+, ($n=0$, $m=1$), and hydrated $Na(H_2O)_k^+$ ($k=1-3$) had the highest abundances by far. While the large aggregate ions produced by ESI are probably due to a charged residue mechanism type process, the abundant Na^+ and Na^+ hydrates must be produced by a different mechanism. Iribarne and Thomson proposed a direct escape of Na^+ charges from the surface of the multiply charged droplets. This process begins to occur after the droplets reach a very small size. Once ion evaporation sets in, there are no more droplet fissions because the excess charges on the droplets are removed by ion evaporation.

The authors also developed theoretical equations for the droplet conditions that will lead to ion evaporation based on Transition State Theory as used in reaction kinetics [34b]. Assuming that the evaporating ion is one of the ionic charges at the surface of the droplet, the leaving ion is *repelled* by the Coulomb repulsion between it and the remaining charges on the droplet at increasing charge separation. But at very short distances from the droplet an *attractive* force between the leaving ion and the droplet is present as a result of the polarization of the droplet induced by the leaving ion. The ion-polarizability attraction is larger at very short distances from the droplet surface, while the repulsion becomes dominant at larger distances. The transition state is located where these two interactions become equal. The graphs in Figure 1.6 show the predicted radius at which droplet fission at the Rayleigh limit is replaced by ion evaporation. It indicates that the charged droplets must reach a radius of approximately 100 Å (10 nm) before ion evaporation replaces droplet fission.

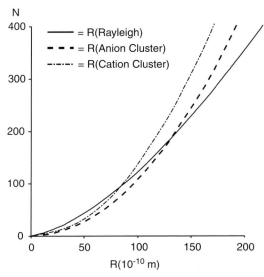

Figure 1.6 Predictions of the ion evaporation theory [34]. The Rayleigh curve provides the droplet radius R and the number of elementary charges N at which a charged water droplet will be at the Rayleigh limit. Solvent evaporation at constant charge to a smaller radius R will cause a Coulomb fission. Similarly, the curves Cation Cluster and Anion Cluster show the threshold of ion evaporation at a given charge N and droplet radius R. For negatively charged droplets, moving at constant charge to a smaller radius R due to solvent evaporation will lead to negative ion evaporation when the radius $R = 140\,\text{Å}$ (1 nm = 10 Å) and for positively charged droplets at $R = 84\,\text{Å}$, where the ion evaporation (Cation Cluster) and Rayleigh curves cross. Below this radius, ion evaporation replaces Coulomb fission. Thus, taking a radius of $R \approx 100\,\text{Å}$ provides a useful benchmark for the region where ion evaporation takes over.

Several research groups have performed experiments to examine the predictions of the theory. Some of the most relevant work is due to Fernandez de la Mora and coworkers [32], who used an interesting approach to provide strong evidence for the qualitative validity of the ion evaporation mechanism. Instead of concentrating on the evaporating droplets, they focused on the sizes and charges of the solid residues formed after complete evaporation of the solvent from the droplets. Since the solid residues had been 'charged droplets' before the last of the solvent evaporated, the size and charge of these residues should represent to a fair approximation the sizes and charges of the final charged droplets. The solid residues representing final droplets frozen in time are amenable to measurement. This approach provided results that were in good agreement with IEM [32].

Theoretical work involving simulations of ion evaporation from charged droplets can also provide valuable insights into IEM. A good example is the work by Vertes and coworkers [35] on the evaporation of H_3O^+ ions from charged water droplets. The authors used classical molecular dynamics simulations to study droplets of 6.5 nm diameter. Checks were made that the parameters used led to predictions of properties (such as the radial distribution function, the enthalpy of evaporation and the self diffusion coefficients) that are in agreement with experimental values.

Droplets of 6.5 nm diameter, which consist of some 4000 water molecules, were charged with H_3O^+ ions; ion pairs corresponding to a dissolved solute were not added. Remarkably, not all H_3O^+ ions were located on the surface of the droplet at equilibrium as is generally assumed for IEM. As H_3O^+ charges were added, fluctuations of water molecules at the droplet surface became much more pronounced. Some of these fluctuations developed into large protuberances that separated as hydrated H_3O^+ ions. Generally the 'solvation shell' of the departing H_3O^+ consisted of some 10 water molecules. Interested readers can observe the simulation of such ion evaporation at the website of Vertes (see http://www.gwu.edu/~vertes/publicat.html).

In summary, the Ion Evaporation Model is experimentally well supported for small ions of the kind that one encounters in inorganic and organic chemistry. However, when the ions become very large, such as polymers, dendrimers or biological supramolecular complexes like proteins and enzymes, the Charged Residue Model (CRM) becomes much more plausible, see Section 1.2.10. Because many applications of ESMS in analytical organometallic and physical organic chemistry involve small ions it is desirable to consider the expected relative sensitivities for these analytes when detected with ESMS.

1.2.9
Observed Relative Ion Intensity of Small Analytes. Dependence on the Nature of the Analyte, its Concentration and Presence of Other Electrolytes in the Solution. High Sensitivities of Surface-Active Analytes

The dependence of the sensitivities of ionic analytes on the nature of the analyte, its concentration, and the presence of other electrolytes in solution is of interest to users of electrospray mass spectrometry. The analytes considered in this section are smaller molecules that most likely enter the gas phase via the Ion Evaporation Model (IEM).

The dependence of the total droplet current produced at the spray capillary on various parameters was given in Eq. (1.7). Relevant to the present discussion is the dependence of the current (I) on the square root of the conductivity of the solution. At the low total electrolyte concentrations generally used in ESI, the conductivity is proportional to the concentration of the electrolyte. Thus, if a single electrolyte (E) was present in the sprayed solution, one would expect that the observed peak intensity I_E will increase with the square root of the concentration of that electrolyte (C_E, see Eq. (1.7)). At flow rates higher than that corresponding to the cone jet mode, the dependence on the concentration is lower than the 0.5 power [36]. Because ESI-MS is a very sensitive method, so that detection of electrolytes down to 10^{-8} M is easily feasible, one seldom works in practice with a single electrolyte system. The presence of electrolyte ions E leads to two concentration regimes for the analyte A:

(a) C_A much higher than C_E. In that case, the I_A is expected to increase with the square root (or slower) of C_A.

(b) C_A much lower than C_E. In that case, I_A is expected to increase with the first power of C_A because now I_A will depend on a statistical competition between A^+ and E^+ for being charges on the droplet surface.

To cover both regions, Tang and Kebarle [36] proposed Eq. (1.11a) for a two-component system in the positive ion mode. Equation (1.11a) predicts that when C_E is much higher than C_A and constant, the observed ion current I_A will be proportional to C_A:

$$\text{Two components;} \quad I_{A^+} = pf \frac{k_A C_A}{k_A C_A + k_E C_E} I \tag{1.11a}$$

$$\text{Three components;} \quad I_{A^+} = pf \frac{k_A C_A}{k_A C_A + k_B C_B + k_E C_E} I \tag{1.11b}$$

$$\text{For } C_A = C_E; \quad I_A = \text{const} \times C_A \quad \text{const} = k_A I_E / k_E C_E \tag{1.11c}$$

Equation (1.11b) is for three components A^+, B^+, and E^+ with solution concentrations C_A, C_B, and C_E; I is the total electrospray current leaving the spray capillary (I can easily be measured, see Figure 1.1) and p and f are proportionally constants (see Ref. [36]). The sensitivity coefficients for A^+, B^+ and E^+ are k_A, k_B, and k_E and depend on the specific chemical ability of the respective ion species to become part of the charge on the surface of the droplet and subsequently enter the gas phase. In the regime where $C_A \ll C_E$, Eq. (1.11a) reduces to Eq. (1.11c).

The experimental results [36] shown in Figure 1.7 give an example of a two-component system where the protonated morphine (MorH$^+$) is the analyte A, used at different concentrations, and the impurity ions NH_4^+ and Na^+, present at constant concentrations, are the electrolyte, E. The observed linear region of MorH$^+$ in the log–log plot used has a slope of unity at low concentrations ($10^{-8} - 10^{-5}$ M), which means that the MorH$^+$ ion is proportional to the morphine concentration. This region is suitable for quantitative determinations of analytes. At about 10^{-5} M the increase in the MorH$^+$ intensity slows down because the MorH$^+$ concentration used comes close to that of the impurity electrolytes. Above that region, where MorH$^+$ becomes the major electrolyte, the peak intensity of MorH$^+$ can increase only in proportion to the square root (or even a lower power) of the electrolyte concentration.

Experimental examination of a three-component system of two analytes (A and B) and the impurity (E) in order to determine the relative sensitivities k_A and k_B leads to an unexpected result (see Figure 1.8). In this experiment the concentrations of the two analytes tetrabutylammonium and cocaine (upper figure) or tetrabutylammonium and codeine (lower figure) are increased together such that $C_A = C_B$. The concentration (C_E) of the impurity is constant. It is easily shown using Eq. (1.11b) that when $C_A = C_B$ and C_E is much larger than C_A and C_B, the relationship $I_A/I_B = k_A/k_B$ holds. In the log–log plot used, the difference, log I_A − log I_B should correspond to the difference log k_A − log k_B = log(k_A/k_B) and should be constant and in general not equal to zero. However, this is not the case (see Figure 1.8). The difference is constant only at high $C_A = C_B$ concentrations and becomes zero at low concentrations.

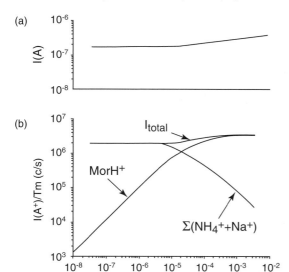

Figure 1.7 (a) Total electrospray current (ampere) with increasing concentration of analyte morphine hydrochloride. Because of the presence of impurity ions (Na$^+$ and NH$_4^+$) at 10^{-5} M, I_{total} remains constant up to the point where the analyte reaches concentrations above 10^{-5} M. (b) Analyte MorH$^+$ ion intensity (corrected for mass-dependent ion transmission, Tm, of quadrupole mass spectrometer used) is proportional to concentration of morphine hydrochloride up to the point where the morphine hydrochloride concentration approaches the concentration of impurity ions. Above that concentration, analyte ion increases much more slowly. (Reprinted from Ref. [36] with permission from the American Chemical Society.)

The tendency of k_A/k_E to approach unity at low C_A and C_B indicates [36] that there is a *depletion* of the ion that has the higher sensitivity k. This is the tetrabutylammonium ion (A) in the present example. At $C_A = C_B \ll C_E \approx 10^{-5}$ M, the current I, the total charge Q of the droplets, and the number of charged droplets are maintained by the presence of the electrolyte E, whose concentration is much higher. Under these conditions, species like A$^+$ and B$^+$ with large coefficients k_A and k_B find plenty of droplet surface to go to, and the ions evaporate rapidly even when present at very low concentrations. This results in a depletion of their concentration. The ion A of higher sensitivity is depleted more than B, and this leads to an apparent finding $k_A = k_B$.

Experimental determination of the coefficient ratios k_A/k_B were performed [36] by working at high concentrations C_A/C_B for a number of analytes in methanol. Under these conditions Eq. (1.11c) holds. It was found that the singly charged inorganic ions, Na$^+$, K$^+$, Rb$^+$, Cs$^+$, and NH$_4^+$ had low sensitivity coefficients, while analyte ions which were expected to be enriched on the droplet surface, that is, which were surface active, had high coefficients that increased with the surface activity of the ions. Thus, assuming that $k_{Cs} = 1$, the relative values k_A for the ions were: Cs$^+ \approx 1$; Et$_4$N$^+ = 3$; Pr$_4$N$^+ = 5$; Bu$_4$N$^+ = 9$; Pen$_4$N$^+ = 16$; HepNH$_3^+ = 8$ (Et is ethyl,

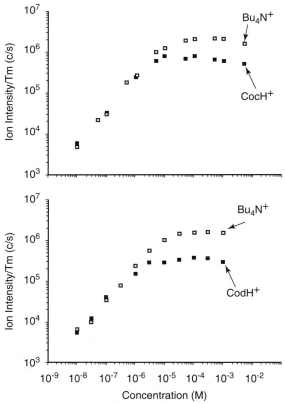

Figure 1.8 Ion intensities (corrected for mass-dependent ion transmission, Tm, of quadrupole mass spectrometer used) for pairs of analytes at equal concentration in solution. The different ESI sensitivities of the analytes are observable only at high analyte concentrations (above 10^{-5} M). (Reprinted from Ref. [36] with permission from the American Chemical Society.)

Pr is *n*-propyl, Bu is *n*-butyl, Pen is *n*-pentyl, Hep is *n*-heptyl) see Table 1.1 in Tang, Kebarle [36]. The tetraalkyl ammonium salts and alkylammonium salts are known surfactants. It is notable that they can be used also as mass calibrants under nonaqueous, anaerobic conditions [37].

Assuming that IEM holds, ions from the droplet surface will leave the droplets and become gas-phase ions. In this case, the gas-phase ion sensitivity coefficient, k_A for ions A^+ will depend on the relative surface population of the droplet surface, that is, on the surface activity of ions A^+ given by a surface activity equilibrium constant K_{SA} and on the rate constant for ion evaporation. The rate constant for ion evaporation is also expected to increase with the surface activity of the ion, because surface active ions have low solvation energies (see Section 1.2.8). A third effect can also be expected. The droplets that lead to ion evaporation will, in general, be first, second or third generation progeny droplets, see Figure 1.5. Because the progeny droplets have higher surface-to-volume ratios relative to the parent droplets, a large enrichment of the surface active ions is expected for the progeny droplets.

More recent work by Enke [38a], starting from somewhat different premises, led to further advances in predicting and understanding relative ion intensities. Instead of working with the ion currents I, I_A, a conversion to mole charge/liter was used. Thus the role of the total ion current I was replaced with the molar concentration of the ionic charges $[Q]$. This can be done on the basis of Eq. (1.12);

$$[Q] = I/F\Gamma \qquad (1.12)$$

where $[Q]$ is the mol L^{-1} electron charges; I the total droplet current (TDC see Figure 1.1) in ampere (coulomb s^{-1}); F is Faraday's constant (96 485 coulomb) and Γ is the flow rate in L s^{-1} through the spray capillary. The same type of relationship is also used in the conversion of the analyte currents I_A, I_B and so on into molar concentrations of charges on the droplets due to the given ion species. Thus, $[A^+]_S$ is the molar concentration of charges on the surface of droplets due to A^+ species. The analyte A^+ was assumed to distribute itself between the interiors of the droplets with a concentration $[A^+]_I$ and as charge at the surface of the droplets $[A^+]_S$. It should be noted that while $[A^+]_I$ in mol L^{-1} is straightforward, $[A^+]_S$ as mol L^{-1} at the surface is not because expressing the surface as volume is unconventional. However, one could imagine that the ions (charges) at the surface are still interacting with a thin layer of solution below the surface, and this corresponds to a volume.

An equilibrium between $[A^+]_S$ and $[A^+]_I$ was assumed. The other electrolytes (E) were treated in the same way. Introduction of equations of charge balance and mass balance for each electrolyte led to an equation which predicts values for $[A^+]_S$, $[E^+]_S$ on the basis of the parameters $[Q]$, which is known (see Eq. (1.12)), the constants K_A, K_E and the concentrations C_A, C_E. The assumption was made that $[A^+]_S$, $[E^+]_S$ will be converted to gas-phase ions and are therefore proportional with the same proportionality constant, pf (see Eq. (1.11)) to the ion currents I_A, I_E. The equation of $[A^+]_S$, $[E^+]_S$ is of the same form as Eq. (1.11) in the high concentration range, but not in the low concentration range. By taking into account, via mass balance, the depletion of the concentration C_A, C_E of the analytes with high coeffients $k_A = K_A$, $k_E = K_E$, the equation of Enke provides an excellent fit of the ion abundance curves, such as shown in Figure 1.8, over the full concentration range, preserving a constant k_A/k_E ratio. Further development by Enke and coworkers has led to a most successful formalism. For other work by Enke and coworkers, dealing with the ESI mechanism and consequences for analytical work with ESI, see Ref. [38].

The preceding discussion deals with analytes that are charged, that is, ions in the solution used. When the analyte is not an ion in solution, charging of the analyte by ions that are present at the surface of the droplet can occur. It should be noted that at the surface the regime is very different from that in the bulk of the solvent. Suppose that the solution sprayed contains ammonium acetate (NH$_4$Ac) as additive and an organic analyte that has an unprotonated basic functional group because the basicity of that group is lower in solution than the basicity of NH$_3$. The ions at the surface of the droplets will be NH$_4^+$. Droplets that through evaporation and fission have reached the size where ion evaporation becomes possible could emit not only NH$_4^+$ but also

Table 1.3 Some gas-phase basicities of bases B for reaction: $BH^+ = B + H^+$.

Base	GP(B)^a (kcal/mol)	Base	GP(B)^a (kcal/mol)
H_2O	157.7	NH_3	195.7
$(H_2O)_2$	181.2	CH_3NH_2	206.6
CH_3OH	173.2	$C_2H_5NH_2$	210.0
$(CH_3OH)_2$	196.3	$(CH_3)_2NH$	214.3
C_2H_5OH	178.0	$n\text{-}(C_3H_7)NH_2$	211.5
$(CH_3)_2O$	179.0	N-Methyl acetamide	205.0
$(C_2H_5)_2O$	182.7	Pyridine	214.8
CH_3CN	191.0		

^{a}GP(B) = Gas-Phase Basicity; GP(B)^a = ΔG^o_{298} All values from NIST Database, http://webbook.nist.gov (Also used are Proton Affinities. They correspond to the ΔH^o value for the gas-phase reaction $BH^+ = B + H^+$).

protonated analytes because the basic group of the analyte might have a higher *gas-phase* basicity than that of NH_3. When the analyte is *at the surface*, the regime is gas-phase like and gas-phase basicites will count. There are large differences between *solution* and *gas-phase* basicities. For values of some gas-phase basicities, see Table 1.3. In the gas phase, in the absence of charge stabilizing solvent, basicities can increase with the size of the compound through dispersal of the charge in the analyte. This is the case for organic analytes where stabilization of the charge by charge dispersal can occur.

The discussion above also illustrates that under certain conditions, ESI may be 'blind' for analytes with low gas-phase basicity. This will be particularly the case when solutes other than the analyte are present at relatively high concentrations and have a higher gas-phase basicity than the analyte. In this case suitable charged groups will have to be introduced on the analyte (see for example Ref. [39]).

1.2.10
Large Analyte Ions such as Dendrimers and Proteins are Most Probably Produced by the Charged Residue Model (CRM)

Although the first experiments of Dole used polystyrene, subsequent experiments by Fenn shifted attention to proteins and protein complexes that routinely can be produced in the gas phase by ESI. As a result, mechanistic studies have focused on these systems although other large synthetic supramolecular systems and polymers are expected to be transferred to the gas phase following the same mechanistic model. In the absence of sufficient data on synthetic macromolecular systems, subsequent sections will discuss mechanistic insights obtained using native proteins.

Native proteins are expected to remain folded when sprayed from neutral aqueous solutions. Under these conditions the folded (nondenatured) proteins lead to mass spectra consisting of a compact series of peaks that correspond to the molecular mass of the protein charged by a narrow range of H^+ ions when the positive ion mode is used. Thus, a small protein-like lysozyme (molecular mass around 15 000 daltons) is observed to lead to three peaks due to three different charge states with $Z = 8$, 9 and 10. Obviously it is of special interest to understand why gas-phase proteins are multiply charged. Is the charge observed related to the protein charge in solution or are other factors involved which are due to the ESI mechanism?

An early study by R. D. Smith and coworkers [40] provided good evidence that proteins are produced via CRM. If CRM holds, one would expect that, when small charged droplets evaporate, there could be one protein but also more than one protein in such droplets, particularly so when high concentrations of the protein are used. Therefore, mass spectra should show the monomers, as originally present in solution, as well as multimers which are due to more than one protein being present in some of the final droplets. The authors observed a preponderance of multiply charged monomers and a rapidly decreasing series of low-intensity dimers, trimers, and higher multimers [40]. All of these results are consistent with CRM and a droplet evolution following a scheme of the type shown in Figure 1.5 [40].

In later work, Smith and coworkers [41a] found an interesting empirical correlation between the molecular mass, M, and the average charge, Z_{av}, see Eq. (1.13), of starburst dendrimers. Starburst dendrimers are multibranched alkylamine polymers that have relatively rigid structures and are close to spherical, that is, with shapes resembling those of globular proteins.

$$Z_{av} = aM^b \tag{1.13}$$

Z_{av} is the observed average charge and M the molecular mass of the dendrimer, while a and b are constants. The value $b = 0.53$ led to the best fit. An identical relationship was observed by Standing and coworkers [41b] for a large number of nondenatured proteins; a value between 0.52 and 0.55 was reported for b.

Independently Fernandez de la Mora [42] was able to show that the empirical relationship (Eq. (1.13)) holds and that this relationship can be derived on the basis of the charged residue mechanism. The plot shown in Figure 1.9 is based on data from the literature used by Fernandez de la Mora, but also includes the data of Standing et al. [41b]

The derivation of Fernandez de la Mora [42] was based on the following arguments. There was theoretical evidence that the evaporating charged droplets (which in the present context are assumed to contain one globular protein molecule) stay close to the Rayleigh limit. This is also supported by more recent experimental results [28] which involve charged evaporating water droplets of 5–35 μm diameter, as well as theoretical studies [43]. These show that the charge is approximately 95% of the Rayleigh limit when the droplets experience a Coulomb fission and approximately 75% of the Rayleigh limit immediately after the Coulomb fission. Thus, the droplets stay at all times within the limits of 95–75% of the Rayleigh limit, and both of these values are close to the Rayleigh limit. Fernandez de la Mora [42] reasoned that when the charged water droplet, containing one protein molecule, evaporates completely, the charges on the droplet will be transferred to the protein. He assumed also that the protein will be neutral when all the water is gone so that the charges on the surface of the droplet become the charge of the protein observed in the ESI mass

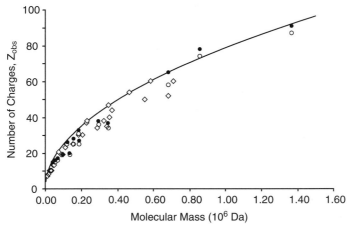

Figure 1.9 Reproduction of a plot used by Fernandez de la Mora [42] and extended to include also data by Standing et al. [41b]. Z_{obs} is the number of charges observed on proteins produced by ESI-MS under nondenaturing conditions, ● highest charge, ○ lowest charge in mass spectrum (Fernandez de la Mora [42]). ◇ average Z_{obs} (Standing et al. [41b]). Solid curve corresponds to charge Z predicted by Eq. (1.14).

spectrum of the protein. Fernandez de la Mora further assumed that the nondenatured proteins have the same density φ as water. Evidence in support of that assumption, based on mobility measurements by Jarrold and Clemmer [44a,c] and Hodgins [44b], is given in Section 1.2.2 of Fernandez de la Mora [42], so that the radius R of any *spherical* protein can be evaluated based on the molecular mass, without reference to the structure of the protein. The radius R of the protein is evaluated with Eq. (1.14);

$$(4/3\pi R^3 \varphi) N_A = M \tag{1.14}$$

where φ is the density of the protein equal to that of water, N_A is Avogadro's number, R the radius of the protein, and M the molecular mass of the protein.

The number of charges Z on the protein is taken to be the same as the number of charges on a water droplet at the Rayleigh limit, with a radius equal to that of the protein. The number of charges Z can be obtained by expressing the charge, $Q = Z \times e$, substituting it in the Rayleigh equation (Eq. (1.10)), and using the relationship between the molecular mass M and the radius of the droplet (Eq. (1.14)). The result is given in Eq. (1.15a);

$$Z = 4(\pi \gamma \varepsilon_0 / e^2 N_A \varphi)^{1/2} \times M^{1/2} \tag{1.15a}$$

$$Z = 0.078 \times M^{1/2} \tag{1.15b}$$

where Z is the number of charges of the protein, γ the surface tension of water, ε_o the electrical permittivity, e the electron charge, N_A is Avogadro number, φ the density of water, and M the molecular mass of the protein. The constant 0.078

in Eq. (1.15b) gives the number of charges on a protein of molecular mass M in megadaltons.

The solid curve in Figure 1.9 gives the predicted charge based on Eq. (1.15a). Good agreement with the experimental results is observed. Notable also is the predicted exponent of M which is 0.5, while the exponent deduced from the experimental data [41a,b] is 0.53.

The agreement of Eq. (1.15) with the observed charges Z can be considered as very strong evidence that globular proteins and protein complexes are produced by the charged residue mechanism. A recent compilation of data by Heck and coworkers [45] has shown that the square root dependence of the charge Z on M (see Eq. (1.15a)) also holds for protein complexes.

The experimental points in Figure 1.9 show considerable scatter and deviations from the theoretical curve. This is most likely because of the measurements being made in different laboratories or deviation from the assumed spherical form. Recent work by Katashov and Mohimen [46] (see Figure 1.10, in which all experimental points were obtained by the authors) and by Nesatyy and Suter [47] provide a very good fit. Kaltashov and Mohimen [46] have shown that in certain rare cases where the shape of the protein deviates strongly from spherical, the charge is determined not by the molecular mass but by the surface area of the protein. However, this finding [46] was based on a single protein (see point with considerable deviation in Figure 1.10), and the authors have not followed up with additional experiments.

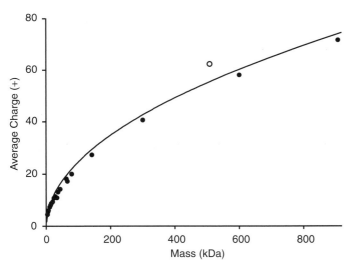

Figure 1.10 Plot of average charge of proteins observed by Kaltashov and Mohimen [46] versus molecular mass of protein. The solid line curve gives average charge predicted by the Fernandez de la Mora equation, Eq. (1.14). A very good fit is observed except for one experimental point, for ferritin with a mass ≈ 510 kDa which has a significantly higher charge Z. This protein is approximately spherical but has a cavity that increases its surface. (The Figure 1.10 plot was kindly provided to the authors by Dr. Justin Benesh.)

Fernandez de la Mora [42] did not consider the actual chemical reactions by which the charging of the protein occurs. These reactions will depend on what additives were present in the solution. Thus, in the presence of 1% of acetic acid in the solution, the charges at the surface of the droplets will be H_3O^+ ions. Charging of the protein will occur by proton transfer from H_3O^+ to functional groups on the surface of the protein that have a higher gas-phase basicity than H_2O. The *gas*-phase basicities are relevant because the solvent will essentially have disappeared. There are plenty of functional groups on the protein that have gas-phase basicities that are higher than that of H_2O. These could be basic residues or amide groups of the peptide backbone at the surface of the protein. Gas-phase basicities of several representative compounds are given in Table 1.3.

In summary, the Charged Residue Mechanism has allowed quantitative predictions of the protein charge state in the gas phase and is well supported for large proteins of widely varying mass. It is likely that it will be important also for the analysis of larger supramolecular and polymeric systems.

1.2.11
Nanospray and Insights into Fundamentals of Electro and Nanospray

Nanospray was developed by Wilm and Mann [48, 49]. Their primary interest was an electrospray device that requires much smaller quantities of analyte. Such a device would be particularly important in situations where only very small amounts of the analyte are available. Most of the analyte is wasted using ESI. The large diameter of the spray tip produces large droplets whose evolution to small droplets requires the presence of a large distance between the spray tip and the sampling orifice or capillary (see Figures 1.1 and 1.2a). As a result, only a small fraction of the generated gas-phase ions enter the sampling orifice/capillary.

With nanospray, the spray tip has a much smaller diameter. Also, for the by far most often used nonviscous solutions, the flow is not a forced flow due to a driven syringe as used in ESI (see Figure 1.2a). Instead, the entrance end of the spray capillary is left open and a 'self-flow' results, which is due to the pull of the applied electric field on the solution at the capillary tip (see Section 1.2.2). This self-flow is controlled by the diameter of the tip of the spray capillary.

In their first effort [48], using what was essentially an electrospray source, Wilm and Mann developed an equation for the radius of the zone at the tip of the Taylor cone from which the charged droplets are ejected. This radius is related to the resulting droplets' radii, see Figure 1.3a. It was found that the radius depends on the 2/3 power of the flow rate. To minimize the radius of the zone, a modified electrospray ion source with a smaller orifice was developed which led to a 'microspray' version of ESI. Further development [49] using capillary orifices as small as 1–2 μm diameter, led to nanospray. Such small orifices could be obtained by pulling small diameter borosilicate capillaries with a microcapillary puller. About 1 μL of solvent is loaded directly into the wide entrance end of the capillary. The droplets produced had a volume that was close to 1000 times smaller than the volume of droplets obtained with conventional ESI. Such small droplets will evaporate very rapidly, so that the capillary tip can be placed very close to the sampling orifice that leads to the mass spectrometer, thereby minimizing sample loss and allowing efficient use of a large fraction of the solution subjected to MS analysis. Therefore, even though the amount of analyte sample used is 10–100 times smaller than that used with ESI, the observed mass spectrum peak intensities are equal if not larger than those in conventional ESI. Nanoelectrospray has proven to be of enormous importance for the analysis of biochemical and biopharmaceutical samples. So far, the use of nanoelectrospray in physical organic or organometallic chemistry has not been reported and remains to be explored (see Section 1.2.12.2).

Nanoelectrospray is also important to research on the fundamentals of electrospray and nanospray. Karas and coworkers [50–52] have been major contributors to this research.

Experimentally it was found that the mass spectra of analytes showed much less dependence on background electrolytes (such as sodium) when nanospray was used for the electrostatic dispersion. Gas-phase ions are produced from charged droplets only when the droplets are very small. This holds both for IEM and CRM. Therefore, if one starts with relatively small initial droplets as generated with nanospray, much less solvent evaporation will be required to reach the final droplet size required for the generation of gas-phase ions. Therefore, in the presence of impurities such as sodium salts the concentration increase of the salt will be much smaller with nanospray.

In an extension and expansion of the above work [50], Schmidt *et al.* [51] studied the effect of different solution flow rates on the analyte signal. A series of analytes was chosen that have decreasing surface activities. The authors found that the ion abundance of the analyte with low surface activity was suppressed at higher 'ESI-like' flow rates, while at the lowest flow rates this suppression disappeared. At high flow rates, the charged droplets emitted from the spray tip are much larger and require extended evaporation and successive fission events before the final droplets are reached. Droplet evaporation as well as each fission increase the surface-to-volume ratio of the droplets. As a result, surface-active analytes will preferentially enter the progeny droplets, leading to a high ion abundance for the surface-active analytes. Considering nanospray and the corresponding low flow rate, very small initial droplets will be obtained and the evolution to the final droplet will be very short. In the extreme, there would be no such evolution and this would lead to minimal discrimination against analytes that are not surface active.

Another well-documented work by Chernushevich, Bahr, and Karas [52] deals with a disadvantage of nanospray relative to conventional electrospray. Using nanospray some analytes were found to appear with delays of tens of minutes; a few analytes were not detected at all. No such suppression was found with ESI. The effect was found to be related to cation exchange on glass surfaces. Glass surfaces are known to be negatively charged and are thus expected to retain positive ions. The surface ion-exchange problem in nanospray could simply be avoided by using pure silica capillaries.

1.2.12
Consequences of the Increase in Concentration Caused by Extensive Evaporation of Solvent in ESI Process. Promotion of Bimolecular Reactions Involving Analyte Ions

As discussed in Section 1.2.7, the formation of the final charged droplets that lead to gas-phase ions is associated with a very large loss of solvent by evaporation. This leads to a large increase in the concentration of the solutes present in the electrosprayed solution. Such a large increase in concentration will promote bimolecular reactions, and these will have an effect on the observed mass spectrum. We will consider two examples. In the first example bimolecular reactions are shown to take place on native proteins, indicating various levels of protein *surface* gas-phase chemistry. In the second example, a comparison of biochemical and physical organic applications of ESMS to the determination of equilibrium constants indicates the limitations on such measurements imposed by the electrospray mechanism.

1.2.12.1 Positive-Negative Ion-Pairing Reactions Involving Impurities such as Na^+

Sodium ions are a common impurity in solvents used for ESI. Analyte samples (especially proteins) are often contaminated by sodium ions in the production process, while sample solution storage in glass vessels also contributes to the sodium contamination. The extensive evaporation of solvent in the ESI process promotes solution ion-pairing reactions. When the analyte is a protein, ion-pairing reactions will involve Na^+ and the ionized acid residues of the protein, see Eq. (1.16a). Similarly, impurity anions such as Cl^- will ion pair with ionized basic residues of the protein (see Eq. (1.16b)).

$$NH_3^+\text{-Protein-}COO^- + Na^+ \rightarrow NH_3^+\text{-Protein-}COO^-Na^+ \quad (1.16a)$$

$$NH_3^+\text{-Protein-}COO^-Na^+ + Cl^- \rightarrow Cl^-NH_3^+\text{-Protein-}COO^-Na^+ \quad (1.16b)$$

The occurrence of these ion pairing reactions was demonstrated by Verkerk and Kebarle [53]. In the collisional activation stage of the ESI-MS dissociation occurs following Eq. (1.17):

$$Cl^-NH_3^+\text{-Protein-}COO^-Na^+ \rightarrow H_2N\text{-Protein-}COO^-Na^+ + HCl \quad (1.17)$$

Thus the Na^+ adduct remains while the Cl^- pulls off a proton from the basic residue of the protein. In general there are at least several acidic residues at the surface of proteins and the addition of sodium to most of these leads to mass spectra that are difficult to interpret. The authors [53] also demonstrated that one can prevent the undesirable sodium addition by adding a millimolar concentration of ammonium acetate to the solution. Because the concentration of ammonium acetate is much higher than that of the impurity ions, ion pairing of NH_4^+ with the acidic residue $CH_3CH_2CO_2^-$ and ion pairing of $CH_3CO_2^-$ with the basic residue occur as shown below:

$$NH_3^+\text{-Protein-}COO^- + NH_4^+ \rightarrow NH_3^+\text{Protein-}COO^-NH_4^+ \quad (1.18a)$$

$$NH_3^+\text{-Protein-}COO^-NH_4^+ + CH_3CO_2^- \rightarrow CH_3CO_2^-NH_3^+\text{-Protein-}COO^-NH_4^+ \quad (1.18b)$$

This result may seem an undesirable change. However in the clean up stages of the ESI-MS these ion pairs are unstable; intramolecular proton transfer occurs followed by dissociation resulting in the products (Eq. (1.19)):

$$CH_3CO_2^- NH_3^+ \text{-Protein-}COO^- NH_4^+ \rightarrow NH_2\text{-Protein-}COOH + NH_3 + CH_3CO_2H \quad (1.19)$$

Because the ammonium acetate concentration is much higher than that of impurities such as sodium ions, reactions, such as Eq. (1.16) are prevented. A clean mass spectrum results and the protein appears at the right mass. In ESI-MS practice, ammonium acetate is routinely used as an additive and the assumption is made that its purpose is to act as a buffer. The above findings [53] demonstrate that the popularity of ammonium acetate may not due to its action as a buffer but on its ability to lead to clean mass spectra.

1.2.12.2 Determination of Equilibrium Constants in Solution via ESI-MS

Electrospray mass spectrometry may appear a very attractive technique for the determination of the equilibrium constant of a reaction occurring in solution because of the soft nature of ESI, the simultaneous determination of the relative ion abundance of the reactants participating in the equilibrium, the high sensitivity, and the low sample consumption.

The determinations of equilibrium constants using ESI-MS can be divided into two categories: (a) equilibria of small positive ions and ligands (such as crown ethers), so that the gas-phase ion creation involves IEM, and (b) equilibria of macro ions (such as proteins) and ligands (which may or may not be large organic molecules) so that gas-phase ion creation involves CRM.

However, uncritical application of ESI-MS can lead to erroneous results due to the complexity of the electrospray process. First of all, ESI-MS spectra are sensitive to instrumental parameters, and these therefore must be chosen in such a way as to minimize such effects. Secondly, during the charged droplets evaporation, the concentration and pH of the sampled solution changes. If the equilibrium rate constants are high enough, the observed ligand-substrate ratio may reflect the concentration changes experienced during the evaporation/ionization step and thus lead to erroneous determinations of the equilibrium constant in solution.

Consider the general reaction, Eq. (1.20a), where P is a ligand and S is a substrate, in a reaction that has reached equilibrium in the solution used. The equilibrium constant K_{AS} is given by Eq. (1.20b) where [P], [S] and [PS] are the concentrations at equilibrium.

$$P + S = PS \quad (1.20a)$$

$$K_{AS} = [PS]/[P] \times [S] = I_{PS}/I_P \times I_S \quad (1.20b)$$

Sampling the solution with ESI, the concentrations can be replaced with the ESI-MS observed peak intensities I_P, I_S and I_{PS} assuming that the individual sensitivity coefficients are equal to one. However this cannot be expected, and different strategies have been proposed to deal with this, see reference [54].

Thus, one can repeat the experiment at several, gradually increasing, concentrations of S and examine whether the association constant, evaluated with Eq. (1.20b), remains constant. This procedure is called the 'Titration Method' and is generally used in ESI-MS determinations of the equilibrium constant. When the differences in molecular mass of S, P or PS are large, erroneous results may be obtained due to m/z-dependent transmission differences of the MS analyzer. While less of a problem for small metal (S)-ligand-(P) interactions, for proteins (P) of very large mass and substrate (S) of much smaller mass it is advantageous to use only the ratio I_{PS}/I_P. Zenobi and coworkers [55] have provided an equation for the determination of K_{AS} in which only the ratio I_{PS}/I_P is used.

From the standpoint of the mechanism of ESI, an agreement of the K_{AS} values determined via ESI-MS with values obtained with conventional in-solution methods may appear surprising. One could expect that the very large increase in the concentration of the solutes in the charged droplets due to evaporation of the solvent in the droplet evolution will lead to an apparent K_{AS} that is much too high. However, this equilibrium shift need not occur if rates of the forward and reverse reactions leading to the equilibrium are slow compared to the time of droplet evaporation.

Peschke et al. [33] evaluated an approximate droplet history scheme for water droplets produced by nanoelectrospray. The early part of this scheme is shown in Figure 1.5, Section 1.2.7. Because the initial droplets produced by nanospray have a small diameter (< 1 μm), their evaporation is very fast, so that they reach the Rayleigh instability condition in just a few μs. It could also be established that the first generation progeny droplets will be the major source of analyte ions [33].

Assuming even the fastest possible reaction rates, that is, the diffusion limit rates for the forward reaction $P + S = PS$, it could be shown that the total droplet evaporation time can be too short for the equilibrium (Eq. 1.20b) to shift in response to the increasing concentration due to solvent evaporation. The rate constant at the diffusion limit decreases with an increase in the substrate size. Substrates of medium size such as erythrohydroxy aspartate, adenosine di- and tri-phosphate, with diffusion-limited rate constants k from 10^6 to $10^7 \, M^{-1} s^{-1}$, are too slow to cause an equilibrium shift that will lead to a significant error in the equilibrium constant (K_{AS}) determination via ESI-MS. Thus, an equilibrium shift for substrates that are not too small is not expected using nanospray. This is in contrast to electrospray, where droplet evaporation takes place in the millisecond range [36]. It is noteworthy that, as far as we are aware, electrospray is the method most used for equilibrium measurements of small metal-ligand systems, while our analysis indicates that the use of nanospray may be less error-prone. For a thorough review on small metal-ligand systems, see Ref. [54]. More recent investigations on small systems are reported in work by Zenobi and coworkers [56].

Because large complexes are most probably transferred to the gas phase via CRM, another question must also be examined. Considering a protein-substrate complex, and assuming close to equal concentrations of protein (P) and substrate (S), an evaluation [33] shows that, for an initial concentration of 10 μM, in most cases there will be one protein and one substrate molecule in the average first generation progeny droplet that has evaporated down to the size of the protein. In that case, protein and substrate will form a nonspecific complex because the substrate makes a random encounter with the protein and has 'no time' to find the site of specific strong bonding. Since the mass of the nonspecific complex is the same as that of the specific complex, the observed peak intensity (I_{SP}) will lead to an apparent K_{AS} that is too high. If however the nonspecific complex is weakly bound, the nonspecific contribution in the measured K_{AS} could be minimized because of dissociation in the clean-up stages of the electrospray process. This *a priori* assumption about weakly bound nonspecific complexes being decomposed in the clean-up stage may not predict the correct outcome in all cases. It neglects to consider the strong ion-neutral bonding in the gas phase and the fact that the protein will be multiply charged. Strong hydrogen bonds can form between charged (protonated) sites of the protein and functional groups of the substrate such as −OH and −NH_2 groups. A recent study by Wang, Kitova, and Klassen [57] exactly described such effects for protein (P)-carbohydrate (S) complexes. In this study, accidental PS and SPS complexes were observed in the gas phase and the nonspecific bond was found to be stronger than the specific bond. This suggested that the nonspecific S-P bond was to a protonated site of the charged protein [57]. These results are not surprising; proton-bridged dimers have long been known to form very strong bonds in the gas phase [58]. Further important examples of the use of ESI-MS in the study of large noncovalent complexes are provided in Ref. [59].

The above examples provide a good illustration of the fact that an understanding of the ESI-MS mechanisms is necessary for the correct interpretation of results obtained with ESI-MS.

References

1 (a) Bakhtiar, R. and Hop, C.E.C.A. (1999) Unravelling seemingly complex chemistry of reactions using electrospray ionization mass spectrometry. *J. Phys. Org. Chem.*, **12**, 511–527; (b) Traeger, J.C. (2000) Electrospray mass spectrometry of organometallic compounds. *Int. J. Mass Spectrom.*, **200**, 387–401; (c) Chen, P. (2003) Electrospray ionization tandem mass spectrometry in high throughput screening of homogeneous catalysis. *Angew. Chem. Int. Ed.*, **42**, 2832–2847; (d) Griep-Raming, J., Meyer, S., Bruhn, T., and Metzger, J.O. (2002) investigation of reactive intermediates of chemical reactions in solution by electrospray ionization mass spectrometry: radical chain reactions. *Angew. Chem. Int. Ed.*, **41**, 2738–2742; (e) Santos, L.S., Knaack, L., and Metzger, J.O. (2005) Investigations of Chemical Reactions in Solution using API MS. *Inter. J. Mass Spectrom.*, **246**, 84–1004.

2 Fenn, J.B. Electrospray Wings for Molecular Elephants. Nobel Lecture available on web: http://nobelprize.org/nobel_prizes/chemistry/laureates/2002/fenn-lecture.pdf.

3 (a) Chapman, S. (1937) Carrier mobility spectra of spray electrified liquids. *Phys. Rev.*, **52**, 184–190; (b) Chapman, S. (1938) Carrier mobility spectra of liquids electrified by bubbling. *Phys. Rev.*, **54**, 520–527.

4 Dole, M. (1989) *My Life in the Golden Age of America*, Vantage Press, New York, p. 69; (b) Dole, M., Mack, L.L., Hines, R.L.,

Mobley, R.C., Ferguson, L.D., and Alice, M.B. (1968) Molecular beams of macroions. *J. Chem. Phys.*, **49**, 2240–2249.

5 (a) Yamashita, M. and Fenn, J.B. (1984) Electrospray ion source. Another variation of the free-jet theme. *J. Phys. Chem.*, **88**, 4451–4459; (b) Yamashita, M. and Fenn, J.B. (1984) Negative ion production with the electrospray ion source. *J. Phys. Chem.*, **88**, 4672–4675.

6 Whitehouse, C.M., Dreyer, R.N., Yamashita, M., and Fenn, J.B. (1985) Electrospray interface for liquid chromatographs and mass spectrometers. *Anal. Chem.*, **57**, 675–679.

7 Smith, R.D., Loo, J.L., Ogorzalek-Loo, R.R., Busman, M., and Udseth, H.R. (1991) Principles and practice of electrospray ionization mass spectrometry for large peptides and proteins. *Mass Spectrom. Reviews*, **10**, 359–451.

8 Electrospray: theory and applications. Special Issue *J. Aerosol Sci.*, **25**, (1994) 1005–1252.

9 (a) Loeb, L., Kip, A.F., Hudson, G.G., and Bennet, W.H. (1941) Pulses in negative point-to-plane corona. *Phys. Rev.*, **60**, 714–722; (b) Pfeifer, R.J. and Hendricks, C.D. (1968) Parametric studies of electrohydrodynamic spraying. *AIAAJ*, **6**, 496–502.

10 (a) Grace, J.M. and Marijnissen, J.C.M. (1994) A review of liquid atomization by electrical means. *J. Aerosol Sci.*, **25**, 1005–1019; (b) Szabo, P.T. and Kele, Z. (2001) Electrospray mass spectrometry of hydrophobic compounds using dimethyl sulfoxide and dimethylformamide as solvents. *Rapid Comm. Mass Spectrom.*, **15**, 2415–2419; (c) Henderson, M.A. and McIndoe, J.S. (2006) Ionic liquids enable electrospray ionisation mass spectrometry in hexane. *Chem. Commun.*, 2872–2874.

11 Taylor, G.I. (1965) The stability of horizontal fluid interface in a vertical electric field. *J. Fluid. Mech.*, **2**, 1–15.

12 Fernandez de la Mora, J. (2007) The fluid dynamics of Taylor cones. *J. Ann. Rev. Fluid. Mech.*, **39**, 217–243.

13 (a) Cloupeau, M. and Prunet-Foch, B. (1994) Electrohydrodynamic spraying functioning modes: a critical review. *J. Aerosol Sci.*, **25**, 1021–1036; (b) Cloupeau, M. (1994) Recipes for use of EHD spraying in cone-jet mode and notes on corona discharge. *J. Aerosol Sci.*, **25**, 1143–1157; (c) Cloupeau, M. and Prunet-Foch, B. (1989) Recipes for use of EHD spraying of liquids in cone-jet mode. *J. Aerosol Sci.*, **22**, 165–184.

14 Tang, K. and Gomez, A. (1994) On the structure of an electrospray of monodisperse droplets. *Phys. Fluids*, **6**, 2317–2322; (b) Tang, K. and Gomez, A. (1994) Generation by electrospray of monodisperse water droplets for targeted drug delivery by inhalation. *J. Aerosol Sci.*, **25**, 1237–1249; (c) Tang, K. and Gomez, A. (1995) Generation of monodisperse water droplets from electrosprays in a corona-assisted cone-jet mode. *J. Colloid Sci.*, **175**, 326–332; (d) Tang, K. and Gomez, A. (1996) Monodisperse electrosprays of low electric conductivity liquids in the cone-jet mode. *J. Colloid Sci.*, **184**, 500–511.

15 Marginean, I., Parvin, L., Heffernan, L., and Vertes, A. (2004) Flexing the electrified meniscus: the birth of a jet in electrosprays. *Anal. Chem.*, **76**, 4202–4207.

16 Blades, A.T., Ikonomou, M.G., and Kebarle, P. (1991) Mechanism of electrospray mass spectrometry. Electrospray as an electrolysis cell. *Anal. Chem.*, **63**, 2109–2114.

17 (a) Van Berkel, G.J. (2000) insights into analyte electrolysis in an electrospray emitter from chronopotentiometry experiments and mass transport calculations. *J. Am. Soc. Mass Spectrom.*, **11**, 951–960; (b) Van Berkel, G.J., Zhou, F., and Aronson, J.T. (1997) Changes in bulk solution pH caused by the inherent controlled-current electrolytic process of an electrospray ion source. *Intern. J. Mass Spectrom. Ion Process.*, **162**, 55–62; (c) Van Berkel, G.J. and Zhou, F. (1996) Observation of gas-phase molecular dications formed from neutral organics in

solution via the controlled-current electrolytic process inherent to electrospray. *J. Am. Soc. Mass Spectrom.*, **7**, 157–162; (d) Zhou, F., Van Berkel, G.J., and Donovan, B.T. (1994) Electron-transfer reactions of fluorofullerene $C_{60}F_{48}$. *J. Am. Chem. Soc.*, **116**, 5485–5486; (e) Van Berkel, G.J., McLuckey, S.A., and Glish, G.L. (1991) Performing ions in solution via charge-transfer complexation for analysis by electrospray ionization mass spectrometry. *Anal. Chem.*, **63**, 2064–2068.

18 Moini, M., Cao, P., and Bard, A.J. (1999) Hydroquinone as a buffer additive for suppression of bubbles formed b electrochemical oxidation of the CE buffer at the outlet electrode in capillary electrophoresis/electrospray ionization-mass spectrometry. *Anal. Chem.*, **71**, 1658–1661.

19 Smith, D.P.H. (1986) The electrohydrodynamic atomization of liquids. *IEEE Trans. Ind. Appl.*, **22**, 527–535.

20 Ikonomou, M.G., Blades, A.T., and Kebarle, P. (1991) Electrospray mass spectrometry of methanol and water solutions. Suppression of electric discharge with SF_6 gas. *J. Am. Soc. Mass Spectrom.*, **2**, 497–505.

21 Wampler, F.W., Blades, A.T., and Kebarle, P. (1993) Negative ion electrospray mass spectrometry of nucleotides: ionization from water solution with SF_6 discharge suppression. *J. Am. Soc. Mass Spectrom.*, **4**, 289–295.

22 (a) Jaworek, A., Czech, T., Rajch, E., and Lackowksi, M. (2005) Spectroscopic studies of electric discharges in electrospraying. *J. Electrost.*, **63**, 635–641; (b) Jaworek, A. and Krupa, A. (1997) Studies of corona discharge in EHD spraying. *J. Electrost.*, **40/41**, 173–178.

23 Fernandez-de la Mora, J. and Locertales, I.G. (1994) The current emitted by highly conducting taylor cones. *J. Fluid. Mech.*, **260**, 155–184.

24 Cherney, L.T. (1999) Structure of the taylor cone jets: limit of low flow rates. *J. Fluid. Mech.*, **378**, 167–196.

25 Chen, D.R. and Pui, D.Y.H. (1997) Experimental investigations of scaling laws for electrospray: dielectric constant effect. *Aerosol Sci. Technology*, **27**, 367–380.

26 Rayleigh, L. (1882) On the equilibrium of liquid conducting masses charged with electricity. *Phil. Mag. Ser. 5*, **14**, 184–186.

27 Gomez, A. and Tang, K. (1994) Charge and fission of droplets in electrostatic sprays. *Phys. Fluids.*, **6**, 404–414.

28 (a) Smith, J.N., Flagan, R.C., and Beauchamp, J.L. (2002) Droplet evaporation and discharge dynamics in electrospray ionization. *J. Phys. Chem. A.*, **106**, 9957–9967; (b) Grimm, R.L. and Beauchamp, J.L. (2002) Evaporation and discharge dynamics of highly charged droplets of heptane, octane and p-xylene generated by electrospray ionization. *Anal. Chem.*, **74**, 6291–6297.

29 Taflin, D.C., Ward, T.L., and Davis, E.J. (1989) Electrified droplet fission and the Rayleigh limit. *Langmuir*, **5**, 376–384.

30 Richardson, C.B., Pigg, A.L., and Hightower, R.L. (1989) On the stability limit of charged droplets. *Proc. Roy. Soc. A.*, **422**, 319–328.

31 Schweitzer, J.W. and Hanson, D.N. (1971) Stability limit of charged drops. *J. Colloid Interface Sci.*, **35**, 417–423; (b) Duft, D., Achtzehn, T., Muller, R., Huber, B.A., and Leisner, T. (2003) Rayleigh jets from levitated microdroplets. *Nature*, **421**, 128.

32 (a) Loscertales, I.G. and Fernandez de la Mora, J. (1995) Experiments on the kinetics of field evaporation of small ions from droplets. *J. Chem. Phys.*, **103**, 5041–5060; (b) Gamero-Castano, M. and Fernadez de la Mora, J. (2000) Kinetics of small ion evaporation from the charge and mass distribution of multiply charged clusters in electrosprays. *J. Mass Spectrom.*, **35**, 790–803; (c) Gamero-Castano, M. and Fernandez de la Mora, J. (2000) Direct measurement of ion evaporation kinetics from electrified liquid surfaces. *J. Chem. Phys.*, **113**, 815–832.

33 Peschke, M., Verkerk, U.H., and Kebarle, P. (2004) Features of the ESI mechanism

that affect the observation of multiply charged noncovalent complexes and the determination of the association constant by the titration method. *J. Am. Soc. Mass Spectrom.*, **15**, 1424–1434.

34 (a) Iribarne, J.V. and Thomson, B.A. (1976) On the evaporation of small ions from charged droplets. *J. Chem. Phys.*, **64**, 2287–2294; (b) Thomson, B.A. and Iribarne, J.V. (1979) Field induced ion evaporation from liquid surfaces at atmospheric pressure. *J. Phys. Chem.*, **71**, 4451–4463.

35 Znamenskiy, V., Marginean, I., and Vertes, A. (2003) Solvated ion evaporation from charged water nanodroplets. *J. Phys. Chem. A.*, **107**, 7406–7412.

36 Tang, L. and Kebarle, P. (1993) Dependence of the ion intensity in electrospray mass spectrometry on the concentration of the analytes in the electrosprayed solution. *Anal. Chem.*, **65**, 3654–3668.

37 Lubben, A.T., McIndoe, J.S., and Weller, A.S. (2008) Coupling an electrospray ionization mass spectrometer with a glovebox: a straightforward, powerful, and convenient combination for analysis of air-sensitive organometallics. *Organometallics*, **27**, 3303–3306.

38 (a) Enke, C.G. (1997) A predictive model for matrix and analyte effects in the electrospray ionization of singly-charged ionic analytes. *Anal. Chem.*, **69**, 4885–4893; (b) Cech, N.B. and Enke, C.G. (2006) Electrospray ionization mass spectrometry: how and when it works, in *Encyclopedia of Mass Spectrometry*, vol. 8 (eds M.L. Gross and R. Caprioli), Pergamon, New York, pp. 171–180; (c) Enke, C.G. (1997) A predictive model for matrix and analyte effects in electrospray ionization of singly-charged ionic analytes. *Anal. Chem.*, **69**, 4885–4893; (d) Cech, N.B. and Enke, C.G. (2001) The effect of affinity for charged droplet surfaces on the fraction of analyte charged in the electrospray process. *Anal. Chem.*, **73**, 4632–4639; (e) Amad, M.H., Cech, N.B., Jackson, G.S., and Enke, C.G. (2000) Importance of gas-phase proton affinities in determining the electrospray ionization response for analytes and solvents. *J. Mass Spectrom.*, **35**, 784–789.

39 Chisholm, D.M. and McIndoe, J.S. (2008) Charged ligands for catalyst immobilisation and analysis. *Dalton Trans.*, 3933–3945.

40 Winger, B.A., Light-Wahl, K.J., Ogorzalek-Loo, R.R., Udseth, H.R., and Smith, R.D. (1993) Observations and implications of high mass-to-charge ratio ions from electrospray ionization mass spectrometery. *J. Am. Soc. Mass. Spectrom.*, **4**, 536–545.

41 (a) Tolic, R.P., Anderson, G.A., Smith, R.D., Brothers, H.M., Spindler, R., and Tomalia, D.A. (1997) Electrospray ionization fourier transform ion cyclotron resonance mass spectrometric characterization of high molecular mass starburst (TM) dendrimers. *Int. J. Mass Spectrom. Ion Proc.*, **165/166**, 405–418; (b) Chernuschevich, I.V., Ens, W., and Standing, K.G. (1998) *New Methods for the Study of Biomolecular Complexes* (eds W. Ens, K.G. Standing, and I.V. Chernuschevich), Kluwer Academic Publishers, Dordrecht, Boston, London, pp. 101–117.

42 Fernandez de la Mora, J. (2000) Electrospray ionization of large multiply charged species proceeds via Dole's charged residue mechanism. *Anal. Chim. Acta*, **406**, 93–104.

43 Shrimpton, J.S. (2005) Dielectric charged drop break-up at sub-rayleigh limit conditions. *IEEE Trans. Dielec. Elec. Ins.*, **12**, 573–578.

44 (a) Valentine, S.J., Anderson, J.G., Ellington, D.E., and Clemmer, D.E. (1997) Disulfide intact and reduced lysozyme in the gas-phase: conformations and pathways of folding and unfolding. *J. Phys. Chem B*, **101**, 3891–3900; (b) Hudgins, R.R. (1997) High resolution ion mobility measurements for gas-phase conformations. *Int. J. Mass Spectrom.*, **165**,

497–507; (c) Shelimov, K.B., Clemmer, D.E., Hudgins, R.R., and Jarrold, M. (1997) Protein structure *in vacuo*: gas-phase conformations of BPTI and cytochrome C. *J. Am. Chem. Soc.*, **119**, 2240–2248.

45 Heck, A.J.R. and van den Heuvel, R.H.H. (2004) Investigation of intact protein complexes by mass spectrometry. *Mass Spectrom. Rev.*, **23**, 368–389.

46 Kaltashov, I.A. and Mohimen, A. (2005) Estimates of protein areas in solution by electrospray ionization mass spectrometry. *Anal. Chem.*, **77**, 5370–5379.

47 Nesatty, V.J. and Suter, M.J.F. (2004) On the conformation-dependent neutralization theory and charging of individual proteins and their non-covalent complexes in the gas phase. *J. Mass Spectrom.*, **39**, 93–97.

48 Wilm, M. and Mann, M. (1994) Electrospray and Taylor-cone theory, Dole's beam of macromolecules at last? *Int. J. Mass Spectrom. Ion Proc.*, **136**, 167–180.

49 Wilm, M. and Mann, M. (1996) Analytical properties of the nanoelectrospray ion source. *Anal. Chem.*, **68**, 1–8.

50 Juraschek, R., Dulks, T., and Karas, M. (1999) Nanoelectrospray - more than just a minimized-flow electrospray ion source. *J. Am. Soc. Mass Spectrom.*, **10**, 300–308.

51 Schmidt, A., Karas, M., and Dulks, T. (2003) Effect of different solution flow rates on analyte signals in nano-ESI-MS, or: when does ESI turn into nano-ESI. *J. Am. Soc. Mass Spectrom.*, **14**, 492–500.

52 Chernushevich, I.V., Bahr, U., and Karas, M. (2004) Nanospray taxation and how to avoid it. *Rapid Commun. Mass Spectrom.*, **18**, 2479–2485.

53 Verkerk, U.H. and Kebarle, P. (2005) Ion-ion and ion-molecule reactions at the surface of proteins produced by nanospray. Information on the number of acidic residues and control of the number of ionized acidic and basic residues. *J. Am. Soc. Mass Spectrom.*, **16**, 1325–1341.

54 Di Marco, V.B. and Bombi, G.G. (2006) Electrospray mass spectrometry (ESI-MS) in the study of metal-ligand solution equilibria. *Mass Spectrom. Rev.*, **25**, 347–379.

55 Daniel, J.M., Friess, S.D., Rajagopalan, S., Wend, S., and Zenobi, R. (2002) Quantitative determination of noncovalent binding interactions using soft ionization mass spectrometry. *Int. J. Mass Spectrom.*, **216**, 1–27.

56 Wortman, A., Kistler-Momotova, A., Zenobi, R., Heine, M.C., Wilhelm, D., and Pratsinis, S.E. (2007) Shrinking droplets in electrospray ionization and their influence on chemical equilibria. *J. Am. Soc. Mass Spectrom.*, **18**, 385–393.

57 Wang, W., Kitova, E.N., and Klassen, J.S. (2003) Bioactive recognition sites may not be energetically preferred in protein-carbohydrate complexes in the gas-phase. *J Am. Chem. Soc.*, **125**, 13630–13861.

58 Lau, Y.K., Saluja, P.S., and Kebarle, P. (1980) The proton in dimethyl sulfoxide and acetone. Results from gas-phase ion equilibria. *J. Amer. Chem. Soc.*, 7429–7433.

59 (a) Loo, J. (1997) Studying noncovalent protein complexes by electrospray ionization mass spectrometry. *Mass Spectrom. Rev.*, **16**, 1–23; (b) Beck, J.L., Colgrave, M.L., Ralph, S.F., and Sheill, M.M. (2001) Electrospray ionization mass spectrometry of oligonucleotide complexes with drugs, metals and proteins. *Mass Spectrom. Rev.*, **20**, 61–87; (c) Daniel, J.M., McCombie, G., Wend, S., and Zenobi, R. (2003) Mass spectrometric determination of association constants of adenylate kinase with two noncovalent inhibitors. *J. Am. Soc. Mass Spectrom.*, **14**, 442–448; (d) Sun, J., Kitova, E.N., Wang, W., and Klassen, J.S. (2006) Method for distinguishing specific from nonspecific protein-ligand complexes in nanoelectrospray ionization mass spectrometry. *Anal. Chem.*, **78**, 3010–3018.

2
Historical Perspectives in the Study of Ion Chemistry by Mass Spectrometry: From the Gas Phase to Solution
Hao Chen

Mass spectrometry (MS), originally developing from nineteenth-century physics, has a history dating back more than a hundred years. Since then, the technique has advanced enormously. The advent of new methods of ion production, novel mass analyzers, and new tools for data processing has made it possible to analyze almost all chemical entities, ranging from small organic compounds to large biological molecules and whole living cells/tissues. Today, mass spectrometry has become one of the most powerful and popular modern physical-chemical methods to study the details of elemental and molecular processes in nature. Because of its unparalleled capability of providing information on molecular weight, chemical structures, isotopic content, and even molecular dynamics, mass spectrometry is playing an increasingly significant role in the chemical and life sciences. As a traditional analytical tool and a booming technology for biomedicine, it has found extensive applications in nearly every discipline involving chemical analysis such as petroleum prospecting and refining, food processing, drug discovery, metablomics, genomics, proteomics and structural biology, as well as the environment, public safety, and industrial process monitoring. Along with the development of mass spectrometric technology, the chemistry of gaseous ions has been explored by the study of unimolecular and collision-induced dissociation (CID) [1–4] and by the study of ion/molecule reactions [2, 5–17] and ion/ion reactions [18–26] in the gas phase. The study of gas-phase ion chemistry has both fundamental and practical significance because it not only provides diverse information on ion reactivity, thermochemistry, and dynamics, but it also forms one basis for chemical analysis using mass spectrometry. In addition, the ion reactivity examined in the isolated gas-phase environment of mass spectrometers has a strong connection with reaction chemistry in solution, as well as contrasting with it. Many important ionic intermediates which are hard to probe in the condensed phase can be easily accessed in the gas phase. Therefore, mass spectrometry has also become an important and unique tool for the mechanistic investigation of organic reactions (particularly organometallic reactions), since soft electrospray ionization (ESI) technology [27, 28] became available. In this regard, there is a large body of work in the literature, which is reviewed in this book.

Reactive Intermediates: MS Investigations in Solution. Edited by Leonardo S. Santos.
Copyright © 2010 WILEY-VCH Verlag GmbH & Co. KGaA, Weinheim
ISBN: 978-3-527-32351-7

This chapter will provide an overview of the history and the current state of the art of mass spectrometry-based study of ion chemistry, specifically of ion/molecule reactions. As this is a broad topic with an immense literature (over 15 000 papers), our emphasis will be placed on the applications of ESI-MS coupled with gas-phase ion/molecule reactions and tandem mass spectrometry in organometallic chemistry, including probing key intermediates and reaction pathways of organometallic reactions in solution. We will start with a brief history, including recent advances in mass spectrometry, which should serve as general background information for readers. Then we will focus on the study of ion/molecule reactions, including their origins, types, and the experimental methods used in their study as well as their role in bridging the gap between gas-phase ion chemistry and condensed-phase organic chemistry. Also, the study of ion/molecule reactions at high pressure, including atmospheric pressure, will be described.

2.1
A Brief History and Recent Advances in Mass Spectrometry

2.1.1
Early Developments

This summary of some of the significant milestones in the development of MS is based on excellent reviews of the long history of mass spectrometry [29–32]. The first known mass spectrometer was an apparatus built by J.J. Thomson in the early 1900s to study and measure the m/e values of the 'corpuscles' which make up 'positive rays' [33], a new type of radiation initially observed by the German physicist Eugen Goldstein [34]. Following the seminal work of Thomson, mass spectrometry has undergone countless improvements in instrumentation, ionization methods, and applications. In 1919, Francis Aston constructed velocity-focusing instruments capable of high resolution and with sufficient mass measurement accuracy to allow the natural abundances and the mass defects of the isotopes to be measured [35–37]. In 1918, Arthur Dempster developed the first direction-focusing mass spectrometer based on a sector-shaped magnet [38]; he also introduced a number of innovations in measuring accurate isotopic abundances. Dempster also devised the electron impact (EI) ion source, improved later by Walker Bleakney [39] and Alfred Nier [40], which became a widely used standard for ionization of volatile organic compounds. In the 1930s, double focusing instruments (velocity and direction focusing) using a combination of sectors for high resolution, developed and designed by Nier, was used in the separation and collection of uranium-235 [41]. Subsequently, Ernest Lawrence converted the Berkeley cyclotron to a mass spectrometer for uranium enrichment and collected 5 µA of U^+ beams in 1941; just a year later, 1 mA beams were being made in Calutrons at Oak Ridge with instruments that were two storeys high and operated in banks 100 yards long. This work ushered in the nuclear age. In 1945, Hipple and Condon [42] discovered the phenomenon of spontaneous dissociation of metastable ions long after collision-induced dissociation had been evidenced in the data of Thomson and Aston and studied systematically by Smythe. In the fifties and sixties, Manfred von Ardenne introduced negative ion MS using 'elektronenanlagerung' (electron-addition) ionization in a special ion source with high vapor density and a very high density of slow electrons [43]. In 1948, Cameron [44] built a time-of-flight (ToF) instrument based on the analysis principle described by Hammer in 1911[45] (i.e., the determination of ion mass by measuring the travel time of the ion over a certain distance). In 1953, Wolfgang Paul and his coworkers invented the quadrupole ion trap mass analyzer [46]. The technique of the ion cyclotron resonance (ICR) mass analyzer was first published in the middle 1950s by Hipple [47] and later greatly improved with the addition of the Fourier Transform method by Alan Marshall and Melvin Comisarow in the early 1970s [48] for high-resolution mass measurement. The introduction of these ion-trap analyzers allowed the sequential decomposition of a mass-selected ion through multiple stages to be studied in a single device.

It is evident that the early development of mass spectrometry was mainly driven by physicists, who used the technique to resolve questions about the fundamental nature of the atom [31]. However, as early as 1913, Thomson urged '... especially chemists, to try this method of analysis. I feel sure that there are many problems in Chemistry which could be solved with far greater ease by this than any other method. The method is surprisingly sensitive – more so than that of spectrum analysis, requires infinitesimal amounts of material, and does not require this to be specially purified: the technique is not difficult...' [33]. Early MS applications in chemistry included bond energies, ionization energies, and other thermochemical determinations. In about 1940, chemical (i.e. molecular) analysis began with the use of mass spectrometric fragmentation patterns to characterize petroleum distillates. Harold Washburn was one of the people responsible for the first practical application in chemistry, the analysis of hydrocarbon types formed in petroleum refining, an MS-based method of quantitative analysis which preceded gas chromatography [49]. This led in turn to the development of commercial 'organic' instruments for high-resolution mass spectrometry which was initiated principally by John Beynon following a visit to Alfred Nier's laboratory [50]. Systematic studies of organic compound fragmentation were begun in the 1950s by Fred McLafferty [51]. Aside from simple carbon-carbon chain cleavage mechanisms responsible for the formation of fragment ions, rearrangement mechanisms were also discovered, among which was the well-known McLafferty rearrangement [52]. Applications of mass spectrometry to natural products, especially to alkaloids, was begun in the late 1950s by Carl Djerassi [53, 54] and Klaus Biemann [55], while peptides were examined by Biemann [56] and McLafferty [57] soon after this. The use of a mass spectrometer as the detector in gas chromatography (or alternatively, the gas chromatograph as the pre-filter for the mass spectrometer) was developed during the 1950s by Roland Gohlke and Fred McLafferty [58]. In the 1970s, Beynon et al. used mass-analyzed ion kinetic energy spectrometry (MIKES) to study metastable ion fragmentations [59]. Later, direct mixture analysis using tandem mass spectrometry was started by Cooks [60].

In order to meet the need to ionize compounds without the occurrence of significant dissociation, much effort went into achieving soft ionization. Chemical ionization (CI) using ion/molecule reactions was developed in the 1960s [61, 62]. Ionization of a sample by CI was achieved by interaction of its molecules with reagent ions. CI allowed ionization without a significant degree of ion fragmentation, but still required vapor phase samples. Also in the 1960s, two instruments of secondary-ion mass spectrometry (SIMS) were developed independently by two groups, one led by Liebel and Herzog [63] and the other by Castaing [64]. These instruments were used to analyze the composition of solid surfaces and thin films by sputtering the surface of the specimen with a focused primary ion beam and collecting ejected secondary ions. Field desorption (FD)/field ionization (FI) was reported by Beckey in 1969 [65], in which electron tunneling triggered by a very high electric field results in ionization of gaseous analyte molecules. In 1976, plasma desorption (PD) ionization, a breakthrough in the analysis of solid samples, was invented by Macfarlane [66]. The method involves ionization of material in a solid sample by bombarding it with ionic or neutral atoms formed as a result of the nuclear fission of a suitable nuclide, typically the Californium isotope ^{252}Cf. In 1975, atmospheric-pressure chemical ionization (APCI) using corona discharge was created [67]. Fast-atom bombardment (FAB) was introduced in 1981 [68], which consisted in focusing on the sample a beam of neutral atoms or molecules. In 1983, Blakely and Vestal [69] introduced the thermospray ionization source (TSP) to produce ions from an aqueous solution sprayed directly into the mass spectrometer. Matrix-assisted laser desorption/ionization (MALDI) [70], first used in 1985 by Franz Hillenkamp, Michael Karas and their colleagues, a culmination of a long series of experiments using desorption ionization, is a soft ionization technique for the analysis of biomolecules and large organic molecules. The breakthrough for large-molecule laser desorption ionization came in 1987 when Koichi Tanaka combined 30-nm cobalt particles in glycerol with a 337-nm nitrogen laser for ionization [71]. In 1984, John Fenn and coworkers started to use electrospray to ionize biomolecules [27], and the first ESI applications of biopolymers such as proteins was published in 1989 [28].

2.1.2
Recent Advances

Mass spectrometry is a rapidly developing field, which has seen tremendous developments in recent years. For instance, a new high-resolution mass analyzer, the Orbitrap, was successfully introduced by Alexander Makarov in 1999 [72, 73]. Field-portable and miniature mass spectrometers [74, 75] have become available and have been commercialized. Many novel MS applications have been revealed, including enzyme reaction monitoring [76], protein conformation change probing [77], the study of noncovalent interactions of host-guest complexes, oligonucleotide ladder sequencing [78], protein mass mapping [79], intact virus analyses [80], quantitative proteomics and metabolomics with isotope labels [81, 82], chemical imaging [83, 84], combination with ion mobility [85–88], and so on.

In addition, new tandem mass spectrometry technologies were also among the important innovations. Apart from traditional collision-induced dissociation (CID) [89–91], a variety of activation methods (used to add energy to mass-selected ions) based on inelastic collisions and photon absorption have been widely utilized. They include IR multiphoton excitation [92, 93], UV laser excitation [94–97], surface-induced dissociation (SID) [98–100], black body radiation [101, 102], thermal dissociation [103], and others. As the fragmentation of peptide/protein ions is a central topic in proteomics, there is strong interest in such novel ion dissociation methods as electron capture dissociation (ECD) [104, 105] and electron transfer dissociation [22]. These new methods can provide structural information that complements that obtained by traditional collisional activation. Also, very recently, ambient ion dissociation methods such as atmospheric pressure thermal dissociation [106] and low temperature plasma assisted ion dissociation [107] have been reported.

Ambient MS is another advance in the field. It allows the analysis of samples with little or no sample preparation. Following the introduction of desorption electrospray ionization (DESI) [108, 109], direct analysis in real time (DART) [110], and desorption atmospheric pressure chemical ionization (DAPCI) [111, 112], a number of ambient ionization methods have been introduced. They include electrospray-assisted laser desorption/ionization (ELDI) [113], matrix-assisted laser desorption electrospray ionization (MALDESI) [114], atmospheric solids analysis probe (ASAP) [115], jet desorption ionization (JeDI) [116], desorption sonic spray ionization (DeSSI) [117], field-induced droplet ionization (FIDI) [118], desorption atmospheric pressure photoionization (DAPPI) [119], plasma-assisted desorption ionization (PADI) [120], dielectric barrier discharge ionization (DBDI) [121], and the liquid microjunction surface sampling probe method (LMJ-SSP) [122], etc. All these techniques have shown that ambient MS can be used as a rapid tool to provide efficient desorption and ionization and hence to allow mass spectrometric characterization of target compounds.

Because of space limitation and the broad scope of the field, it is impractical to attempt to provide a comprehensive list of all recent advances in mass spectrometry. Nevertheless, there is no doubt that we will continue to see rapid development in all areas of mass spectrometry in the near future.

2.2
Overview of the Study of Ion/Molecule Reactions in the Gas Phase by Mass Spectrometry

2.2.1
Brief History

The origins of much of mass spectrometry lie in the work of the great English physicist, J.J. Thomson; the observation (if not the full understanding) of the first ion/molecule reactions is no exception. In 1913, Thomson discovered that operating his positive-ray parabola apparatus in a hydrogen atmosphere produced a signal at a mass-to-charge ratio of 3, which he correctly attributed to the species H_3 [33]. Three years later, Dempster confirmed that the formation of the species H_3^+ is from the reaction between H_2^+ and H_2 [38]. In 1928, Hogness and Harkness reported the formation of I_3^+ and I_3^- in iodine vapor subjected to EI for ionization [123]. However, the arena of ion/molecule reaction chemistry was largely overlooked during the following decades when the main interests in mass spectrometry lay in the physics and chemistry of ionization and dissociation. The modern era of ion/molecule reaction studies began in the early 1950s following the work of Tal'roze [124], Stevenson [125], and Field et al. [126]. The discovery of methionium ion CH_5^+ generated from the ion/molecule reaction of CH_4^+ with a neutral CH_4 molecule in the gas phase as a stable species [124–126] and later the indication of its existence in superacid media by Olah [127] aroused the interest of chemists concerned with structure and bonding. This penta-coordinated carbocation was long regarded with some suspicion by organic chemists grounded in tetravalent carbon. As a result, a number of studies of gas-phase ion/molecule reactions were undertaken. Subsequently, Munson and Field realized the promise of ion/molecule reactions (e.g., proton transfer) as a soft ionization technique and termed the experiment chemical ionization in 1966, as we mentioned before [61, 62]. In the CI experiments, analyte molecules are ionized via reactions with reagent ions generated by EI of an appropriate reagent gas. Since the ionization depends on chemical reactions, it can be used to control energy deposited and to achieve selective ionization, and so on. Unlike EI, which involves unimolecular ion dissociation, CI represents the first break into bimolecular chemistry of ions. Munson and Field recognized CI as an excellent method for acquiring molecular-weight information for analytes that would undergo extensive fragmentation upon EI. Their work, along with early fundamental studies [128–133], led to the study of the products, rates [15, 134, 135], thermochemistry [136–138], and dynamics [139–143] of ion/molecule reactions, which became a field of scientific activity with applications in many diverse fields such as interstellar space [144], semiconductor processing [145], flames [146], plasma [147], combustion [148], chemical and pharmaceutical manufacturing [149], and in many vital biosystems.

2.2.2
Basic Types of Ion/Molecule Reactions

The common reactions between cations and neutral molecules include charge exchange (Scheme 2.1, Eq. (2.1)), proton transfer (Scheme 2.1, Eq. (2.2)), clustering (Scheme 2.1, Eq. (2.3)) and condensation reactions (with elimination of water or some other small neutral molecules, Scheme 2.1, Eq. (2.4)). A charge exchange reaction is exothermic if the recombination energy (RE) of the reactant ion is greater than the ionization energy (IE) of the neutral molecule M. Proton transfer can occur if the reaction is exothermic, that is, if the proton affinity of M_2 is larger than that of M_1. In the clustering reaction, the cluster formed must be stabilized by collisions or emission of light. If collisional stabilization is not

$X^{+\cdot} + M \rightarrow M^{+\cdot} + X$	(2.1)	charge exchange
$M_1H^+ + M_2 \rightarrow M_2H^+ + M_1$	(2.2)	proton transfer
$M_1^+ + M_2 \rightarrow M_1M_2^+$	(2.3)	clustering
$M_1^+ + M_2 \rightarrow [M_1M_2^+]^* \rightarrow M_3^+ + M_4$	(2.4)	condensation
$X^- + MY \rightarrow MX + Y^-$	(2.5)	displacement
$X^- + HY \rightarrow HX + Y^-$	(2.6)	proton transfer
$X^{-\cdot} + Y \rightarrow X + Y^{-\cdot}$	(2.7)	charge exchange
$X^- + Y \rightarrow [XY]^-$	(2.8)	association

Scheme 2.1 Basic types of ion/molecule reactions.

available or there is not enough time for emission of light, the cluster decomposes to products different from the reactants (Scheme 2.1, Eq. (2.4)) or dissociates back to reactants. Many ion/molecule reactions proceed without an activation barrier, and their cross sections are governed by the long-range attractive potential of the approaching reactants [17]. Models based on long-range potential capture [15] predict reaction rates in excess of $10^{-9}\,\text{cm}^3\,\text{molecule}^{-1}\,\text{s}^{-1}$. Thermochemical effects play a significant role in these reactions, and exothermicity is usually required to drive low-energy ion/molecule reactions. For example, exothermic proton transfer reactions are usually highly efficient, with reaction occurring essentially on every collision. As the proton transfer reactions becomes thermoneutral, the reaction efficiency drops and becomes very low for endothermic or endoergic reactions [150, 151].

Typical ion/molecule reactions between anions and neutral molecules [7, 9, 152–155] can be classified as displacement (Scheme 2.1, Eq. (2.5)), proton transfer (Scheme 2.1, Eq. (2.6)), charge exchange (Scheme 2.1, Eq. (2.7)) and association (Scheme 2.1, Eq. (2.8)). Among these, the displacement reaction has been studied extensively in the gas phase [156, 157], and the prototypical example is an anionic S_N2 reaction studied by Brauman [156]. In addition, interactions between a neutral molecule and an electron involving electron capture [158] and dissociative electron capture [159], are also important types of ion/molecule reactions in the gas phase. A molecule M with a positive electron affinity can form a stable anion M^- by capturing a thermal electron. In the case of dissociative electron capture, capture of an electron by a compound MX leads to a repulsive state of MX^-, which dissociates to form M and X^- with excess internal and/or translational energy.

2.2.3
Relationship to Reaction Analogies in Solution

2.2.3.1 Mechanism Elucidation of Classical Organic Reactions

Ion/molecule reactions traditionally performed in the low-pressure, very dilute environment of mass spectrometers are long regarded as a powerful tool to probe key intrinsic properties and intrinsic reactivities of gaseous ions. In the absence of solvation and ion-pairing interactions, gas-phase studies reveal the subtle details of reaction mechanisms and unambiguously characterize ion reactivity. These diverse ionic species have been formed, isolated, measured, and then allowed to react in the gas-phase environment by a variety of MS techniques able to determine the masses, structures (i.e. the connectivities) of the ionic reaction products and a number of physico-chemical properties.

Many classical organic reactions have been studied in the gas phase using MS methods. They include nucleophilic substitution reactions [9, 157, 160–162], the Friedel–Crafts acylation [163], the Diels–Alder cycloaddition [164, 165], the Wittig reaction [166, 167], the Reformatsky reaction [168], Michael addition [169, 170], the Kolbe reaction [171], the Cannizzaro reaction [172], Claisen-Schmidt reactions [173], the Meerwein-Ponndorf-Verley reduction/Oppenauer oxidation [174, 175], the Haber-Weiss reaction [176], the pinacol [177], Claisen [178, 179], Hofmann [180, 181], Cope [182, 183], Lossen [184, 185], and Wolff rearrangements [186], and the Fischer indole [177] and cumulene syntheses [187]. A wide range of analogies to condensed phase in reaction pathways are found in the gas phase. Also, comparison of gas-phase and condensed-phase data provides a powerful means for understanding the role that solvation and ion pairing play in determining the outcomes of ionic reactions [7].

2.2.3.2 Mechanism Elucidation of Organometallic Reactions

Because of the important catalytic role of some organometallic compounds in synthetic and biological processes, the study of metal ions and ionic transition metal compounds as well as their reactivities started more than 30 years ago [188–193]. Such have been the advances in ion/molecule reaction techniques that it has been possible to study single, elementary reaction steps or catalytic cycles mediated by transition metal ions on both the qualitative and quantitative level [189, 194–196].

In the mid-1970s, the mass spectrometric study of ion reactivity only focused on singly and multiply charged 'naked' ions and very simple organometallics of the type $[M(L)_n]^+$, with L representing some simple ligands (e.g., L = hydride, O, alkyl, CO, H_2O) hardly considered to be relevant to authentic catalysts in the condensed phase by synthetic chemists. The reason for this was the limitation in available ionization methods at that time. For example, EI is not amenable to many thermally fragile organometallic systems, as it requires volatility and generates extensive fragmentation, even at low ionizing electron energies. FD [115, 197, 198] and FAB [192, 199] have addressed this problem with some success, but the experimental conditions required for practical implementation are often difficult. The ionization of organometallic compounds using the more recent ionization technique of MALDI, however, faces the issue of analyte aggregation and the inability to directly analyze reaction solutions.

ESI has proven to be a most convenient means to transfer typical organometallics from solution to the gas phase, especially when the species of interest is present in the ionic form in solution. The first study of using ESI-MS for the analysis of ionic transition-metal complexes was made by Chait in 1994 [200], in which intact principal ions for ruthenium(II) dipyridyl complexes was observed. This represents the beginning of mass spectrometric characterization of known and defined solution-phase organometallic species [201–204]. Since its development, ESI-MS has been most elaborated for the ionization of large biomolecules such as proteins as a biochemical tool. However, when coupled with ion/molecule reactions and tandem mass spectrometry, ESI-MS is rapidly becoming the technique of choice for solution mechanistic studies such as probing transient ionic intermediates and confirming proposed reaction pathways.

There are a number of practical reasons and motivations for inorganic and organic chemists to embrace ESI-MS for reaction mechanism elucidation [195, 201, 203, 205, 206]. Firstly, because of its softness, ESI permits the transfer of ionic complexes into the gas phase directly from solution. An important question arose about whether or not the structures of ions observed in the gas phase truly reflect the species in solution. Previous studies had shown that for almost every case where the identity of ions in solution was established by some techniques other than MS such as NMR spectroscopy or electrochemistry, the intact ions observed by ESI-MS agreed with the prior identification [203, 207]. Secondly, once ions of interest are transferred into the gas phase, they can be singled out by ion selection according to their mass-to-charge ratio without any perturbation of the other compounds in solution. Thus, isolated ions are available for structural analysis by tandem mass spectrometry using techniques such as CID, providing information about organometallic complex stoichiometries. In addition, unlike the highly charged bipolymers often studied by ESI-MS, the charges on most organometallic ions are small. Consequently, the characteristic isotope patterns of many commonly encountered metals can be readily resolved with typical low-resolution mass analyzers and used to provide a positive identification, even in the presence of other ions with the same nominal m/z [201, 208]. Thirdly, crucial intermediates in catalytic processes are often, by their very nature, extremely short lived; transferring them to the gas phase (or, alternatively, transforming a suitable precursor in the gas phase) provides reactive intermediates which are almost infinitely stable because of the high-vacuum conditions; their reactivity can then be probed by directed collision with reaction partners (i.e., using ion/molecule reactions). Fourthly, in contrast to NMR, ESI-MS has advantages in observation of ionic species in solution that are paramagnetic or may not have a suitable nucleus [201]. Also, because ESI-MS is capable of detecting all ionic components of a rapidly exchanging labile system [209, 210], it provides a good alternative to low-temperature NMR studies. Furthermore, one notes the ability of ESI-MS to detect reaction intermediates in solution, often at very low concentrations. As a result, ESI-MS is ideal for the rapid screening of microscale reactions (thus minimizing wastage) and directing subsequent synthetic chemistry on the macroscopic scale [205]. The high sensitivity, the rapid analysis speed, and the capability to provide structural information on intermediates makes ESI-MS an excellent tool complementary and sometimes superior to electrochemistry and NMR for reaction mechanism investigation.

Table 2.1 lists some of the mechanistic studies of organic and organometallic reactions reported in the literature by ESI-MS. All sorts of reactions have been successfully explored in the gas phase, such as the Baylis-Hillman reaction [211–213], C–H or N–H activation [214–219], cyclopropanation reaction [220], Diels-Alder reactions [221], displacement reactions [222], electrophilic fluorination [223, 224], Fischer indole synthesis [225], Gilman reaction [226, 227], Grubbs metathesis reaction [228–231], Heck reaction [194], methylenation [232], oxidation [233, 234], Petasis olefination reaction [235], Raney Nickel-catalyzed coupling [236], ruthenium

Table 2.1 List of the mechanistic studies of organic and organometallic reactions by ESI-MS.

Year	Research Groups	Reaction description and discovery	References
1993	S.R. Wilson	Detection of transient ionic intermediates in the *Wittig*, *Mitsunobu*, and *Staudinger* reactions directly from solution	[241]
1993	S.R. Wilson	*Raney Nickel-catalyzed coupling* of 2-bromo-6-methylpyridine; dimeric Ni(II) complex intermediates were found.	[236]
1994	J.W. Canary	Observation of catalytic intermediates in the *Suzuki reaction* of Pd-catalyzed coupling of aryl boronic acid and aryl halides	[240]
1996	A. Hallberg	Palladium-catalyzed intramolecular *arylation* of enamidines	[253]
1997	D.A. Platterner and P. Chen	*C-H activation* reaction of $[Cp^*Ir(PMe_3)(CH_3)]^+$; the true 3-member-ring intermediate of iridium complex $[Cp^*Ir(\eta^2\text{-}CH_2PMe_2)]^+$ was found.	[214, 215]
1997	D.A. Platterner	*Oxygen transfer* mechanism of $[(salen)Mn^{III}]^+$; the elusive Mn^{III} and Mn^{V}-oxo species were detected.	[254, 255]
1997	G. Licini	Titanium-alkoxide mediated enantioselective *sulfoxidation*; the reactive chiral intermediates was characterized.	[239]
1998	D.A. Platterner and P. Chen	*Ziegler-Natta-like olefin oligomerization*; higher reaction rate constant for the addition of $[Cp_2ZrCH_3]^+$ to 1-butene in the gas phase than in-solution was observed due to 'isolated ions'. Product ions with multiple butene insertions up to $[Cp_2Zr(CH_2CHEt)_4CH_3]^+$ were detected.	[243]
1998	P. Chen	The *olefin metathesis* reaction of the Grubbs ruthenium carbene complexes; it confirmed that the metalacyclobutane structure is a transition state rather than an intermediate.	[228–231]
2000	W. Handerson	*Silver(I) oxide-mediated synthesis* of a platinalactam complex; platinum(II)–amidate complex cis-$[PtCl\{N(CO_2Et)C(O)CH_2CN\}(PPh_3)_2]$ was isolated.	[256]
2000	M. Wills	*Ruthenium(II)-catalyzed asymmetric reduction* of ketones to alcohols; the speculated precatalyst $[RuCl(\eta^6\text{-cymene})]_2$ was identified.	[237]
2001	W. Handerson and T.S.A. Hor	*Displacement reactions* of $[Pt_2(\mu\text{-Se})_2(PPh_3)_4]$ on tin substrates; the reaction process was monitored and ionic product structures were identified.	[222]
2001	C. Kutal	ESI-MS was used to probe the solution photochemical behavior of $[CpFebz]PF_6$, a photoinitiator for *polymerization of epoxides*	[257]
2003	J.L. Beauchamp	*Wolff rearrangement*; gas-phase synthesis of charged copper and silver Fischer carbenes from diazomalonates	[242]

2.2 Overview of the Study of Ion/Molecule Reactions in the Gas Phase by Mass Spectrometry

Table 2.1 (Continued)

Year	Research Groups	Reaction description and discovery	References
2003	R.A.J. O'Hair	Gas-phase catalytic cycles for the two-electron *oxidation* of primary and secondary alcohols to aldehydes and ketones by molybdenum complexes; its process was found to be analogous with industrial oxidation of methanol to formaldehyde using MoO_3.	[234]
2003	M.N. Eberlin	*Petasis olefination*; oxatitanacycle intermediates were transferred from solution to the gas phase and detected; the CID behavior of the intermediates fully supports the Hughes mechanism.	[235]
2004	J.O. Metzger	*Electron transfer-initiated Diels-Alder reactions*; the transient radical cations were detected.	[221]
2004	R.A.J. O'Hair	*Gilman reaction*; dimethyl cuprate undergoes C-C bond coupling with methyliodide in the Gas Phase; the preferred mechanism for Cu involves the formation of a T-shaped Cu transition state.	[226, 227]
2004	M.N. Eberlin	*Heck reaction*; key ionic intermediates of the Heck reaction of arene diazonium salts were intercepted and structurally characterized.	[194]
2004	M.N. Eberlin	*Baylis-Hillman reaction*; protonated intermediates in the catalytic cycle of the Baylis-Hillman reaction between an activated alkene and an electrophile were intercepted and characterized, providing direct evidence for the currently accepted mechanism.	[211–213]
2004	M.N. Eberlin	*Pd-catalyzed coupling* of vinylic tellurides with alkynes; Pd-Te cationic intermediates were, for the first time, intercepted and transferred to the gas phase for structural characterization.	[244]
2004	M.N. Eberlin	*Nucleophilic substitution reactions* of methylated hydroxylamines with bis(2,4-dinitrophenyl) phosphate	[250, 251]
2005	M.N. Eberlin	*Oxidation* of caffeine in water; continuous on-line and real-time monitoring of the oxidation process were performed.	[233]
2005	Y. Guo and S. Ma	*Palladium-catalyzed addition* of organoboronic acids to allenes	[248]
2005	S. Ma and Y. Guo	*Pd(0)-catalyzed three-component tandem double addition-cyclization reaction*; four cationic palladium intermediates were characterized by high-resolution ESI-FTMS	[249]
2005	Y. Guo	*Electrophilic fluorination* – detection and characterization of the key radical cationic intermediates fully supported the SET mechanism.	[223, 224]
2006	D. Schroeder and J. Roithova	*Activation of CH_4* by $[MgO]^+$	[217]

(Continued)

Table 2.1 (Continued)

Year	Research Groups	Reaction description and discovery	References
2007	D. Schroeder and H. Schwarz	Nickel-mediated thermal *activation of CH₄*; pronounced ligand effects and the role of formal oxidation states were observed.	[218]
2007	H. Schwarz	Thermal *activation of CH₄* by Group 10 metal hydrides MH⁺	[219]
2007	M.N. Eberlin	*Tröger's base formation*	[121]
2007	M.N. Eberlin	*Stille reaction*; ESI(+)-MS was able to transfer, directly from solution to the gas phase, the species involved in all main steps of a Stille reaction.	[238]
2007	M.N. Eberlin	Probing the mechanism of direct Mannich-type α-*methylenation* of ketoesters	[232]
2008	A. Pfaltz	*Pd-catalyzed allylic substitution and Diels-Alder reactions*; fast MS screening of effective chiral ligands or catalysts for stereoselective reactions was achieved.	[245–247]
2008	R.G. Cooks	*Fischer indole synthesis, Borsche-Drechsel cyclization*, and the *pinacol rearrangement*; organic reactions of ionic intermediates were promoted by atmospheric pressure thermal activation.	[225]
2008	P. Chen	Gas-phase synthesis and reactivity of a gold carbene complex; a gold benzylidene complex ion was observed, corresponding to the presumed reactive intermediates of a *gold-catalyzed cyclopropanation reaction*.	[220]
2008	Z. Wang and K. Ding	*Asymmetric ring-opening aminolysis* of 4,4-dimethyl-3,5,8-trioxabicyclo[5.1.0]octane catalyzed by titanium/BINOLate/water system; evidence for the Ti(BINOLate)₂-bearing active catalyst entities was obtained and the role of water for the reaction was examined.	[258]
2008	D. Schroeder	Gas-phase *C-H and N-H bond activation* by a high valent nitrido-iron dication	[216]

(II)-catalyzed asymmetric reduction [237], Stille reaction [238], sulfoxidation [239], Suzuki reaction [240], Wittig reaction [241], Wolff rearrangement [242], Ziegler-Natta olefin oligomerization [243], Pd-catalyzed C—C bond formation [244–249], and nucleophilic substitution reactions [250, 251]. These studies have helped the confirmation of hypothesized reaction mechanisms and identification of true reaction intermediates and pathways, even to the point of suggesting novel and superior catalysts in solution [195]. An elegant example is the study of C–H activation reaction by $[Cp^*Ir(PMe_3)(CH_3)]^+$, which showed that the reaction proceeds by a dissociative mechanism via the 3-member-ring intermediate of iridium complex $[Cp^*Ir(\eta^2-CH_2PMe_2]^+$ as opposed to 18-electron iridium(V) species [214, 215]. Less than one year later, the catalyst of Cp*Ir(η^2-CH$_2$PMe$_2$)(OTf), designed and synthesized according to the ESI-MS results, was found to have much better reactivity toward benzene than its acyclic analog [252].

2.2.3.3 Catalyst Screening

The traditional trial-and-error approach for the discovery of new catalysts or new ligands is laborious. Recently, high-throughput catalyst screening has been undertaken using ESI-MS and gas-phase ion/molecule reactions [259]. One of the most interesting methodologies using ESI is the so-called 'ion-fishing' technique [260, 261]. The suspected catalytic ionic species is 'fished' from solution (i.e., mass selection of ionized species of interest by mass spectrometry) and transferred to the collision cell of the mass spectrometer, where the gas-phase catalytic reaction is studied. For example, in 2001, Chen and coworkers used this technique to study important homogeneous transition metal-catalyzed reactions such as ring-opening metathesis polymerization (ROMP) [262]. In this screen, solution-phase synthesis of library compounds was followed by electrospray ionization of the mixture of the synthesized compounds, gas-phase ion selection of desired monomeric cations, and investigation of their reactivities by ion/molecule reactions. The high throughput nature of this methodology was well demonstrated by examination of 180 reactions, each in 10-fold replicates, in just two weeks. In 2003, Finn and his coworkers discovered new rhodium catalysts for the asymmetric hydrosilylation of ketones by mass spectrometry screening [263]. In their experiments, a family of chiral phosphate P,N-ligands were synthesized and converted into their rhodium complexes by reaction with [RhCl(cod)$_2$]. These complexes were then screened for catalytic activity using mass spectrometry-based enantiomeric excess determination. More recently, fast MS screening of effective chiral ligands or catalysts for stereoselective reactions such as allylic substitution and Diels–Alder reactions was also achieved by Pfaltz, in which a double mass-labeling strategy (i.e., use of mixtures of quasi-enantiomeric substrates) was employed so that the selectivity of chiral catalyst toward the substrates could be directly determined by mass spectrometry [245–247].

2.2.3.4 Synthesis of Elusive Ionic Species

Ion/molecule reactions also play an important role in the preparation of elusive ionic species which are not easily accessible in solution or from direct ionization or dissociation of neutral molecules [264–268]. A historic but still remarkable example is the readily preparation of the methionium ion, CH_5^+, by the gas-phase ion/molecule reaction of CH_4^+ with CH_4 [269] mentioned above. Exotic examples include nonclassical distonic ions (a radical cation or anion in which the charge site and the spin charge are not formally located in the same atom or group of atoms) [265, 268]. Gas-phase synthesis of transition metal ion complex/solvent clusters by electrospray ionization was also shown also by Posey using ion/molecule reactions of [Fe(bpy)$_3$]$^{2+}$ with a variety of solvents [270]. Other related examples are the synthesis of naked metal clusters using a ligand-stripping process of ESI-generated [CoRu$_3$(CO)$_{13}$]$^-$ to get [CoRu$_3$]$^-$ from McIndoe's group [271] and the gas-phase preparation of organomagnesates [196] and organocuprates [227] by O'Hair.

2.2.3.5 Probing Reactivity of Microsolvated Cluster Ions

With the aim of understanding bulk solution chemistry from gas-phase studies, a considerable amount of effort using high-pressure mass spectrometry [272] and ESI-MS has already addressed the reactivity of microsolvated cluster ions [273]. The partially solvated ions studied include core ions (hydroxide [274], acylium [275], halides [276], alkali metals [277], transitional metals [102], and nucleic acid ions [278], etc.) solvated with single or a few molecules of water, alcohols, and other solvent species of organic and atmospheric interest. All these species are much closer to those of interest in solution phase chemistry than the radical cations of traditional mass spectrometry. These studies provide information on ion solvation energy as well as the rate constants, isotope effects, and product distributions as a function of the cluster size and composition, increasing our knowledge of solvation at the molecular level.

2.2.4
Experimental Methods for the Study of Ion/Molecule Reactions

2.2.4.1 Low-Pressure Ion/Molecule Reactions

Ion/molecule reactions are mainly performed in the low-pressure environment of the mass spectrometer. Many mass spectrometers and methods for their operation have been developed to study ion/molecule reactions under controlled conditions. They include flow instruments such as flowing afterglow mass spectrometers [152, 193, 279–282] and selected-ion flow tubes (SIFT) [283–285], ion trapping instruments including Fourier transform ion cyclotron resonance (FT-ICR) [286–289] and quadrupole ion trap (QIT) mass spectrometers [290, 291], tandem mass spectrometers such as multiple-sector tandem mass spectrometers [292], guided ion beam tandem mass spectrometers [293–296], triple quadrupoles (QqQ, Q signifies the first or third quadrupole and q refers to the collision/reaction quadrupole cell) [297, 298] and pentaquadrupoles [299–301]. In addition, more specialized instruments have been developed for special purposes such as crossed-beam or merged-beam instruments for the study of thermal collisions [302, 303] and pulsed-electron high-pressure mass spectrometers [272, 304–306] for the evaluation of ion/molecule equilibria used to provide thermochemical information.

Tandem mass spectrometers are particularly well suited, almost necessary, for the study of ion/molecule reactions. For instance, triple quadrupoles are extraordinarily convenient and efficient for MS^2 experiments in a *tandem-in-space* mode, can be operated over a range of ion kinetic energies (from zero to a few hundred eV) to permit access to reactive and dissociative collisions, and give unit mass resolution for the pre- and post-collision ions as well as high ion transmission in the 'rf-only' quadrupole reaction cell ('rf-only' means no DC voltage applied on the quadrupole). These instruments also display high tolerance to poor vacuum and ease of operation and software control. In addition to inheriting the advantages of triple quadrupoles, the more complex but considerably more powerful pentaquadrupoles can be used to carry out triple-stage mass spectrometry (MS^3) experiments and therefore overcome an important limitation of triple quadrupoles for ion/molecule reaction studies, that is, they provide the ability to obtain mass and hence structural information on the products of the reactions of mass-selected ions. Pentaquadrupole mass spectrometers are particularly well suited to such studies as they allow the full range (all 21 types) [307] of MS^0 to MS^3 experiments to be easily accessed, with easy on-line mass-selection and control of the reactant and product ions [308]. A detailed description of the instrumental aspects and the applications of pentaquadrupoles in performing ion/molecule reactions has been published [309].

Recently, O'Hair also modified a 3D quadrupole ion trap mass spectrometer equipped with an ESI ion source as a complete chemical laboratory for fundamental gas-phase studies of metal-mediated chemistry. Resolving to the multistage capabilities of the quadrupole ion trap mass spectrometer for collision-induced dissociation and ion/molecule reactions, metal-mediated chemistry relevant to catalysis, C—C bond coupling, bioinorganic, and supramolecular chemistry were examined [310].

2.2.4.2 High-Pressure Ion/Molecule Reactions

Gas-phase ion chemistry under high-pressure conditions established decades ago by Kebarle *et al.* and used to make important thermochemical measurements is now a subject of growing interest [159, 311–313] for two reasons. (i) All chemical species encountered along the reaction coordinate are thermally equilibrated with the buffer gas, which allows the observed rate constants of ion/molecule reactions to be interpreted rigorously in terms of candidate mechanisms and potential energy surfaces for that reaction [311, 314]; an excellent illustration of this point is provided by the extensively studied S_N2 nucleophilic displacement reaction between chloride and methyl bromide [315, 316]. (ii) The structural and stereochemical information derived from gas-phase ion chemistry under high-pressure conditions, including atmospheric pressure, should be more fully comparable with data from solution chemistry than low-pressure data, thus allowing meaningful correlations between gas-phase and condensed-phase ionic reactivity [313].

Previously, owing to limitations in the available instrumentation for high-pressure ionization, high-pressure ion/molecule reactions were not extensively studied [317–324] and the neutral reactants were limited to volatile small molecules. Only a limited number of investigations under high pressure (0.01–10 atm) in instruments such as ion mobility spectrometers [322], pulsed high-pressure mass spectrometers [321], and time-resolved atmospheric pressure ionization mass spectrometers [325] as well as those integrated with radiolytic technique [317] were reported. Fortunately, the advent and recent development of atmospheric pressure ionization methods such as ESI, APCI, DESI, FIDI and atmospheric pressure thermal desorption ionization (APTDI) have provided a platform for the general study of ion/molecule reactions under ambient conditions [14, 118, 326–332]. In addition to the investigation of ionic reactivity at atmospheric pressure, studies using these ionization methods coupled with appropriate mass analyzers allow selective and rapid product detection and ion structural determination, encompassing the study of ion/molecule reactions of nonvolatile, polar, and heavy biomolecules. Also, because the ionization process involves a variety of ionic reactions such as protonation, dissociation, and adduct formation, the study of ambient ion/molecule reactions could help in understanding the ionization mechanism for these atmospheric-pressure ionization methods in return.

An early example of an ion/molecule reaction carried out in an atmospheric-pressure capillary inlet reactor based on an ESI interface to a quadrupole mass spectrometer involved the study of proton transfer between amines and multiply protonated proteins by Smith [333, 334]. Smith also observed protonated adenosine 5'-monophosphate (AMP) as the product of charge inversion in the reaction of [AMP-H]$^-$ with multiply-charged positive macroions using merged gas streams at near-atmospheric pressure [335]. In addition, the formation of TNT Meisenheimer complex [336] and the interaction between piperidine and multiply-charged lysozyme ions generated by electrosonic spray ionization (ESSI) [337] under atmospheric pressure has been reported. Eberlin reactions (i.e., the gas-phase transacetalization reaction) of acylium ions have also been observed to occur at atmospheric pressure under in-source ion/molecule reaction conditions [338].

Recently, ion/molecule reactions have been applied to DESI experiments for rapid chemical analysis [339–342]. In the reactive DESI experiments, a DESI spray solution is doped with specific reagents intended to allow particular ion/molecule reactions during the sampling process. Such ion/molecule reactions differ in several important characteristics from conventional ion/molecule reactions: (i) they take place under heterogeneous not homogenous conditions with respect to physical phase, namely at the solution/solid interface; (ii) they occur under ambient conditions, specifically at atmospheric pressure instead of vacuum; (iii) the neutral analyte molecule as well as the molecular precursor to the reactant ion can be heavy, polar, and nonvolatile; (iv) *in-situ* derivatization of the analyte on the surface can occur through the reaction and offer a fast and selective method of detection of the analyte; (v) standard solution-phase reagents can be used to produce finely controlled reactions based on well-known solution chemistry. These characteristics have the potential to extend the range and increase the richness of the ion chemistry accessible to mass spectrometry and also to provide new analytical advantages. Besides simple cationization/adducting reactions employed in DESI experiments [225, 341, 343, 344], other examples of ion/molecule reactions involving bond formation are transacetalization of acylium cations [345], cyclization of phenylboronic acid with *cis*-diols [339], and oximation of neutral steroids or anabolic steroid glucuronides with hydroxylamine [340].

2.3
Future Perspectives

On a final note, MS-based ion chemistry studied in the gas phase has already brought a wealth of new insights into organic and organometallic chemistry, such as capturing elusive reaction intermediates, increasing understanding of the catalytic process, and rapidly screening new catalysts. The further development of mass spectrometric technology may open up new and previously unexplored areas of chemistry. For instance, probing the chemistry of transition metal complexes that are highly reactive to air and moisture or metal-catalyzed biological reactions may become routine using mass spectrometry. The new ion chemistry discovered in the gas phase may be also possibly transferred to the condensed phase. Gas-phase synthesis on a large scale using ion selection and ion soft landing technology [346] is also possible. When complemented by ion spectroscopy, mass spectrometry can be a very powerful and ideal technology for the investigation of condensed-phase chemistry.

References

1 Lifshitz, C. (1987) *Int. Rev. Phys. Chem.*, **6**, 35.
2 Comita, P.B. (1985) *Science*, **227**, 863.
3 Shukla, A.K. and Futrell, J.H. (2000) *J. Mass. Spectrom.*, **35**, 1069.
4 Hayes, R.N. and Gross, M.L. (1990) *Methods Enzymol.*, **193**, 237.
5 Futrell, J.H. (1986) *Gaseous Ion Chemistry and Mass Spectrometry*, John Wiley & Sons, Inc., New York.
6 Nibbering, N.M.M. (1990) *Acc. Chem. Res.*, **23**, 279.
7 Gronert, S. (2001) *Chem. Rev.*, **101**, 329.
8 Bowers, M.T., Marshall, A.G., and McLafferty, F.W. (1996) *J. Phys. Chem.*, **100**, 12897.
9 Nibbering, N.M.M. (1988) *Adv. Phys. Org. Chem.*, **24**, 1.
10 Damrauer, R. and Hankin, J.A. (1995) *Chem. Rev.*, **95**, 1137.
11 DePuy, C.H., Grabowski, J.J., and Bierbaum, V.M. (1982) *Science*, **218**, 955.
12 Squires, R.R. (1992) *Acc. Chem. Res.*, **25**, 461.
13 Born, M., Ingemann, S., and Nibbering, N.M.M. (1997) *Mass Spectrom Rev.*, **16**, 181.
14 Cooks, R.G., Chen, H., Eberlin, M.N., Zheng, X., and Tao, W.A. (2006) *Chem. Rev.*, **106**, 188.
15 Bowers, M.T. (1979) *Gas Phase Ion Chemistry*, Academic Press, New York.
16 Futrell, J.H. and Tiernan, T.O. (1979) *Ion-Molecule Reactions*, vol. 2 (ed. J.L. S Franklin), Plenum Press, New York, p. 485.
17 Farrar, J.M. (2006) *Springer Handbook of Atomic, Molecular, and Optical Physics* (ed. G.W.F. S Drake), Springer, New York, p. 983.
18 McLuckey, S.A. and Stephenson, J.L.J. (1998) *Mass Spectrom. Rev.*, **17**, 369.
19 Chrisman, P.A., Pitteri, S.J., and McLuckey, S.A. (2005) *Anal. Chem.*, **77**, 3411.
20 McLuckey, S.A. (2006) *Principles of Mass Spectrometry Applied to Biomolecules* (eds J. S Laskin and C. S Lifshitz), John Wiley & Sons, Inc., p. 519.
21 Xia, Y. and McLuckey, S.A. (2008) *J. Am. Soc. Mass. Spectrom.*, **19**, 173.
22 Syka, J.E.P., Coon, J.J., Schroeder, M.J., Shabanowitz, J., and Hunt, D.F. (2004) *Proc. Natl. Acad. Sci. USA*, **101**, 9528.
23 Gunawardena, H.P., He, M., Chrisman, P.A., Pitteri, S.J., Hogan, J.M., Hodges, B.D.M., and McLuckey, S.A. (2005) *J. Am. Chem. Soc.*, **127**, 12627.
24 Huang, T.-Y., Emory, J.F., O'Hair, R.A.J., and McLuckey, S.A. (2006) *Anal. Chem.*, **78**, 7387.
25 Ogorzalek-Loo, R.R., Udseth, H.R., and Smith, R.D. (1991) *J. Phy. Chem.*, **95**, 6412.
26 Scalf, M., Westphall, M.S., Krause, J., Kaufman, S.L., and Smith, L.M. (1999) *Science*, **283**, 194.
27 Yamashita, M. and Fenn, J.B. (1984) *J. Phys. Chem.*, **88**, 4451.
28 Fenn, J.B., Mann, M., Meng, C.K., Wong, S.F., and Whitehouse, C.M. (1989) *Science*, **246**, 64.

29. Grayson, M.A. (ed.) (2002) *Measuring Mass: From Positive Rays to Proteins*, Chemical Heritage Press, Philadelphia.
30. Hoffmann, E.D., Charette, J., and Stroobant, V. (1996) *Mass Spectrometry: Principles and Applications*, John Wiley & Sons, Chichester.
31. Griffiths, J. (2008) *Anal. Chem.*, **80**, 5678.
32. Nier, K.A. (1999) *Instruments of Science: An Historical Encyclopedia* (eds R. Bud and D.J. Warner), The Science Museum, London & The National Museum of American History, Smithsonian Institution.
33. Thomson, J.J. (1913) *Rays of Positive Electricity and Their Application to Chemical Analysis*, Longmans, London.
34. Goldstein, E. (1886) *Berl. Ber.*, **39**, 691.
35. Aston, F.W. (1919) *Philos. Mag.*, **38**, 707.
36. Aston, F.W. (1942) *Mass Spectra and Isotopes*, Arnold, London.
37. Aston, F.W. (1920) *Proc. Cambridge Philos. Soc.*, **19**, 317.
38. Dempster, A.J. (1918) *Phys. Rev.*, **11**, 316.
39. Bleakney, W. (1929) *Phys. Rev.*, **34**, 157.
40. Nier, A.O. (1947) *Rev. Sci. Instrum.*, **18**, 398.
41. Nier, A.O., Booth, E.T., Dunning, J.R., and Grosse, A.V. (1940) *Phys. Rev.*, **57**, 546.
42. Hipple, J.A. and Condon, E.U. (1945) *Phys. Rev.*, **68**, 54.
43. von Ardenne, M. (1959) *Z. angew. Phys.*, **11**, 121.
44. Cameron, A.E. and Eggers, D.F. (1948) *Rev. Sci. Instrum.*, **19**, 605.
45. Hammer, W. (1911) *Phys. Zeitschr.*, **12**, 1077.
46. Paul, W. and Steinwedel, H. (1953) *Z. Naturforsch.* **8a**, 448.
47. Hipple, J.A., Sommer, H., and Thomas, H.A. (1949) *Phys. Rev.*, **76**, 1877.
48. Comisarow, M.B. and Marshall, A.G. (1974) *Chem. Phys. Lett.*, **25**, 282.
49. Hoover, H.C. and Washburn, H.W. (1940) *Petrol. Technol.*, **3**, 1.
50. Beynon, J.H. (1960) *Mass Spectrometry and Its Applications to Organic Chemistry*, Elsevier, Amsterdam.
51. McLafferty, F.W. (1956) *Anal. Chem.*, **28**, 306.
52. McLafferty, F.W. (1959) *Anal. Chem.*, **31**, 82.
53. Djerassi, C., Mills, J.S., and Villotti, R. (1958) *J. Am. Chem. Soc.*, **80**, 1005.
54. Djerassi, C., Gilbert, B., Shoolery, J.N., Johnson, L.F., and Biemann, K. (1961) *Experientia*, **17**, 162.
55. Biemann, K. (1964) *Pure Appl. Chem.*, **9**, 95.
56. Biemann, K., Gapp, F., and Seibl, J. (1959) *J. Am. Chem. Soc.*, **81**, 2274.
57. Senn, M. and McLafferty, F.W. (1966) *Biochem. Biophys. Res. Commun.*, **23**, 381.
58. Gohlke, R.S. and McLafferty, F.W. (1993) *J. Am. Soc. Mass Spectrom.*, **4**, 367.
59. Beynon, J.H., Cooks, R.G., Amy, J.W., Baitinger, W.E., and Ridely, T.Y. (1973) *Anal. Chem.*, **45**, 1023.
60. Kruger, T.L., Litton, J.F., Kondrat, R.W., and Cooks, R.G. (1976) *Anal. Chem.*, **48**, 2113.
61. Munson, M.S.B., and Field, F.H. (1966) *J. Am. Chem. Soc.*, **88**, 2621.
62. Harrison, A.G. (1992) *Chemical Ionization Mass Spectrometry*, 2nd edn, CRC Press, Boca Raton.
63. Liebl, H.J. (1967) *J. Appl. Phys.*, **38**, 5277.
64. Castaing, R. and Slodzian, G.J. (1962) *Microscopie*, **1**, 395.
65. Beckey, H.D. (1969) *Research/Development*, **20**, 26.
66. Macfarlane, R.D. and Torgerson, D.F. (1976) *Science*, **191**, 920.
67. Carroll, D.I., Dzidic, I., Stillwell, R.N., Haegele, K.D., and Horning, E.C. (1975) *Anal. Chem.*, **47**, 2369.
68. Morris, H.R., Panico, M., Barber, M., Bordoli, R.S., Sedgwick, R.D., and Tyler, A. (1981) *Biochem. Biophys. Res. Commun.*, **101**, 623.
69. Blakely, C.R. and Vestal, M.L. (1983) *Anal. Chem.*, **55**, 750.
70. Karas, M., Bachmann, D., and Hillenkamp, F. (1985) *Anal. Chem.*, **57**, 2935.

71 Tanaka, K., Waki, H., Ido, Y., Akita, S., Yoshida, Y., and Yoshida, T. (1988) *Rapid Commun. Mass Spectrom.*, **2**, 151.

72 Hu, Q., Noll, R.J., Li, H., Makarov, A., Hardman, M., and Cooks, R.G. (2005) *J. Mass Spectrom.*, **40**, 430.

73 Perry, R.H., Cooks, R.G., and Noll, R.J. (2008) *Mass Spec. Rev.*, **27**, 661.

74 Badman, E.R. and Cooks, R.G. (2000) *J. Mass Spectrom.*, **35**, 659.

75 Chaudhary, A., van Amerom, F.H.W., Short, R.T., and Bhansali, S. (2006) *Int. J. Mass Spectrom.*, **251**, 32.

76 Lee, E.D., Mueck, W., Henion, J.D., and Covey, T.R. (1989) *J. Am. Chem. Soc.*, **111**, 460.

77 Chowdhury, S.K., Katta, V., and Chait, B.T. (1990) *J. Am. Chem. Soc.*, **112**, 9012.

78 Pieles, U., Zurcher, W., Schoer, M., and Moser, H.E. (1993) *Nucleic Acids Res.*, **21**, 3191.

79 Henzel, W.J., Billeci, T.M., Stults, J.T., Wong, S.C., Grimley, C., and Watanabe, C. (1993) *Proc. Natl. Acad. Sci. USA*, **90**, 50.

80 Siuzdak, G., Bothner, B., Yeager, M., Brugidou, C., Fauquet, C.M., Hoey, K., and Chang, C.M. (1996) *Chem. Biol.*, **3**, 45.

81 Gygi, S.P., Rist, B., Gerber, S.A., Turecek, F., Gelb, M.H., and Aebersold, R. (1999) *Nat. Biotech.*, **17**, 994.

82 Oda, Y., Huang, K., Cross, F.R., Cowburn, D., and Chait, B.T. (1999) *Proc. Natl. Acad Sci. USA*, **96**, 6591.

83 Chaurand, P., Schwartz, S.A., and Caprioli, R.M. (2002) *Curr. Opin. Chem. Biol.*, **6**, 676.

84 Ifa, D.R., Manicke, N.E., Dill, A.L., and Cooks, R.G. (2008) *Science*, **321**, 805.

85 Von Helden, G., Kemper, P.R., Gotts, N.G., and Bowers, M.T. (1993) *Science*, **259**, 1300.

86 Kanu, A.B., Dwivedi, P., Tam, M., Matz, L., and Hill, H.H. (2008) *J. Mass. Spectrom.*, **43**, 1.

87 Valentine, S.J., Plasencia, M.D., Liu, X.Y., Krishnan, M., Naylor, S., Udseth, H.R., Smith, R.D., and Clemmer, D.E. (2006) *J. Proteome Res.*, **5**, 2977.

88 McLean, J.A., Ruotolo, B.T., Gillig, K.J., and Russell, D.H. (2005) *Int. J. Mass Spectrom.*, **240**, 301.

89 Downard, K.M. and Biemann, K. (1994) *J. Am. Soc. Mass Spectrom.*, **5**, 966.

90 Alexander, A.J., Thibault, P., Boyd, R.K., Curtis, J.M., and Rinehart, K.L. (1990) *Int. J. Mass Spectrom. Ion Processes*, **98**, 107.

91 Doroshenko, V.M. and Cotter, R.J. (1995) *Anal. Chem.*, **67**, 2180.

92 Zimmermann, U., Naeher, U., Frank, S., Martin, T.P., and Malinowski, N. (1996) *Large Clusters of Atoms and Molecules* (ed. T.P. S Martin), Kluwer Academic, Dordrecht, The Netherlands, p. 510.

93 Laskin, J. and Futrell, J.H. (2000) *J. Phys. Chem. A*, **104**, 8829.

94 Bowers, W.D., Delbert, S.S., Hunter, R.L., and McIver, R.T. (1984) *J. Am. Chem. Soc.*, **106**, 7288.

95 Martin, S.A., Hill, J.A., Kittrell, C., and Biemann, K. (1990) *J. Am. Soc. Mass Spectrom.*, **1**, 107.

96 Thompson, M.S., Cui, W., and Reilly, J.P. (2004) *Angew. Chem. Int. Ed.*, **43**, 4791.

97 Barbacci, D.C. and Russell, D.H. (1999) *J. Am. Soc. Mass Spectrom.*, **10**, 1038.

98 Wysocki, V.H., Joyce, K.E., Jones, C.M., and Beardsley, R.L. (2008) *J. Am. Soc. Mass Spectrom.*, **19**, 190.

99 Williams, E.R., Henry, K.D., McLafferty, F.W., Shabanowitz, J., and Hunt, D.F. (1990) *J. Am. Soc. Mass Spectrom.*, **1**, 413.

100 Bouchoux, G., Nguyen, M.T., and Salpin, J.-Y. (2000) *J. Phy. Chem. A*, **104**, 5778.

101 Price, W.D., Schnier, P.D., and Williams, E.R. (1996) *Anal. Chem.*, **68**, 859.

102 Stevens, S.M., Dunbar, R.C., Price, W.D., Sena, M., Watson, C.H., Nichols, L.S., Riveros, J.M., Richardson, D.E., and Eyler, J.R. (2004) *J. Phys. Chem., A*, **108**, 9892.

103 Rockwood, A.L., Busman, M., Udseth, H.R., and Smith, R.D. (1991) *Rapid Commun. Mass Spectrom.*, **5**, 582.

104 Zubarev, R.A., Kelleher, N.L., and McLafferty, F.W. (1998) *J. Am. Chem. Soc.*, **120**, 3265.

105 Zubarev, R.A. (2004) *Curr. Opin. Biotech.*, **15**, 12.

106 Chen, H., Eberlin, L.S., and Cooks, R.G. (2007) *J. Am. Chem. Soc.*, **129**, 5880.

107 Xia, Y., Ouyang, Z., and Cooks, R.G. (2008) *Angew. Chem. Int. Ed.*, **47**, 8646.

108 Takats, Z.W., Wiseman, J.M., Gologan, B., and Cooks, R.G. (2004) *Science*, **306**, 471.

109 Takats, Z., Wiseman, J.M., and Cooks, R.G. (2005) *J. Mass Spectrom.*, **40**, 1261.

110 Cody, R.B., Laramee, J.A., and Durst, H.D. (2005) *Anal. Chem.*, **77**, 2297.

111 Williams, J.P., Patel, V.J., Holland, R., and Scrivens, J.H. (2006) *Rapid Commun. Mass Spectrom.*, **20**, 1447.

112 Song, Y. and Cooks, R.G. (2006) *Rapid Commun. Mass Spectrom.*, **20**, 3130.

113 Huang, M.-W., , Chei, H.-L., Huang, J.-P., and Shiea, J. (1999) *Anal. Chem.*, **71**, 2901.

114 Sampson, J.S., Hawkridge, A.M., and Muddiman, D.C. (2007) *Rapid Commun. Mass Spectrom.*, **21**, 1150.

115 McEwen, C.N. and Ittel, S.D. (1980) *Org. Mass Spectrom.*, **15**, 35.

116 Takats, Z., Katona, M., Czuczy, N., and Skoumal, R. (2006) Proceedings of the 54th ASMS Conference on Mass Spectrometry and Allied Topics, Seattle, WA, May 28-June 1.

117 Haddad, R., Sparrapan, R., and Eberlin, M.N. (2006) *Rapid Commun. Mass Spectrom.*, **20**, 2901.

118 Grimm, R.L. and Beauchamp, J.L. (2005) *J. Phys. Chem. B*, **109**, 8244.

119 Haapala, M., Pol, J., Saarela, V., Arvola, V., Kotiaho, T., Ketola, R.A., Franssila, S., Kauppila, T.J., and Kostiainen, R. (2007) *Anal. Chem.*, **79**, 7867.

120 Ratcliffe, L.V., Rutten, F.J.M., Barrett, D.A., Whitmore, T., Seymour, D., Greenwood, C., Aranda-Gonzalvo, Y., Robinson, S., and McCoustra, M. (2007) *Anal. Chem.*, **79**, 6094.

121 Abella, C.A.M., Benassi, M., Santos, L.S., Eberlin, M.N., and Coelho, F. (2007) *J. Org. Chem.*, **72**, 4048.

122 Van Berkel, G.J., Kertesz, V., Koeplinger, K.A., Vavrek, M., and Kong, A.T. (2008) *J. Mass Spectrom.*, **43**, 500.

123 Hogness, T.R. and Harkness, R.W. (1928) *Phys. Rev.*, **32**, 784.

124 Tal'roze, V.L. and Lyubimova, A.K. (1952) *Dokl. Akad. Nauk. SSSR.*, **86**, 969.

125 Stevenson, D.P. and Schissler, D.O. (1955) *J. Chem. Phys.*, **23**, 1353.

126 Field, F.H., Franklin, J.L., and Lampe, F.W. (1957) *J. Am. Chem. Soc.*, **79**, 2419.

127 Olah, G.A. and Schlosberg, R.H. (1968) *J. Am. Chem. Soc.*, **90**, 2726.

128 Ryan, K.R., Sieck, L.W., and Futrell, J.H. (1964) *J. Chem. Phys.*, **41**, 111.

129 Haynes, R.M. and Kebarle, P. (1966) *J. Chem. Phys.*, **45**, 3899.

130 Kebarle, P., Searles, S.K., Zolla, A., Scarborough, J., and Arshadi, M. (1967) *J. Am. Chem. Soc.*, **89**, 6393.

131 Ferguson, E.E., Fehsenfeld, F.C., and Schmeltekopf, A.L. (1967) *Dev. Ind. Microbiol.*, **11**, 41.

132 Su, T. and Bowers, M.T. (1973) *Int. J. Mass Spectrom. Ion Phys.*, **12**, 347.

133 Olmstead, W.N. and Brauman, J.I. (1977) *J. Am. Chem. Soc.*, **99**, 4219.

134 Armentrout, P.B. (2004) *J. Anal. Atomic Spectrom.*, **19**, 571.

135 Futrell, J.H. and Tiernan, T.O. (1968) *Science*, **162**, 415.

136 Cooks, R.G., Ast, T., Pradeep, T., and Wysocki, V. (1994) *Acc. Chem. Res.*, **27**, 316.

137 Kuck, D. (2000) *Angew. Chem., Int. Ed.*, **39**, 125.

138 Aubry, C. and Holmes, J.L. (2000) *Int. J. Mass Spectrom.*, **200**, 277.

139 Armentrout, P.B. and Baer, T. (1996) *J. Phys. Chem.*, **100**, 12866.

140 Lifshitz, C. (1989) *Mass Spectrom.*, **10**, 1.

141 Powis, I. (1985) *Mass Spectrom.*, **8**, 1.

142 Lifshitz, C. (1987) *Mass Spectrom.*, **9**, 1.

143 Ng, C.-Y., Baer, T., and Powis, I. (1994) *Unimolecular and Bimolecular Ion-Molecule Reaction Dynamics*, John Wiley & Sons, Chichester.

144 Herbst, E. (1998) *Adv. Gas Phase Ion Chem.*, **3**, 1.

145 Olich, J. (2000) Ger. Offen.: Germany, DE 1985 5164 A1.

146 Calcote, H.F. (1972) *Ion-Mol. React.*, **2**, 673.
147 Vestal, M.L. (2001) *Chem. Rev.*, **101**, 361.
148 Semo, N.M. and Koski, W.S. (1984) *J. Phy. Chem.*, **88**, 5320.
149 Dash, A.K. (1997) *Process Contr. Qual.*, **10**, 229.
150 Solka, B.H. and Harrison, A.G. (1975) *Int. J. Mass Spectrom. Ion Phys.*, **17**, 379.
151 Bohme, D.K., Mackay, G.I., and Schiff, H.I. (1980) *J. Chem. Phys.*, **73**, 4976.
152 DePuy, C.H. and Bierbaum, V.M. (1981) *Acc. Chem. Res.*, **14**, 146.
153 DePuy, C.H. (2002) *J. Org. Chem.*, **67**, 2393.
154 Bowie, J.H., Trenerry, V.C., and Klass, G. (1981) *Mass Spectrom.*, **6**, 233.
155 O'Hair, R.A.J. (1991) *Mass Spec. Rev*, **10**, 133.
156 Lieder, C.A. and Brauman, J.I. (1974) *J. Am. Chem. Soc.*, **96**, 4028.
157 DePuy, C.H., Gronert, S., Mullin, A., and Bierbaum, V.M. (1990) *J. Am. Chem. Soc.*, **112**, 8650.
158 Christophoru, L.G. and Stockdale, J.A.D. (1968) *J. Chem. Phys.*, **48**, 1956.
159 Knighton, W.B. and Grimsrud, E.P. (1992) *J. Am. Chem. Soc.*, **114**, 2336.
160 Bowie, J.H. (1980) *Acc. Chem. Res.*, **13**, 76.
161 Riveros, J.M., Jose, S.M., and Takashima, K. (1985) *Adv. Gas Phase Ion Chem.*, **21**, 197.
162 Gozzo, F.C., Ifa, D.R., and Eberlin, M.N. (2000) *J. Org. Chem.*, **65**, 3920.
163 Speranza, M. and Sparapani, C. (1981) *Radiochim. Acta*, **28**, 87.
164 Eberlin, M.N. and Cooks, R.G. (1993) *J. Am. Chem. Soc.*, **115**, 9226.
165 Eberlin, M.N. (2004) *Int. J. Mass Spectrom.*, **235**, 263.
166 Johlman, C.L., Ijames, C.F., Wilkins, C.L., and Morton, T.H. (1983) *J. Org. Chem.*, **48**, 2628.
167 Lum, R.C. and Grabowski, J.J. (1993) *J. Am. Chem. Soc.*, **115**, 7823.
168 Castle, L.W., Hayes, R.N., and Gross, M.L. (1990) *J. Chem. Soc. Perkin Trans II*, **2**, 267.
169 McDonald, R.N., Chowdhury, A.K., and Setser, D.W. (1980) *J. Am. Chem. Soc.*, **102**, 6491.
170 DePuy, C.H., Van Doren, J.M., Gronert, S., Kass, S.R., Motell, E.L., Ellison, G.B., and Bierbaum, V.M. (1989) *J. Org. Chem.*, **54**, 1846.
171 Shen, J., Evans, C., Wade, N., and Cooks, R.G. (1999) *J. Am. Chem. Soc.*, **121**, 9762.
172 Sheldon, J.C., Bowie, J.H., Dua, S., Smith, J.D., and O'Hair, R.A.J. (1997) *J. Org. Chem.*, **62**, 3931.
173 Hass, G.W. and Gross, M.L. (1996) *J. Am Soc. Mass Spectrom.*, **7**, 82.
174 Schroeder, D. and Schwarz, H. (1990) *Angew. Chem.*, **102**, 925.
175 Van Der Waal, J.C., Kunkeler, P.J., Tan, K., and Van Bekkum, H. (1998) *J. Catalysis*, **173**, 74.
176 Blanksby, S.J., Bierbaum, V.M., Ellison, G.B., and Kato, S. (2007) *Angew. Chem. Int. Ed.*, **46**, 4948.
177 Glish, G.L. and Cooks, R.G. (1978) *J. Am. Chem. Soc.*, **100**, 6720.
178 Van der Wel, H., Nibbering, N.M.M., Kingston, E.E., and Beynon, J.H. (1985) *Org. Mass Spectrom.*, **20**, 535.
179 Eichinger, P.C.H. and Bowie, J.H. (1990) *Aust. J. Chem.*, **43**, 1479.
180 Hammerum, S. (1981) *Tetrahed. Lett.*, **22**, 157.
181 Yates, B.F. and Radom, L. (1987) *J. Am. Chem. Soc.*, **109**, 2910.
182 Rozeboom, M.D., Kiplinger, J.P., and Bartmess, J.E. (1984) *J. Am. Chem. Soc.*, **106**, 1025.
183 Schulze, S.M., Santella, N., Grabowski, J.J., and Lee, J.K. (2001) *J. Org. Chem.*, **66**, 7247.
184 Adams, G.W., Bowie, J.H., and Hayes, R.N. (1991) *J. Chem. Soc. Perkin Trans II*, **5**, 689.
185 Eichinger, P.C.H., Dua, S., and Bowie, J.H. (1994) *Int. J. Mass Spectrom. Ion Processes*, **133**, 1.
186 Lebedev, A.T., Hayes, R.N., and Bowie, J.H. (1991) *J. Chem. Soc. Perkin Trans II*, **8**, 1127.

187 Blanksby, S.J. and Bowie, J.H. (1999) *Mass Spectrom. Rev.*, **18**, 131.
188 Litzow, M.R. and Spalding, T.R. (1973) *Physical Inorganic Chemistry, Monograph No. 2: Mass Spectrometry of Inorganic and Organometallic Compounds*, Elsevier, New York, N.Y.
189 Freiser, B.S. (1996) *Organometallic Ion Chemistry*, Kluwer, Dordrecht, The Netherlands.
190 Armentrout, P.B. (1995) *Acc. Chem. Res.*, **28**, 430.
191 Eller, K. (1993) *Coordination Chem. Rev.*, **126**, 93.
192 Miller, J.M. (1989) *Mass Spec. Rev.*, **9**, 319.
193 Damrauer, R. (2004) *Organometallics*, **23**, 1462.
194 Sabino, A.A., Machado, A.H.L., Correia, C.R.D., and Eberlin, M.N. (2004) *Angew. Chem., Int. Ed.*, **43**, 2514.
195 Plattner, D.A. (2001) *Int. J. Mass Spectrom.*, **207**, 125.
196 O'Hair, R.A.J., Vrkic, A.K., and James, P.F. (2004) *J. Am. Chem. Soc.*, **126**, 12173.
197 Games, D.E., Jackson, A.H., Kane-Maguire, L.A.P., and Taylor, K. (1975) *J. Organomet. Chem.*, **88**, 345.
198 Gross, J.H., Nieth, N., Linden, H.B., Blumbach, U., Richter, F.J., Tauchert, M.E., Tompers, R., and Hofmann, P. (2006) *Anal. Bioanal. Chem.*, **386**, 52.
199 Bruce, M.I. and Liddell, M.J. (1987) *Appl. Organomet. Chem.*, **1**, 191.
200 Katta, V., Chowdhury, S.K., and Chait, B.T. (1990) *J. Am. Chem. Soc.*, **112**, 5348.
201 Traeger, J.C. (2000) *Int. J. Mass Spectrom.*, **200**, 387.
202 Colton, R. and Traeger, J.C. (1992) *Inorganica Chim. Acta*, **201**, 153.
203 Colton, R., D'Agostino, A., and Traeger, J.C. (1995) *Mass Spec. Rev*, **14**, 79.
204 Gatlin, C.L. and Turecek, F. (1997) *Electrospray Ionization Mass Spectrometry* (ed. R.B. S Cole), John Wiley & Sons, Inc., New York, p. 527.
205 Henderson, W. and McIndoe, J.S. (2004) *Comprehensive Coordination Chemistry II*, vol. 2 (eds J.A. S McCleverty and T.J. S Meyer), Elsevier, Oxford, p. 387.
206 Santos, L.S., Knaack, L., and Metzger, J.O. (2005) *Int. J. Mass Spectrom.*, **246**, 84.
207 Cheng, Z.L., Siu, K.W.M., Guevremont, R., and Berman, S.S. (1992) *J. Am. Soc. Mass Spectrom.*, **3**, 281.
208 Henderson, W. and Sabat, M. (1996) *Polyhedron*, **16**, 1663.
209 Colton, R., James, B.D., Potter, I.D., and Traeger, J.C. (1993) *Inorg. Chem.*, **32**, 2626.
210 Bond, A.M., Colton, R., D'Agostino, A., Harvey, J., and Traeger, J.C. (1993) *Inorg. Chem.*, **32**, 3952.
211 Santos, L.S., Pavam, C.H., Almeida, W.P., Coelho, F., and Eberlin, M.N. (2004) *Angew. Chem., Int. Ed.*, **43**, 4330.
212 Santos, L.S., Da Silveira Neto, B.A., Consorti, C.S., Pavam, C.H., Almeida, W.P., Coelho, F., Dupont, J., and Eberlin, M.N. (2006) *J. Phys. Org. Chem.*, **19**, 731.
213 Amarante, G.W., Benassi, M., Sabino, A.A., Esteves, P.M., Coelho, F., and Eberlin, M.N. (2006) *Tetrahedron Lett.*, **47**, 8427.
214 Hinderling, C., Plattner, D.A., and Chen, P. (1997) *Angew. Chem. Int. Ed.*, **36**, 243.
215 Hinderling, C., Feichtinger, D., Plattner, D.A., and Chen, P. (1997) *J. Am. Chem. Soc.*, **119**, 10793.
216 Schlangen, M., Neugebauer, J., Reiher, M., Schroeder, D., Lopez, J.P., Haryono, M., Heinemann, F.W., Grohmann, A., and Schwarz, H. (2008) *J. Am. Chem. Soc.*, **130**, 4285.
217 Schroeder, D. and Roithova, J. (2006) *Angew. Chem. Int. Ed.*, **45**, 5705.
218 Schlangen, M., Schroeder, D., and Schwarz, H. (2007) *Angew. Chem. Int. Ed.*, **46**, 1641.
219 Schlangen, M. and Schwarz, H. (2007) *Angew. Chem. Int. Ed.*, **46**, 5614.
220 Fedorov, A., Moret, M.-E., and Chen, P. (2008) *J. Am. Chem. Soc.*, **130**, 8880.
221 Fuermeier, S. and Metzger, J.O. (2004) *J. Am. Chem. Soc.*, **126**, 14485.

222 Yeo, J.S.L., Vittal, J.J., Henderson, W., and Hor, T.S.A. (2001) *J. Chem. Soc., Dalton Trans.*, 315.
223 Zhang, X., Liao, Y., Qian, R., Wang, H., and Guo, Y. (2005) *Org. Lett.*, **7**, 3877.
224 Zhang, X., Wang, H., and Guo, Y. (2006) *Rapid Commun. Mass Spectrom.*, **20**, 1877.
225 Chen, H., Eberlin, L.S., Nefliu, M., Augusti, R., and Cooks, R.G. (2008) *Angew. Chem. Int. Ed.*, **47**, 3422.
226 James, P.F. and O'Hair, R.A.J. (2004) *Org. Lett.*, **6**, 2761.
227 Rijs, N., Khairallah, G.N., Waters, T., and O'Hair, R.A.J. (2008) *J. Am. Chem. Soc.*, **130**, 1069.
228 Hinderling, C., Adlhart, C., and Chen, P. (1998) *Angew. Chem. Int. Ed.*, **37**, 2685.
229 Adlhart, C., Hinderling, C., Baumann, H., and Chen, P. (2000) *J. Am. Chem. Soc.*, **122**, 8204.
230 Adlhart, C. and Chen, P. (2003) *Helv. Chim. Acta*, **86**, 941.
231 Adlhart, C., Volland, M.A.O., Hofmann, P., and Chen, P. (2000) *Helv. Chim. Acta*, **83**, 3306.
232 Milagre, C.D.F., Milagre, H.M.S., Santos, L.S., Lopes, M.L.A., Moran, P.J.S., Eberlin, M.N., and Rodrigues, J.A.R. (2007) *J. Mass Spectrom.*, **42**, 10.
233 Dalmazio, I., Santos, L.S., Lopes, R.P., Eberlin, M.N., and Augusti, R. (2005) *Environ. Sci. Tech.*, **39**, 5982.
234 Waters, T., O'Hair, R.A.J., and Wedd, A.G. (2003) *J. Am. Chem. Soc.*, **125**, 3384.
235 Meurer, E.C., Santos, L.S., Pilli, R.A., and Eberlin, M.N. (2003) *Org. Lett.*, **5**, 1391.
236 Wilson, S.R. and Wu, Y. (1993) *Organometallics*, **12**, 1478.
237 Kenny, J.A., Versluis, K., Heck, A.J.R., Walsgrove, T., and Wills, M. (2000) *J. Chem. Soc. Chem. Commun.*, 99.
238 Santos, L.S., Rosso, G.B., Pilli, R.A., and Eberlin, M.N. (2007) *J. Org. Chem.*, **72**, 5809.
239 Bonchio, M., Licini, G., Modena, G., Moro, S., Bortolini, O., Traldi, P., and Nugent, W. (1997) *Chem. Commun.*, 869.
240 Aliprantis, A.O. and Canary, J.W. (1994) *J. Am. Chem. Soc.*, **116**, 6985.
241 Wilson, S.R., Perez, J., and Pasternak, A. (1993) *J. Am. Chem. Soc.*, **115**, 1994.
242 Julian, R.R., May, J.A., Stoltz, B.M., and Beauchamp, J.L. (2003) *J. Am. Chem. Soc.*, **125**, 4478.
243 Feichtinger, D., Plattner, D.A., and Chen, P. (1998) *J. Am. Chem. Soc.*, **120**, 7125.
244 Raminelli, C., Prechtl, M.H.G., Santos, L.S., Eberlin, M.N., and Comasseto, J.V. (2004) *Organometallics*, **23**, 3990.
245 Muller, C.A. and Pfaltz, A. (2008) *Angew. Chem. Int. Ed.*, **47**, 3363.
246 Markert, C., Roesel, P., and Pfaltz, A. (2008) *J. Am. Chem. Soc.*, **130**, 3234.
247 Teichert, A. and Pfaltz, A. (2008) *Angew. Chem. Int. Ed.*, **47**, 3360.
248 Qian, R., Guo, H., Liao, Y., Guo, Y., and Ma, S. (2005) *Angew. Chem. Int. Ed.*, **44**, 4771.
249 Guo, H., Qian, R., Liao, Y., Ma, S., and Guo, Y. (2005) *J. Am. Chem. Soc.*, **127**, 13060.
250 Domingos, J.B., Longhinotti, E., Brandão, T.A.S., Santos, L.S., Eberlin, M.N., Bunton, C.A., and Nome, F. (2004) *J. Org. Chem.*, **69**, 7898.
251 Domingos, J.B., Longhinotti, E., Brandão, T.A.S., Bunton, C.A., Santos, L.S., Eberlin, M.N., and Nome, F. (2004) *J. Org. Chem.*, **69**, 6024.
252 Luecke, H.F. and Bergman, R.G. (1997) *J. Am. Chem. Soc.*, **119**, 11538.
253 Ripa, L. and Hallberg, A. (1996) *J. Org. Chem.*, **61**, 7147.
254 Feichtinger, D. and Plattner, D.A. (1997) *Angew. Chem. Int. Ed.*, **36**, 1718.
255 Feichtinger, D. and Plattner, D.A. (2000) *J. Chem. Soc., Perkin Trans.*, **2**, **5**, 1023.
256 Henderson, W., Oliver, A.G., and Nicholson, B.K. (2000) *Inorganica Chimica Acta*, **298**, 84.
257 Ding, W., Johnson, K.A., Amster, I.J., and Kutal, C. (2001) *Inorg. Chem.*, **40**, 6865.
258 Bao, H., Zhou, J., Wang, Z., Guo, Y., You, T., and Ding, K. (2008) *J. Am. Chem. Soc.*, **130**, 10116.
259 Henderson, W. and McIndole, J.S. (2005) *Mass Spectrometry of Inorganic,*

Coordination and Organometallic Compounds: Tools-Techniques-Tips, John Wiley & Sons, Ltd., Chichester

260 Chen, P. (2003) *Angew. Chem. Int. Ed.*, **42**, 2832.

261 Adlhart, C. and Chen, P. (2000) *Helv. Chim. Acta*, **83**, 2192.

262 Volland, M.A.O., Adlhart, C., Kiener, C.A., Chen, P., and Hofmann, P. (2001) *Chem. Eur. J.*, **7**, 4621.

263 Yao, S., Meng, J.-C., Siuzdak, G., and Finn, M.G. (2003) *J. Org. Chem.*, **68**, 2540.

264 Vairamani, M., Mirza, U.A., and Srinivas, R. (1990) *Mass Spectrom. Rev.*, **9**, 235.

265 Stirk, K.M., Kiminkinen, L.K.M., and Kenttamaa, H.I. (1992) *Chem. Rev.*, **92**, 1649.

266 Moraes, L.A.B., Eberlin, M.N., and Laali, K.K. (2001) *Organometallics*, **20**, 4863.

267 Nagao, S., Kato, A., Nakajima, A., and Kaya, K. (2000) *Trans. Mater. Res. Soc. Japan*, **25**, 959.

268 Gozzo, F.C., Moraes, L.A.B., Eberlin, M.N., and Laali, K.K. (2000) *J. Am. Chem. Soc.*, **122**, 7776.

269 Meisels, G.G., Hamill, W.H., and Williams, R.R. (1956) *J. Chem. Phys.*, **25**, 790.

270 Spence, T.G., Burns, T.D., and Posey, L.A. (1997) *J. Phys. Chem. A*, **101**, 139.

271 Butcher, C.P.G., Dyson, A.D.P.J., Johnson, B.F.G., Langridge-Smith, P.R.R., and McIndoe, J.S. (2003) *Angew. Chem. Int. Ed.*, **42**, 5752.

272 Kebarle, P. (1988) *Techniques of Chemistry: Techniques for the Study of Ion-Molecule Reactions, vol. 20* (eds J.M. S Farrar and W. H. S Saunders JJr), John Wiley & Sons, New York, p. 221.

273 Takashima, K. and Riveros, J.M. (1999) *Mass Spectrom. Rev.*, **17**, 409.

274 Henchman, M., Paulson, J.F., and Hierl, P.M. (1983) *J. Am. Chem. Soc.*, **105**, 5509.

275 Guan, Z. and Liesch, J.M. (2001) *J. Mass Spectrom.*, **36**, 264.

276 Kato, S., Hacaloglu, J., Davico, G.E., DePuy, C.H., and Bierbaum, V.M. (2004) *J. Phy. Chem., A*, **108**, 9887.

277 Wang, G. and Cole, R.B. (1998) *Anal. Chem.*, **70**, 873.

278 Null, A.P., Nepomuceno, A.I., and Muddiman, D.C. (2003) *Anal. Chem.*, **75**, 1331.

279 Adams, N.G. and Smith, D. (1988) *Techniques of Chemistry: Techniques for the Study of Ion-Molecule Reactions*, vol. 20 (eds J.M. S Farrar and W.H. S Saunders JJr), John Wiley & Sons, New York, 165.

280 Bierbaum, V.M., Ellison, G.B., and Leone, S.R. (1984) *Gas Phase Ion Chem.*, **3**, 1.

281 Goebbert, D.J., Chen, H., and Wenthold, P.G. (2006) *J. Mass Spectrom.*, **41**, 242.

282 Chacko, S.A. and Wenthold, P.G. (2007) *J. Org. Chem.*, **72**, 494.

283 Squires, R.R. (1992) *Int. J. Mass Spectrom*, **118–119**, 503.

284 Fishman, V.N., Graul, S.T., and Grabowski, J.J. (1999) *Int. J. Mass Spectrom.*, **185/186/187**, 477.

285 Graul, S.T. and Squires, R.R. (1988) *Mass Spectrom. Rev.*, **7**, 263.

286 Comisarow, M.B. and Marshall, A.G. (1996) *J. Mass Spectrom.*, **31**, 581.

287 McLafferty, F.W. (1994) *Acc. Chem. Res.*, **27**, 379.

288 Stone, J.A. (1997) *Mass Spectrom. Rev.*, **16**, 25.

289 Dearden, D.V., Liang, Y., Nicoll, J.B., and Kellersberger, K.A. (2001) *J. Mass Spectrom.*, **36**, 989.

290 March, R.E. and Hughes, R.J. (eds) (1989) *Quadrupole Storage Mass Spectrometry*, John Wiley & Sons, New York.

291 March, R.E. and Todd, J.F.J. (eds) (1995) *Practical Aspects of Ion Trap Mass Spectrometry, Vol. III: Chemical, Environmental, and Biomedical Applications*, CRC Press, Boca Raton, FL.

292 Futrell, J.H. and Miller, C.D. (1966) *Rev. Sci. Instrum.*, **37**, 1521.

293 Armentrout, P.B. (1999) *Topics Organomet. Chem.*, **4**, 1.

294 Armentrout, P.B. (2002) *J. Am. Soc. Mass Spectrom.*, **13**, 419.

295 Angel, L.A. and Ervin, K.M. (2004) *J. Phys. Chem. A*, **108**, 9827.

296 Yamaguchi, S., Kudoh, S., Kawai, Y., Okada, Y., Orii, T., and Takeuchi, K. (2003) *Chem. Phys. Lett.*, **377**, 37.

297 Batey, J.H. and Tedder, J.M. (1983) *J. Chem. Soc., Perkin Trans.*, 2, **8**, 1263.

298 Kinter, M.T. and Bursey, M.M. (1986) *J. Am. Chem. Soc.*, **108**, 1797.

299 Eberlin, M.N., Moraes, L.A.B., Gozzo, F.C., Carvalho, M.C., Mendes, M.A., and Sparrapan, R. (1998) *Advances Mass Spectrom.*, **14**, A011640/1.

300 Juliano, V.F., Gozzo, F.C., Eberlin, M.N., Kascheres, C., and Lago, C.L. (1996) *Anal. Chem.*, **68**, 1328.

301 Kotiaho, T., Shay, B.J., Cooks, R.G., and Eberlin, M.N. (1993) *J. Am. Chem. Soc.*, **115**, 1004.

302 Vestal, M.L., Blakley, C.R., Ryan, P.W., and Futrell, J.H. (1974) *Chem. Phys. Lett.*, **27**, 490.

303 Futrell, J.H. (1992) *Adv. Chem. Phys.*, **82**, 501.

304 Williamson, D.H., Knighton, W.B., and Grimsrud, E.P. (1996) *Int. J. Mass Spectrom. Ion Processes*, **154**, 15.

305 Hvistendahl, G., Saastad, O.W., and Uggerud, E. (1990) *Int. J. Mas Spectrom. Ion Processes*, **98**, 167.

306 Paul, G. and Kebarle, P. (1991) *J. Am. Chem. Soc.*, **113**, 1148.

307 Schwartz, J.C., Wade, A.P., Enke, C.G., and Cooks, R.G. (1990) *Anal. Chem*, **62**, 1809.

308 Schwartz, J.C., Schey, K.L., and Cooks, R.G. (1990) *Int. J. Mass Spectrom. Ion Processes*, **101**, 1.

309 Eberlin, M.N. (1997) *Mass Spectrom. Rev.*, **16**, 113.

310 O'Hair, R.A.J. (2006) *Chem. Commun.*, 1469.

311 Speranza, M. (1992) *Mass Spectrom. Rev.*, **11**, 73.

312 Knighton, W.B. and Grimsrud, E.P. (1996) *Advances Gas Phase Ion Chem.*, **2**, 219.

313 Dillow, G.W. and Kebarle, P. (1988) *J. Am. Chem. Soc.*, **110**, 4877.

314 Wang, H., Peslherbe, G.H., and Hase, W.L. (1994) *J. Am. Chem. Soc.*, **116**, 9644.

315 Viggiano, A.A., Morris, R.A., Paschkewitz, J.S., and Paulson, J.F. (1992) *J. Am. Chem. Soc.*, **114**, 10477.

316 Giles, K. and Grimsrud, E.P. (1992) *J. Phys. Chem.*, **96**, 6680.

317 Cacace, F. (1988) *Acc. Chem. Res.*, **21**, 215.

318 Cacace, F. and Speranza, M. (1988) *Techniques for the Study of Ion Molecule Reactions*, John Wiley & Sons, Inc., New York.

319 Comisarow, M.B. (1978) *Transform Techniques in Chemistry*, Plenum, New York.

320 Ferguson, E.E., Fehsenfeld, F.C., and Albritton, D.L. (1979) *Gas Phase Ion Chemistry*, Academic, New York.

321 Kebarle, P. (1988) *Techniques for the Study of Ion Molecule Reactions*, John Wiley & Sons, Inc., New York.

322 Knighton, W.B., Bognar, J.A., O'Connor, P.M., and Grimsrud, E.P. (1993) *J. Am. Chem. Soc.*, **115**, 12079.

323 Arnold, S.T., Seeley, J.V., Williamson, J.S., Mundis, P.L., and Viggiano, A.A. (2000) *J. Phys. Chem. A*, **104**, 5511.

324 Bell, A.J., Giles, K., Moody, S., and Watts, P. (1998) *Int. J. Mass Spectrom. Ion Processes*, **173**, 65.

325 Matsuoka, S., Nakamura, H., and Tamura, T. (1981) *J. Chem. Phys.*, **75**, 681.

326 Tam, M. and Hill, H.H. (2004) *Anal. Chem.*, **76**, 2741.

327 Wentworth, W.E., Batten, C.F., D'sa, E.D., and Chen, E.C.M. (1987) *J. Chromatogr.*, **390**, 249.

328 Hirabayashi, A., Takada, Y., Kambara, F H., Umemura, Y., Ohta, H., Ito, H., and Kuchitsu, K. (1992) *Int. J. Mass Spectrom. Ion Processes*, **120**, 207.

329 Ikonomou, M.G. and Kebarle, P. (1992) *Int. J. Mass Spectrom. Ion Processes*, **117**, 283.

330 Meurer, E.C., Sabino, A.A., and Eberlin, M.N. (2003) *Anal. Chem.*, **75**, 4701.

331 Augusti, R., Chen, H., Eberlin, L.S., Nefliu, M., and Cooks, R.G. (2006) *Int. J. Mass Spectrom.*, **253**, 281.

332 Sparrapan, R., Eberlin, L.S., Haddad, R., Cooks, R.G., Eberlin, M.N., and Augusti, R. (2006) *J. Mass Spectrom.*, **41**, 1242.

333 Loo, R.R., Loo, J.A., Udseth, H.R., Fulton, J.L., and Smith, R.D. (1992) *Rapid Comm. Mass Spectrom.*, **6**, 159.

334 Loo, R.R. and Smith, R.D. (1994) *J. Am. Soc. Mass Spectrom.*, **5**, 207.

335 Loo, R.R.O., Udseth, H.R., and Smith, R.D. (1991) *J. Phys. Chem.*, **95**, 6412.

336 Popov, I.A., Chen, H., Kharybin, O.N., Nikolaevb, E.N., and Cooks, R.G. (1953) *Chem. Comm.*, 2005.

337 Takáts, Z., Wiseman, J.M., Gologan, B., and Cooks, R.G. (2004) *Anal. Chem.*, **76**, 4050.

338 Meurer, E.C., Sabino, A.A., and Eberlin, M.N. (2003) *Anal. Chem.*, **75**, 4701.

339 Chen, H., Cotte-Rodriguez, I., and Cooks, R.G. (2006) *Chem. Commun.*, 597.

340 Kushnir, M.M., Rockwood, A.L., Roberts, W.L., Pattison, E.G., Bunker, A.M., Fitzgerald, R.L., and Meikle, A.W. (2006) *Clin. Chem.*, **52**, 120.

341 Nyadong, L., Green, M.D., De Jesus, V.R., Newton, P.N., and Fernandez, F.M. (2007) *Anal. Chem.*, **79**, 2150.

342 Chen, H., Talaty, N.N., Takats, Z., and Cooks, R.G. (2005) *Anal. Chem.*, **77**, 6915.

343 Takats, Z., Wiseman, J.M., Gologan, B., and Cooks, R.G. (2004) *Science*, **306**, 471.

344 Meurer, E.C., Chen, H., Riter, L.S., Cotte-Rodriguez, I., Eberlin, M.N., and Cooks, R.G. (2004) *Chem. Commun.*, **1**, 40.

345 Sparrapan, R., Eberlin, L.S., Haddad, R., Cooks, R.G., Eberlin, M.N., and Augusti, R. (2006) *J. Mass Spectrom.*, **41**, 1242.

346 Ouyang, Z., Takáts, Z., Blake, T.A., Gologan, B., Guymon, A.J., Wiseman, J.M., Oliver, J.C., Davisson, V.J., and Cooks, R.G. (2003) *Science*, **301**, 1351.

3
Organic Reaction Studies by ESI-MS
Fabiane M. Nachtigall and Marcos N. Eberlin

3.1
Introduction

Reactions are the heart and soul of chemistry. Therefore chemists, to effectively control and improve their chemical reactions, need to know the mechanistic details by which these usually sophisticated, multistep and multicomponent processes with many intermediates and transient species proceed. The tremendous success and rapid broad use of electrospray ionization mass spectrometry (ESI-MS) [1] results mainly from its ability to transfer, in a gentle and efficient way, ions of many types, charge states and nearly unlimited masses from the 'real-world' environment of reaction solutions directly to the diluted gas phase and to characterize these intact and isolated gaseous ions according to their masses and connectivities with the outstanding speed, sensitivity and selectivity that only mass spectrometry is able to provide. ESI-MS (for ion detection and mass and isotopic pattern determination) as well as its tandem version ESI-MS(/MS) (for structural investigation) has therefore established itself as the principal technique to study reaction mechanisms of organic reactions in solution. ESI can 'fish,' rapidly and efficiently, reactants, intermediates and products (either ionic species or molecules in ionic forms) and transfer them to mass spectrometers (leaving behind the solvent molecules and counter ions) to measure their masses and access their structures and intrinsic reactivities using tandem MS/MS experiments. Continuous on-line monitoring of organic reactions of many types by ESI-MS(/MS) 'ion fishing' (Figure 3.1) provides instantaneous and comprehensive snapshots of the ionic composition and thus permits organic chemists to follow how these reactions progress as a function of time and reaction conditions, whereas the high sensitivity and speed of ESI-MS allow even transient (low concentrations for short times) intermediates to be detected and characterized. The intrinsic reactivity of each key gaseous species fished by ESI can also be further investigated via ESI-MS(/MS) experiments in the search for the most intrinsically active species via gas-phase ion/molecule reactions in which solvent and counter-ion effects are absent. Therefore, the ability to 'fish' ions (or neutral, zwitterionic, or radical species in ionic forms such as protonated, deprotonated, or cationized or anionized molecules) provides a detailed

Reactive Intermediates: MS Investigations in Solution. Edited by Leonardo S. Santos.
Copyright © 2010 WILEY-VCH Verlag GmbH & Co. KGaA, Weinheim
ISBN: 978-3-527-32351-7

Figure 3.1 A cartoon of the ESI-MS 'ion fishing' process for monitoring organic reactions in solution. Ions functioning as reactants, intermediates and products or neutral species in ionic forms are fished continuously from the reaction solution into the gas phase environment of mass spectrometers. In viewing its mechanisms, cations or anions are ejected ('explode out') from the tiny charged droplets during ESI, thus flying directly to the gas phase. Adapted from a cartoon available at http://www.thefishingline.com/ and signed by R. Stubler.

overview of the reaction steps via the interception, characterization and reactivity investigation of their key players.

Recent reviews have summarized the use of ESI-MS(/MS) to monitor chemical reactions of different types in solution, discussing in detail the basic concepts, advantages, and applications of on-line and off-line reaction screening [2]. In this chapter, we have selected some representative examples, not exclusively but mostly from our own experiences with this emerging and exciting field, in the hope of exemplifying the diversity of aspects that can be evaluated and the important information on the mechanisms of various key organic reactions that such studies can provide, and the strategies used during ESI-MS(/MS) monitoring to investigate various organic reactions.

3.2
Reaction Mechanisms

3.2.1
Morita-Baylis-Hillman Reaction

The Morita-Baylis-Hillman (MBH) reaction can be broadly defined as a coupling reaction between an alkene activated by an electron-withdrawing group and an electrophile that occurs under Lewis base catalysis, 1,4-diazabicyclo[2.2.2]octane (DABCO) normally being used as the base (Scheme 3.1) [3].

In the mechanism initially proposed by Hill and Isaacs [4] and later refined by others [5] (Scheme 3.2), step I consists of the 1,4-addition of the catalytic tertiary

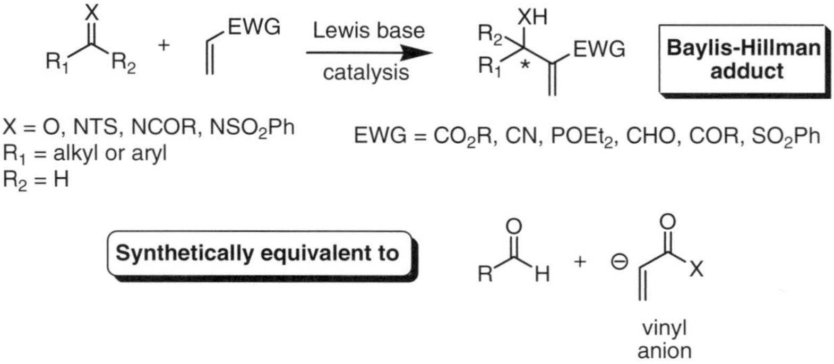

$X = O, NTS, NCOR, NSO_2Ph$
$R_1 =$ alkyl or aryl
$R_2 = H$

$EWG = CO_2R, CN, POEt_2, CHO, COR, SO_2Ph$

Scheme 3.1 General scope of the MBH reaction.

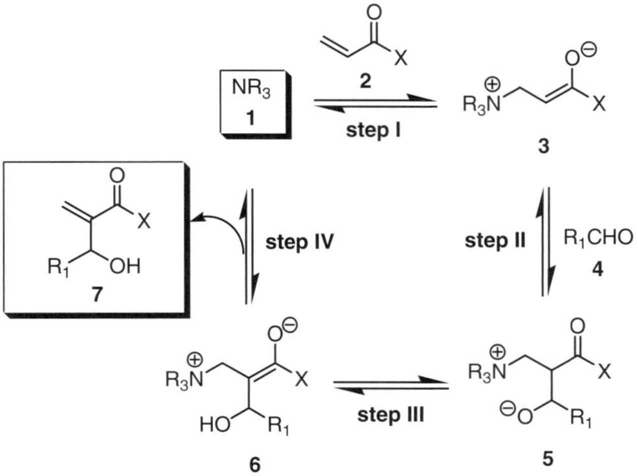

Scheme 3.2 Currently accepted catalytic cycle for the MBH reaction.

amine **1** to the activated alkene **2** (α,β-unsaturated carbonyl compounds), which generates the zwitterion intermediate **3**. In step II, **3** adds to aldehyde **4** by an aldolic addition reaction to yield intermediate **5**. Step III involves an intramolecular prototropic shift within **5** to form the zwitterionic intermediate **6**, which, in step IV, forms the final MBH adduct **7** through E2 or E1cb elimination in the presence of a Lewis base.

Several neutral zwitterionic intermediates are therefore involved in the currently accepted mechanism for the MBH reaction of acrylates with aldehydes catalyzed by DABCO (Scheme 3.2). However, these neutral species (undetectable by ESI-MS) are expected to be in equilibrium with their protonated or cationized forms such as [M + Na]$^+$ in methanolic solutions and might therefore be fished and detected as such by ESI-MS.

Using ESI-MS in both the positive and negative ion modes, Eberlin and coworkers [6] monitored on-line a few MBH reactions. The proposed intermediates for the catalytic cycle of the MBH reaction (**1–7**, Scheme 3.3) were for the first time successfully intercepted and structurally characterized.

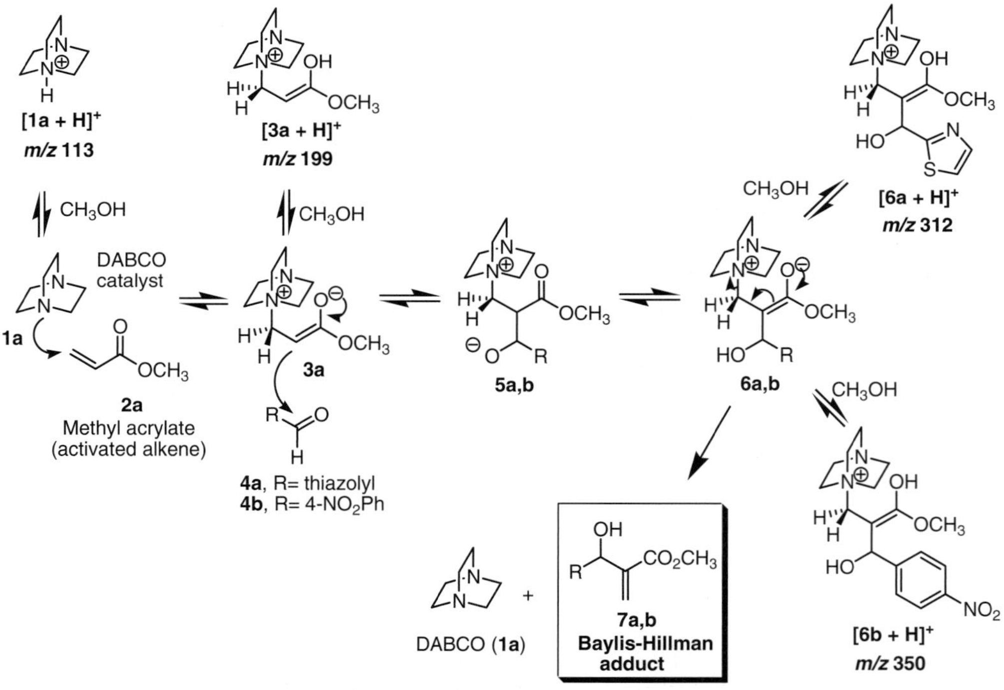

Scheme 3.3 Mechanism of MBH reaction of methyl acrylate (**2a**) and aldehydes (**4a,b**) catalyzed by DABCO (**1a**) showing the species in their protonated forms with their respective m/z values that were intercepted and characterized by ESI(+)-MS(/MS).

3.2.2
Morita-Baylis-Hillman Reaction Co-catalyzed by Ionic Liquids

In a similar manner, we applied ESI-MS(/MS) on-line monitoring to MBH reactions co-catalyzed by ionic liquids to screen for the supramolecular species that could be responsible for the co-catalysis (Scheme 3.4) [7]. Loosely bonded supramolecular species formed by coordination of neutral reagents, products, and the protonated forms of zwitterionic MBH intermediates and final product with cations and anions of ionic liquids have been gently and efficiently 'fished' directly from the solution to the gas phase (Figure 3.2). Mass measurements and ESI-MS(/MS) characterization of these unprecedented species via collision-induced dissociation (CID) were also performed.

By using competitive experiments, the relative efficiency of different co-catalysts was determined to be BMI·CF_3CO_2 > BMI·BF_4 > BMI·PF_6, which was the opposite of that observed by Afonso *et al.* [8] in solution. Based on the interception of these unprecedented supramolecular species, Eberlin and coworkers propose (Scheme 3.4) that 1,3-dialkylimidazolium ionic liquids function as efficient co-catalysts for the MBH reaction by: (i) activating the aldehyde toward nucleophilic attack by BMI coordination (species 8^+) and (ii) stabilizing the zwitterionic species that act as the main MBH intermediates through supramolecular coordination (species 9^+, 10^+, 11^+, and 13^+).

Recently [9], ESI-MS(/MS) monitoring was again used to intercept and characterize new key intermediates for the rate-determining step of MBH reactions. These ESI-MS data provide evidence supporting recent suggestions, based on kinetic experiments and theoretical calculations, for the dualist nature of the elimination step of the MBH reaction mechanism. McQuade and coworkers [10] and Aggarwal and coworkers [11] have recently re-evaluated the MBH reaction mechanism from kinetic and theoretical data, focusing on the elimination step. According to McQuade, the MBH reaction is second order relative to the aldehyde and shows a significant kinetic isotopic effect (KIE): $k_H/k_D = 5.2 \pm 0.6$ in DMSO. Interestingly, regardless of the solvent (DMF, MeCN, THF, $CHCl_3$), the KIEs were found to be greater than 2, indicating the relevance of proton abstraction in the rate-determining step (RDS). Based on these new data, McQuade and coworkers proposed a new mechanistic view for the elimination step (Scheme 3.5), suggesting **IV** as the RDS. Soon after, Aggarwal and coworkers based also on kinetic studies, proposed that the reaction kinetics is second order in relation to the aldehyde but only at its beginning ($\leq$ 20% of conversion), then becoming autocatalytic. Apparently, the MBH adduct **7** may act as a proton donor and therefore assists the elimination step via a six-membered intermediate **16** (Scheme 3.5).

The new kinetic evidence just discussed has also stimulated further theoretical studies on the MBH reaction mechanism [10, 11]. These studies suggest that step **IV** can occur via two pathways (Scheme 3.5): (a) in the absence of a proton source, elimination is assisted by a second molecule of aldehyde (**14**), as proposed by McQuade, or (b) in the presence of a proton source, such as an alcohol, the elimination proceeds via intermediate **16** or a similar species.

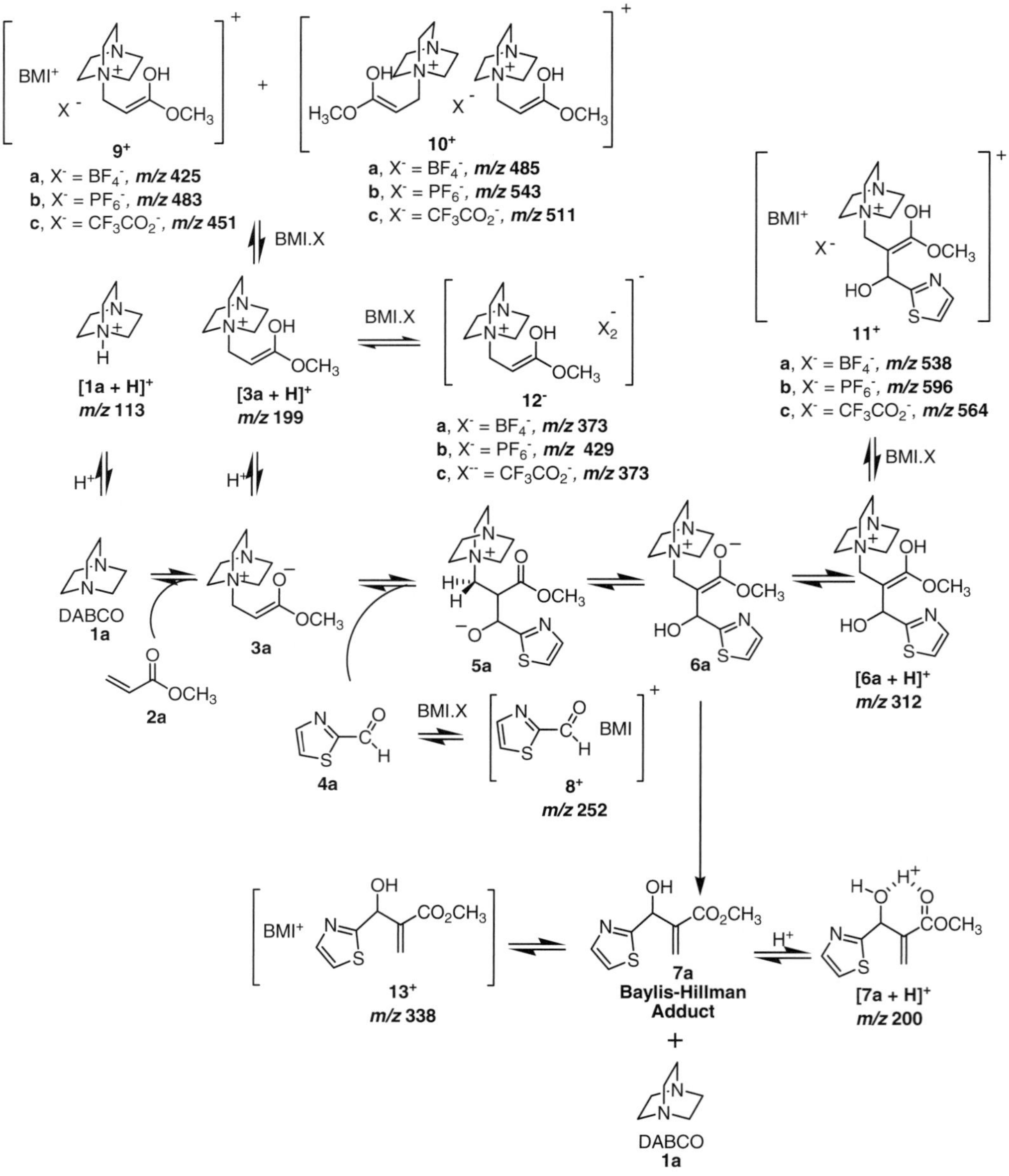

Scheme 3.4 MBH reaction of methyl acrylate (**2a**) and 2-thiazolecarboxaldehyde (**4a**) co-catalyzed by both DABCO (**1a**) and an ionic liquid BMI·X (X = BF_4^-, PF_6^-, $CF_3CO_2^-$).

These mechanistically important new propositions about the elimination step of the MBH reaction have therefore stimulated us to perform complementary ESI-MS(/MS) investigations aiming to intercept and characterize the new MBH reaction intermediates postulated for the dualist nature of the key RDS elimination step. After

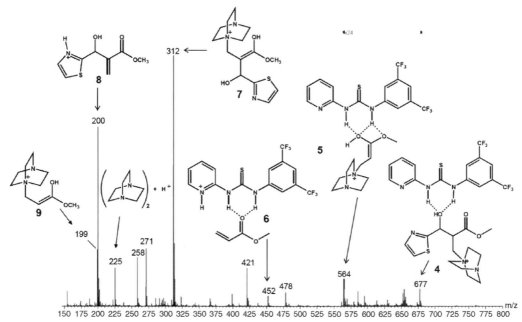

Figure 3.2 ESI(+)-MS of the MBH reaction solution co-catalyzed by a thiourea.

10 min of reaction, an aliquot of an MBH reaction solution was diluted in acetonitrile and its ESI-MS collected. Intermediate **14a** (R = Ph), proposed by McQuade [10] from the nucleophilic attack of the MBH alkoxide on the aldehyde (Scheme 3.6), was intercepted as [**14a** + Na]$^+$ of m/z 433. Considering Aggarwal's proposal for proton

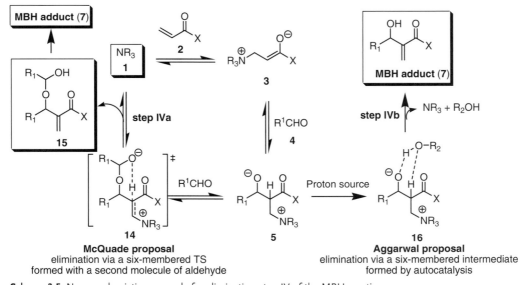

Scheme 3.5 New mechanistic proposals for elimination step IV of the MBH reaction.

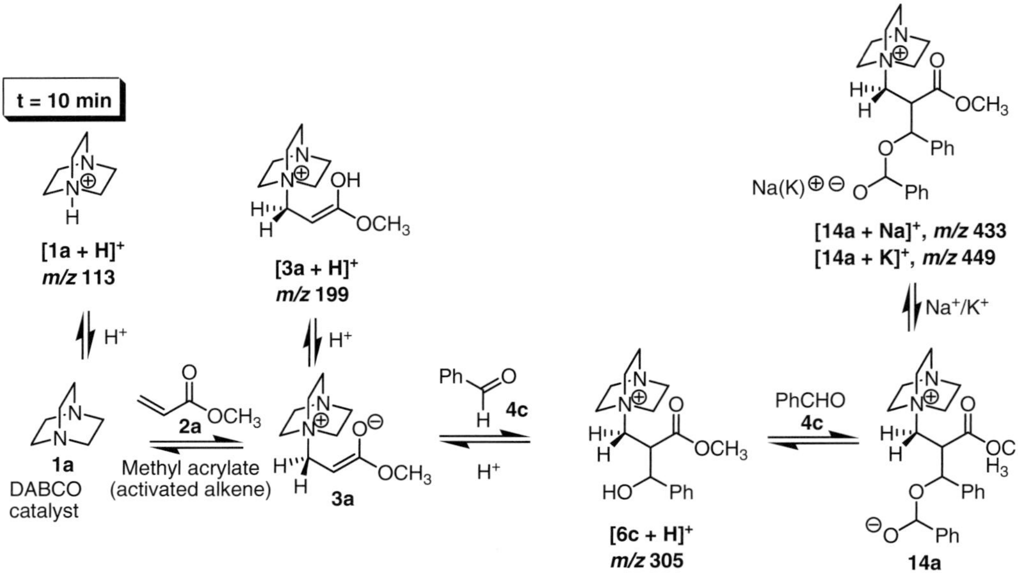

Scheme 3.6 Species characterized by ESI-MS(/MS) for an MBH reaction monitored after 10 min of reaction.

sources [11], we monitored the MBH reaction performed with the same experimental protocol but now adding an additional 3 equivalents of, for instance, β-naphthol (an external proton source). Immediately after mixing the reagents, an aliquot of the reaction solution was subject to ESI-MS, and again a new species corresponding to [**16a** + H]$^+$ (R$_2$OH = β-naphthol) of m/z 449 was detected. The interception and characterization of [**16a** + H]$^+$ agrees with Aggarwal's proposition that a proton source participates in the elimination step by assisting the removal of the base. The 'fishing' and structural characterization of these key intermediates exemplify the complex equilibrations occurring during MBH reactions and the dualistic nature of the RDS elimination step. These findings may also help to develop general asymmetric versions of MBH reactions, an endeavor that must consider all major equilibria and use a fast and efficient proton transfer promoter.

Connon et al. [12] recently described the use of (thio)ureas as efficient organocatalysts for MBH reactions, giving significant increases in rate and yield even for deactivated aromatic aldehydes. This attractive solution stimulated studies on organocatalysis in MBH reactions. Initially, the efforts concentrated on increasing MBH reaction rates but, soon after, asymmetric approaches using chiral ureas as catalyst have became the main targets. However, the catalytic role of (thio)ureas on the mechanism of the MBH reaction leading to the impressive improvements in yield, rate and enantioselectivity was still unclear. Based on ESI-MS(/MS) monitoring and DFT calculations, a new mechanism for the role of (thio)ureas on the organocatalysis of MBH reactions was proposed [13]. An MBH reaction catalyzed by (thio)urea was monitored and key intermediates were intercepted and characterized (Figure 3.2). These intermediates point to the participation of the (thio)urea

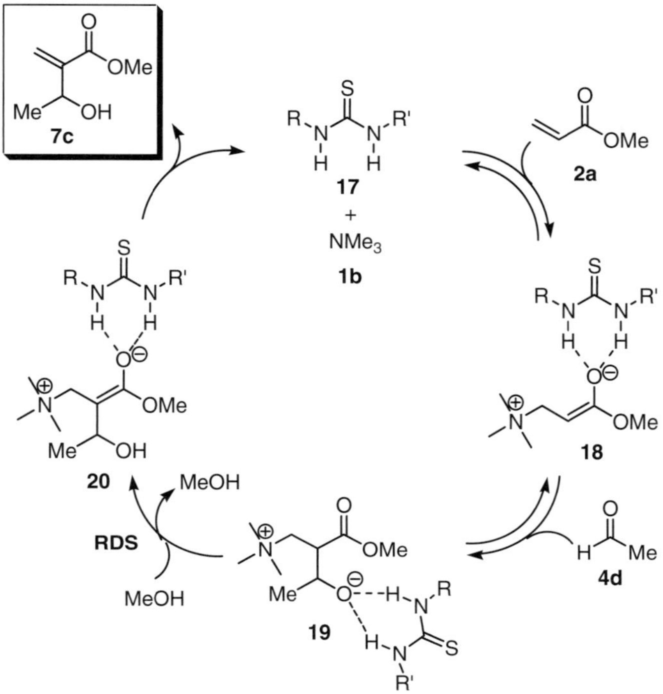

Scheme 3.7 Proposal based on the ESI-MS(/MS) data for the catalytic role of (thio)urea in MBH reactions.

as an organocatalyst in virtually all steps of the MBH cycle, and this key ESI-MS(/MS) finding therefore indicates that (thio)urea participates in the RDS, that is, in the proton transfer step. DFT calculations, performed for a model MBH reaction between formaldehyde and acrolein with trimethylamine as base, either in the presence or the absence of (thio)urea, suggest that (thio)urea accelerates MBH reactions by significantly decreasing the TS energies of all reaction steps in the catalytic cycle, including therefore the rate-limiting proton transfer step. Based on the combined ESI-MS(/MS) and DFT results, a new proposal for the catalytic role of (thio)urea in the MBH reaction is presented (Scheme 3.7).

3.2.3
α-Methylenation of Ketoesters

Many efforts have been made to find optimal synthetic routes of α-methylene carbonyl compounds. These simple but interesting molecules are used in a variety of applications, such as synthetic intermediates [14], to mimic biologically active natural products [15] and as potential antitumor drugs [16].

Direct infusion ESI(+)-MS(/MS) was used to monitor Mannich-type α-methylenation of α-, β-, and γ-ketoesters, and the results guided development of a convenient one-pot method for the efficient preparation of their α-methylene

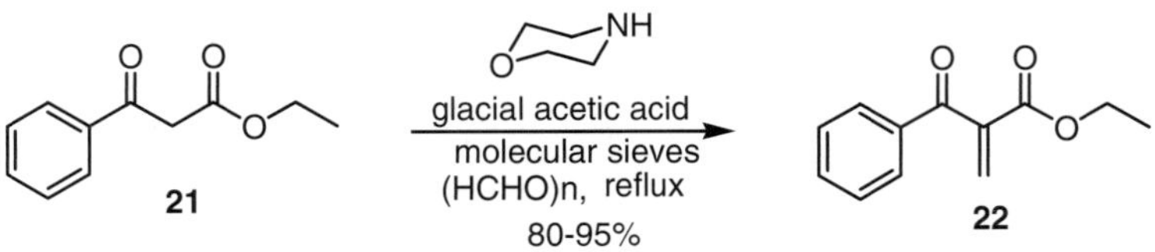

Scheme 3.8 Direct Mannich-type α-methylenation of ethyl benzoylacetate. Optimized conditions: ethyl benzoylacetate (1.0 mmol), morpholine (0.3 mmol), paraformaldehyde (9.0 mmol) and glacial acetic acid (5.0 mL) under reflux, dry nitrogen atmosphere and addition of molecular sieves.

derivatives [17]. More specifically, we monitored the reaction of a ketoester with paraformaldehyde catalyzed by morpholine in an acetic acid/acetonitrile solution (Scheme 3.8).

As Figure 3.3 shows, the ESI(+)-MS collected for the reaction solution was relatively clean and mechanistically enlightening. As soon as the reaction mixture was formed and electrosprayed, ESI(+)-MS detected the two reactants: protonated

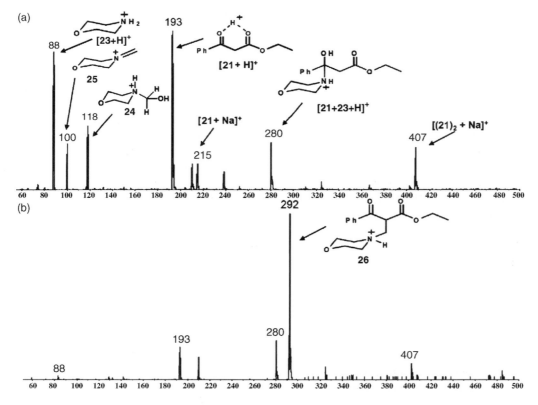

Figure 3.3 ESI(+)-MS for the α-methylenation reaction of **21** after (a) $t_R = 0.1$ min and (b) $t_R = 12$ min. Note the changes in the ion composition of the reaction solution.

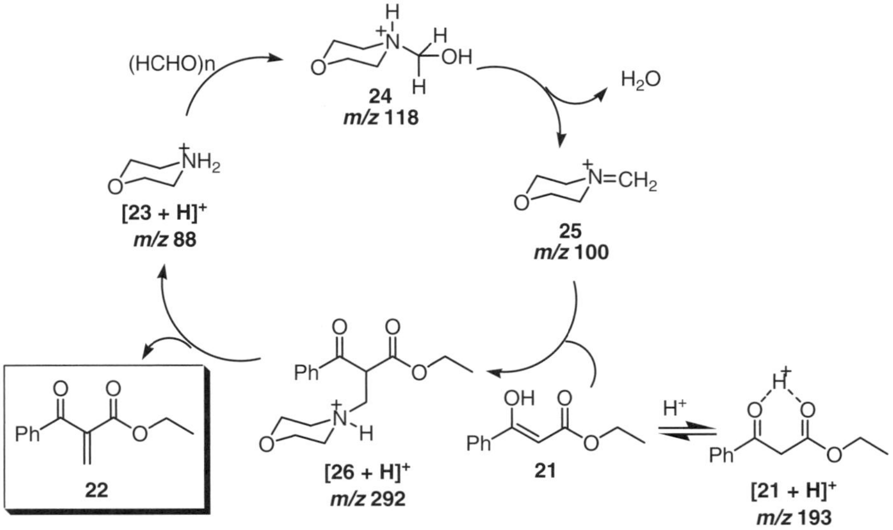

Scheme 3.9 Catalytic cycle for α-methylenation of β-ketoesters in glacial acetic acid medium based on data from ESI-MS(/MS) monitoring.

morpholine, [23 + H]⁺, of m/z 88 and protonated and sodiated ethyl benzoylacetate 21, [21 + H]⁺ of m/z 193 and [21 + Na]⁺ of m/z 215, as well as its proton dimer form, [(21)₂ + H]⁺ of m/z 280. Besides the reactants, two key cationic intermediates were detected: 24 of m/z 118 and 25 of m/z 100.

Previously, only kinetic evidence was available to support the intermediacy of iminium ions such as in Mannich-type reactions [18]. Another key piece of information was provided by continuous monitoring; after 12 min of reaction, the ESI-MS changed considerably and another key intermediate was detected: the aldol 26 in its protonated form of m/z 292 (Figure 3.2b). The abundance of the starting reagents [21 + H]⁺ of m/z 193, [23 + H]⁺ of m/z 88, and intermediates 25 of m/z 100 and 24 of m/z 118 had decreased accordingly. For characterization, the gaseous cationic intermediates were subjected to CID via ESI(+)-MS(/MS) experiments.

Scheme 3.9 summarizes the catalytic cycle proposed from the ESI-MS(/MS) data for direct Mannich-type α-methylenation of ketoester 21, which forms the α-methylene ketoester 22 in the presence of 23. This cycle is based on previous mechanistic interpretations [19] but now shows authentic cationic intermediates (24, 25 and 26) that have been properly detected and characterized via ESI(+)-MS(/MS) monitoring.

3.2.4
Unexpected Synthesis of Conformationally Restricted Analogs of γ-Amino Butyric Acid (GABA) via a Ring Contraction Reaction

γ-Amino butyric acid (GABA) is the major inhibitory neurotransmitter in the central nervous system (CNS) and is likely to be present in about 60–70% of all CNS

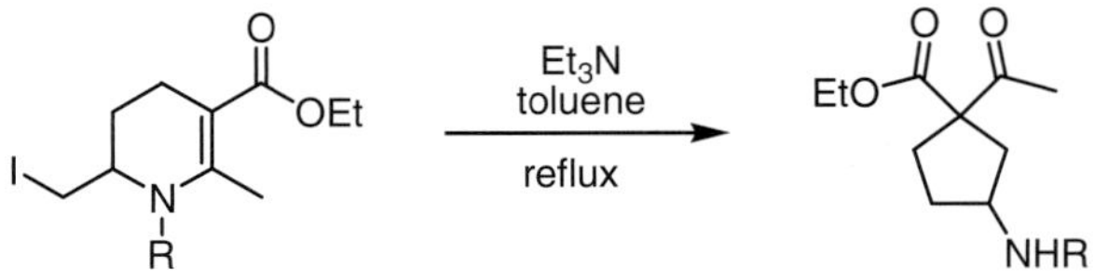

Scheme 3.10 Unexpected ring contraction reaction investigated by ESI-MS(/MS).

27a (R = Bn, 24h)
27b (R = Ph, 72h)
27c (R = p-MeOPh, 72h)

28a (85%)
28b (73%)
28c (77%)

synapses [20]. The synthesis of conformationally restricted GABA analogs is therefore an important current target in organic synthesis, and many attempts have been made to produce GABA model compounds [21] containing chiral centers.

From previous results with lower homologs, dehydroiodination of the three alkenyl-α-enaminoesters (Scheme 3.10) was expected to provide six-membered N-heterocyclic products. The reactions of these enaminoesters with triethylamine were found to lead, however, to the unexpected stereoselective synthesis of trisubstituted cyclopentane derivatives. The cyclopentanes bear two chiral centers and a α-amino ester moiety, and are therefore conformationally restricted analogs of α-amino butyric acid (GABA). Looking for experimental support to validate the mechanism of ring contraction for the reaction outlined in Scheme 3.10, Eberlin and co-workers [22] monitored the reaction by ESI-MS(/MS). A bicyclic iminium ion intermediate 29^+ was proposed to participate in the reaction (Scheme 3.11); both reactant **27** and product **28** are neutral molecules, which could be detected by ESI(+)-MS in their protonated forms $[M + H]^+$. Note that even an unfavorable equilibrium for protonation could be helpful considering the exceptionally high sensitivity of ESI-MS. We first performed the ESI-MS monitoring of the reaction of iodo-β-enamino-esters **27a-c** with Et$_3$N as the base. Iodo-β-enamino-esters **27** (1.0 equiv.) and Et$_3$N (1.0 equiv.) were mixed in 1:1 toluene/methanol (2 mL) at 25 °C, and the reaction was monitored by ESI-MS. Note that even when anhydrous HPLC-grade solvents were used, the traces of water needed for the last hydrolysis step that yields **28** were likely present.

Major ions were clearly detected by ESI-MS, and they corresponded to the protonated reactant $[27a + H]^+$ of m/z 386, intermediate **29a**$^+$ of m/z 258, and the protonated product $[28a + H]^+$ of m/z 276. Likewise for **27b** and **27c**, the protonated reactant $[27b + H]^+$ of m/z 400 and $[27c + H]^+$ of m/z 416, the intermediates **29b**$^+$ of m/z 272, and **29c**$^+$ of m/z 306, and the final protonated products $[28b + H]^+$ of m/z 290 and $[28c + H]^+$ of m/z 306 were also detected. The detected cations, as shown by continuous ESI-MS monitoring, were the same from 1 to 60 min of reaction. Scheme 3.11 summarizes a general reaction pathway showing neutral and protonated reactants and protonated products, as well as the key bicyclic iminium ion intermediates **29a–c**$^+$ that have been intercepted and structurally characterized by ESI-MS(/MS).

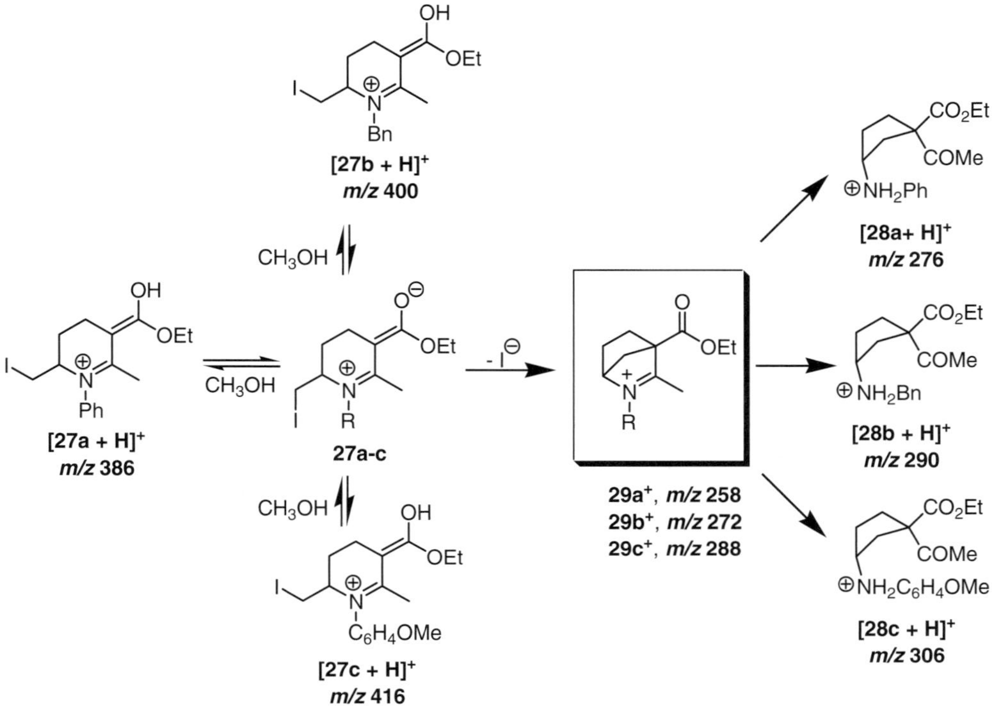

Scheme 3.11 Mechanism for the ring contraction reaction of iodo-β-enamino-esters based on data from ESI-MS(/MS) monitoring. The screening was performed using **27a–c** (1 mmol), Et$_3$N (1 mmol) in toluene/methanol (1 : 1, 2.0 mL) at a flow rate of 0.01 mL min^{-1}.

3.2.5
The Heck Reaction

The Heck reaction, with its many creative and effective variations, finds a prominent place among the synthetic tools available in organic synthesis for the construction of C—C bonds [23]. The Heck reaction has found growing applications mainly because of its rather mild conditions and versatility, and some highly enantioselective versions have been reported [24]. Despite some impressive demonstrations of its synthetic utility, the Heck reaction still has a number of unaccountable features and unverified mechanistic details. The phosphine-free version of the Heck reaction also holds great synthetic potential, and is usually more economical, practical, and experimentally simpler than the phosphine version [25]. One example of the phosphine-free protocol is that in which arene diazonium salts are used as the arylating partner instead of the traditional aryl halides and triflates. This version of the Heck arylation, initially used by Heck and further developed by Matsuda and coworkers [26], has been finding increasing application in the synthesis of natural

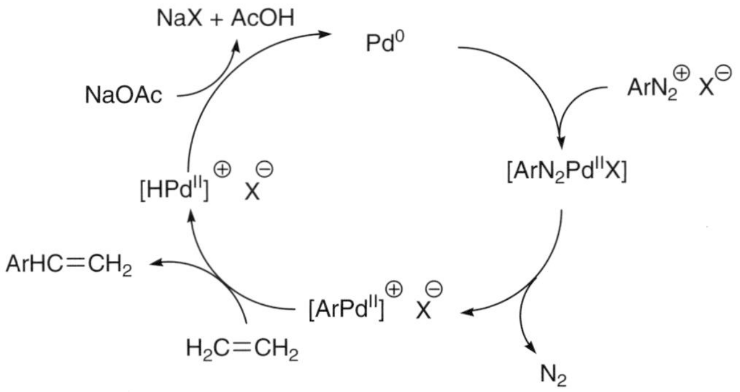

Scheme 3.12 The catalytic cycle proposed by Matsuda and coworkers for the Heck reaction with arene diazonium salts.

product analogs [27] and biologically active compounds [28] but, in contrast to the main-stream halides and triflates, less attention has been given to understanding of its mechanistic aspects. In their initial work [26], Matsuda and coworkers proposed the catalytic cycle outlined in Scheme 3.12 for the Heck reaction with arene diazonium salts.

To 'fish' for the catalytic species in Heck reactions, Eberlin and coworkers [29] performed a study starting with an acetonitrile solution of the arene diazonium salt 4-MeOPhN$_2$$^+BF_4$$^-$ using [Pd$_2$(dba)$_3$]·dba as the source of palladium. The reactant diazonium ion of m/z 135, together with four ionic species, was detected by ESI-MS after 1–5 min of reaction (Figure 3.4a) **30** (m/z 295; all m/z values are reported for ^{106}Pd), **31** (m/z 336), **32** (m/z 488), and **33** (m/z 681), all displaying the characteristic isotopic distributions of Pd-species. Interestingly, the composition of the cationic intermediates before olefin addition changes drastically with time but stabilizes after 90 min, with ions **31** and **30** converting almost completely into **32** (Figure 3.4b). All these potentially catalytic species were selected for ESI-MS(/MS) characterization via CID and to investigate their intrinsic gas-phase reactivity toward model olefins via ion/molecule reactions [30]. These olefins were then added to the reaction solution, and indeed, after 90 min, olefin insertion products were clearly detected, such as those of m/z 324, 517 and 558 formed by the addition of the catalytically active species of m/z 488 to 2,3-dihydrofuran (Figure 3.5).

For the first time, therefore, several cationic intermediates of the oxidative addition step of the Heck reaction involving arene diazonium salts were detected and characterized by ESI-MS(/MS). A dynamic, time-dependent process with ligand equilibria between several ionic intermediates was observed for the oxidative addition step. The most reactive intermediate for olefin addition (the ion of m/z 488 in Figure 3.5) was also detected and characterized, and this species predominates after mixing the arene diazonium salt and [Pd$_2$(dba)$_3$]·dba after about 90 min. Therefore, a novel protocol for the Heck reaction with a delay of 90 min for olefin addition was established for maximum yield. A detailed catalytic cycle for the Heck

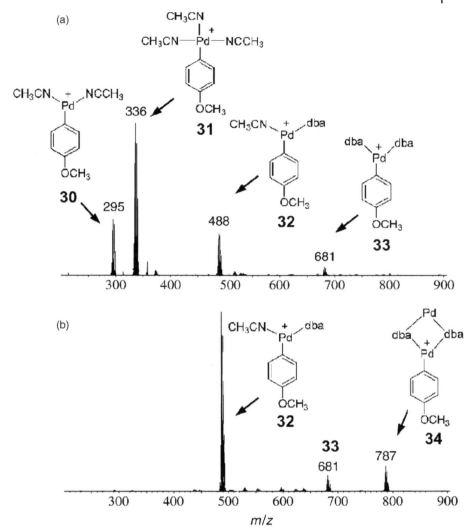

Figure 3.4 ESI(+)-MS of the reaction solution (2 nmol L^{-1}) of the arene diazonium salt 4-MeOPhN$_2^+$ BF$_4^-$ and [Pd$_2$(dba)$_3$]·dba in acetonitrile after mixing for: (a) 5 min and (b) 90 min.

reaction with arene diazonium salts was proposed (Scheme 3.13). The rich set of mechanistic information obtained for the Heck reaction nicely illustrates the power of ESI-MS(/MS) ion fishing for the study of reaction mechanisms both in solution and in the gas phase.

The Heck reaction has been extensively investigated, but less attention has been given to oxa-Heck reactions, and only a few protocols are described for this interesting transformation. In the first examples of Pd-catalyzed oxyarylation reported by Horino and Inoue [31], reactions of styrene-like olefins such as **39** with

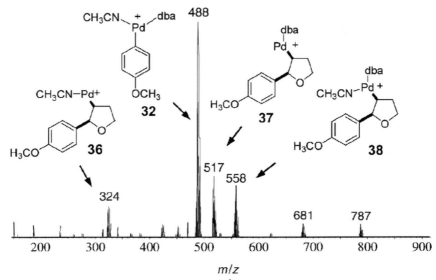

Figure 3.5 ESI(+)-MS for the solution (2 nmol L^{-1}) of the arene diazonium salt 4-MeOPhN$_2$$^+BF_4$$^-$ and [Pd$_2$(dba)$_3$]·dba in acetonitrile after 90 min of mixing with subsequent addition of the olefin **35**.

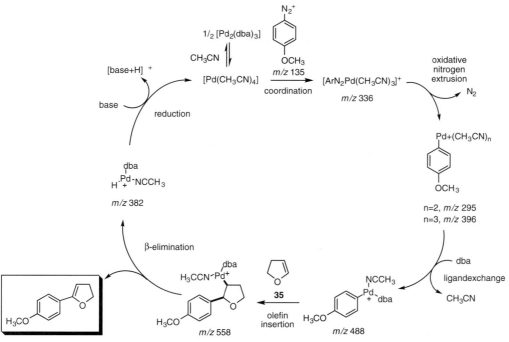

Scheme 3.13 Proposed catalytic cycle for the Heck reaction with arene diazonium salts.

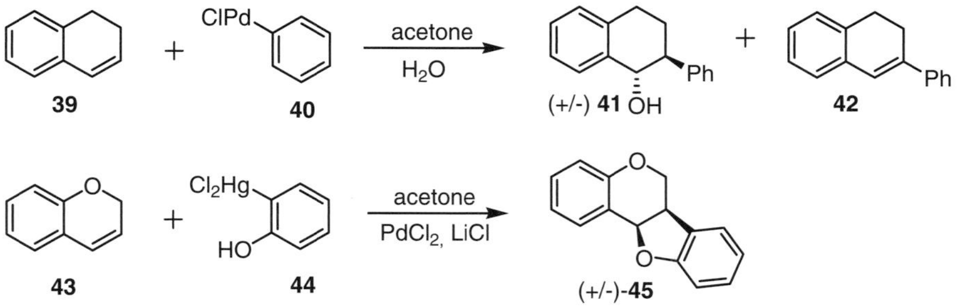

3.2 Reaction Mechanisms | 79

Scheme 3.14 Inter- and intramolecular Heck oxyarylations.

phenylpalladium chloride (**40**) yielded stereoselectively *trans*-**41** in 60% yield along with the Heck adduct **42** (Scheme 3.14). A few years later, the same authors [32] described an intramolecular version of oxyarylation that in contrast, was found to be *cis*-stereoselective, as shown by the reaction of **43** with **44** leading to **45** in 78% yield.

More recently, Emrich and Larock [33] reported the first catalytic oxyarylations of chromens by *ortho*-iodophenols in DMF in the presence of 5 mol% of Pd(OAc)$_2$/Na$_2$CO$_3$, later extending this methodology to other cyclic olefins. Kiss *et al.* [34] also reported the Pd(OAc)$_2$-catalyzed oxyarylation of *O*-6-benzyloxy-chromen by *ortho*-iodophenol in acetone in the presence of Ag$_2$CO$_3$ and PPh$_3$, but the scope of this reaction with other olefins and *ortho*-iodophenols has not yet been investigated.

Recently [35] ESI-MS(/MS) was used to investigate the mechanism of the Pd-catalyzed oxyarylation of **46** by **47a** leading to (+/−)-**48a** (Scheme 3.15). In the presence of PPh$_3$, the arylpalladium ion **51a** and the key intermediate cyclo-palladate **52a** were intercepted and characterized, as Figure 3.6 illustrates. Analogous cationic intermediates with MeCN ligands were also intercepted and characterized in the absence of PPh$_3$. The scope of this synthetically useful reaction was also studied using

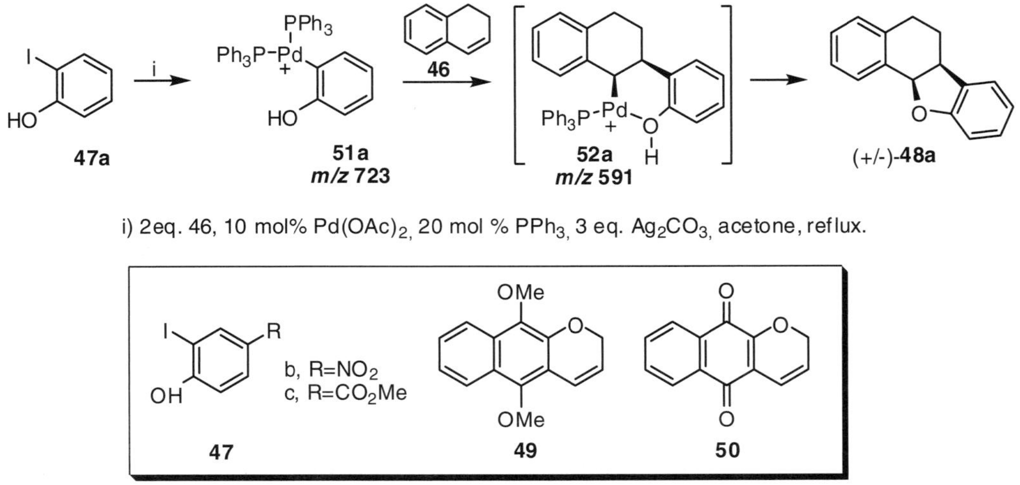

i) 2 eq. **46**, 10 mol% Pd(OAc)$_2$, 20 mol % PPh$_3$, 3 eq. Ag$_2$CO$_3$, acetone, reflux.

Scheme 3.15 Pd-catalyzed oxyarylation of **46** by **47a** leading to (+/−)-**48a**.

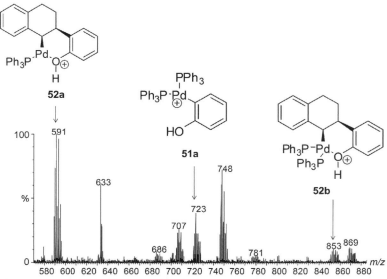

Figure 3.6 ESI(+)-MS of the oxa-Heck arylation reaction solution of **47a** with **46** after dilution with MeCN.

olefins **49** and **50** and *ortho*-iodophenols **47b,c**, and similar intermediates were intercepted. Scheme 3.16 summarizes, using the **46** → **48a** transformation as an example, a mechanistic view for Pd-catalyzed oxyarylation in the presence of PPh$_3$, as indicated by ESI-MS(/MS) monitoring [36, 37].

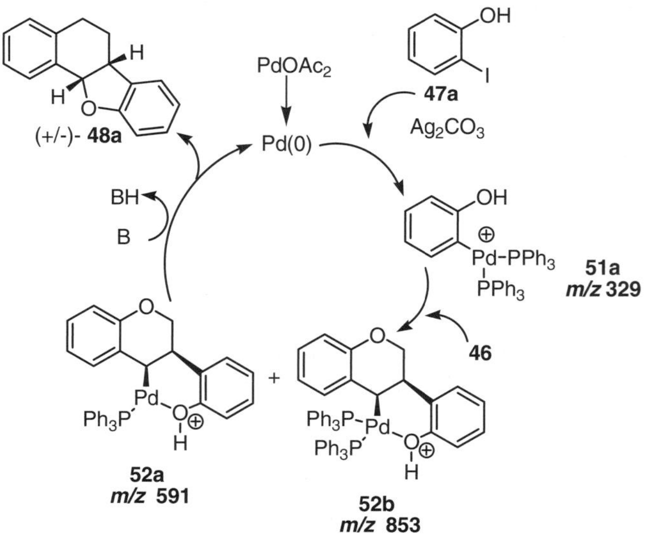

Scheme 3.16 Mechanistic view based on data from the ESI-MS (/MS) monitoring of oxa-Heck reactions of **46** with **47a** in the presence of PPh$_3$.

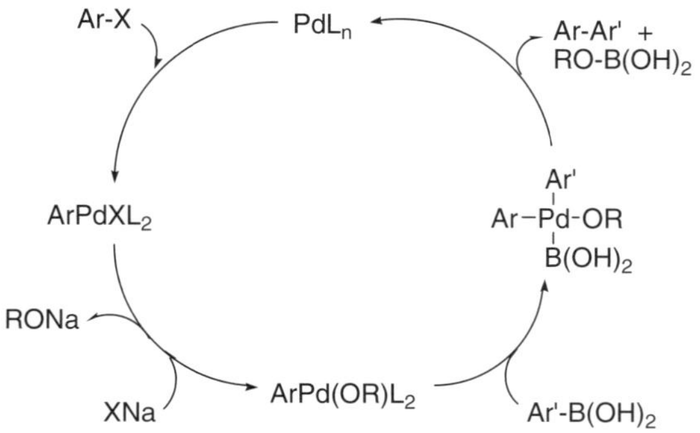

Scheme 3.17 Catalytic cycle for Suzuki-type cross-coupling.

3.2.6
Suzuki Reaction

Palladium-catalyzed cross-coupling between a formal electrophile C−X (X mainly Br, I, OTf) and an organometallic species C−M (M mainly Mg, Zn, Sn, and B) is a versatile synthetic method for making C−C bonds. The cycle proposed by Suzuki and coworkers [38] (Scheme 3.17) is initiated by the oxidative addition of the organic halide to the stabilized Pd(0) species. The transmetalation step transfers the Ar′ group from the boron to the palladium to generate an intermediate containing Ar, Ar′, B(OH)$_2$, and RO in the coordination sphere of palladium. Two reductive eliminations from this intermediate produce the product of Ar-Ar′ coupling and the final boric acid derivative.

Aliprantis and Canary studied the Suzuki coupling reaction of pyridyl halide **53** with phenylboronic acids **55a–c** (Scheme 3.18) [39]. They performed an off-line ESI-MS investigation, identifying pyridylpalladium(II) complexes as the cation species [(pyrH)Pd(PPh$_3$)$_2$Br]$^+$ (**54a**) and [(pyr)Pd(PPh$_3$)$_2$]$^+$ (**54b**). Additionally, diarylPd(II) [(pyrH)(R$_1$R$_2$C$_6$H$_3$)Pd(PPh$_3$)$_2$]$^+$ (**56**) and arylPd(II) [(R$_1$R$_2$C$_6$H$_3$)Pd(PPh$_3$)$_2$]$^+$ were also detected. A few years later, some new results were published by Roglans and coworkers on an ESI-MS investigation of Pd-catalyzed self-coupling reaction of areneboronic acids [40].

3.2.7
Stille Reaction

The Stille reaction [41] is a general, selective, and multifaceted palladium-catalyzed reaction used to construct C−C bonds [42]. It proceeds via Pd-catalyzed coupling of organic electrophiles such as unsaturated halides, sulfonates, or triflates with functionalized organostannanes. Although it is nowadays considered a standard method in organic synthesis, recent modifications and variants of the Stille reaction have opened up a multitude of new and highly attractive synthetic possibilities.

Eberlin and coworkers, [43] performed a ESI(+)-MS(/MS) investigation in which key Pd intermediates involved in the major steps of a real catalytic cycle of a Stille reaction were detected. The well-known working-model mechanism of the Stille [44]

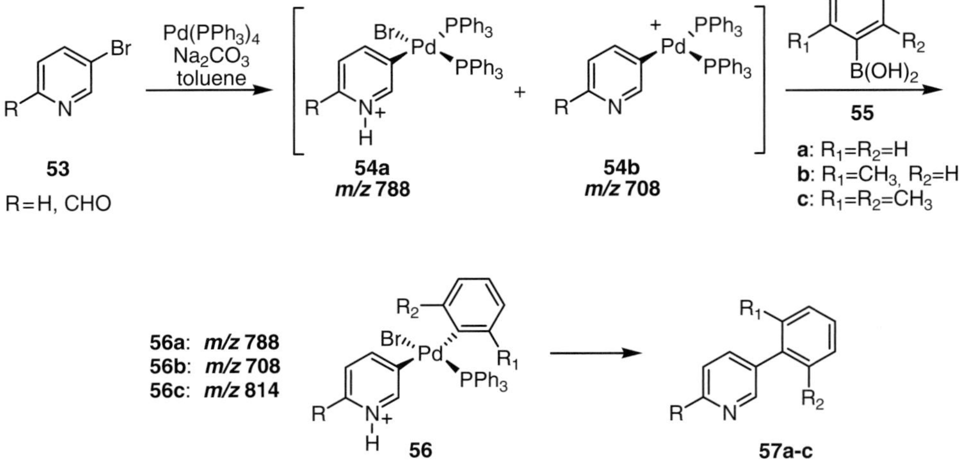

Scheme 3.18 Pd-catalyzed reaction of pyridyl halides **53** with phenylboronic acids **55a–c** after dilution with MeOH for ESI-MS monitoring.

reaction is based on three reaction steps: (i) oxidative addition, (ii) transmetalation, and (iii) reductive elimination (Scheme 3.19).

The investigation was initiated by the Stille coupling reaction of vinyltributyltin (**61**) with 3,4-dichloroiodobenzene (**59**) promoted by Pd(0) precursors, expecting to form styrene derivative **66**. Scheme 3.20 depicts a detailed catalytic cycle for the Stille reaction that emerged from this study. Reaction steps and intermediates are basically similar to the working model proposed in the Stille reaction (Scheme 3.19), but now the reaction intermediates **60a**, **62a**, **63a**, and **65** including the previously elusive

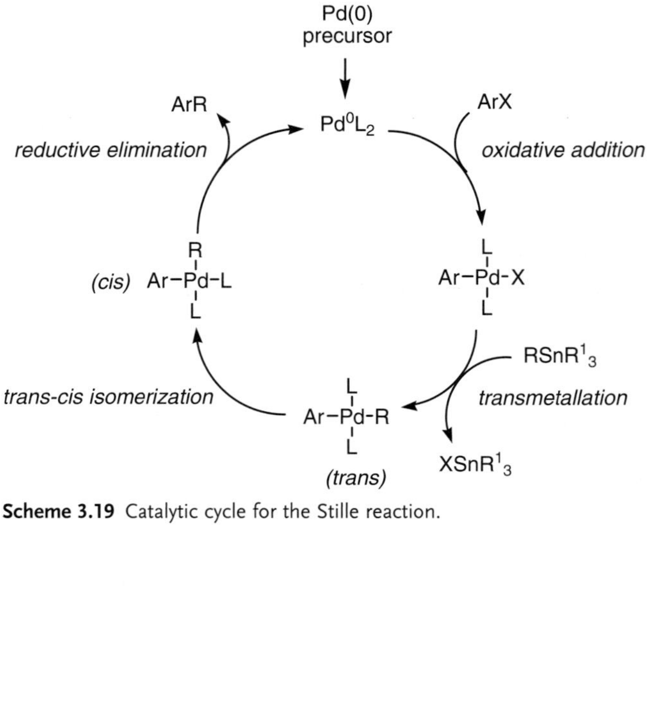

Scheme 3.19 Catalytic cycle for the Stille reaction.

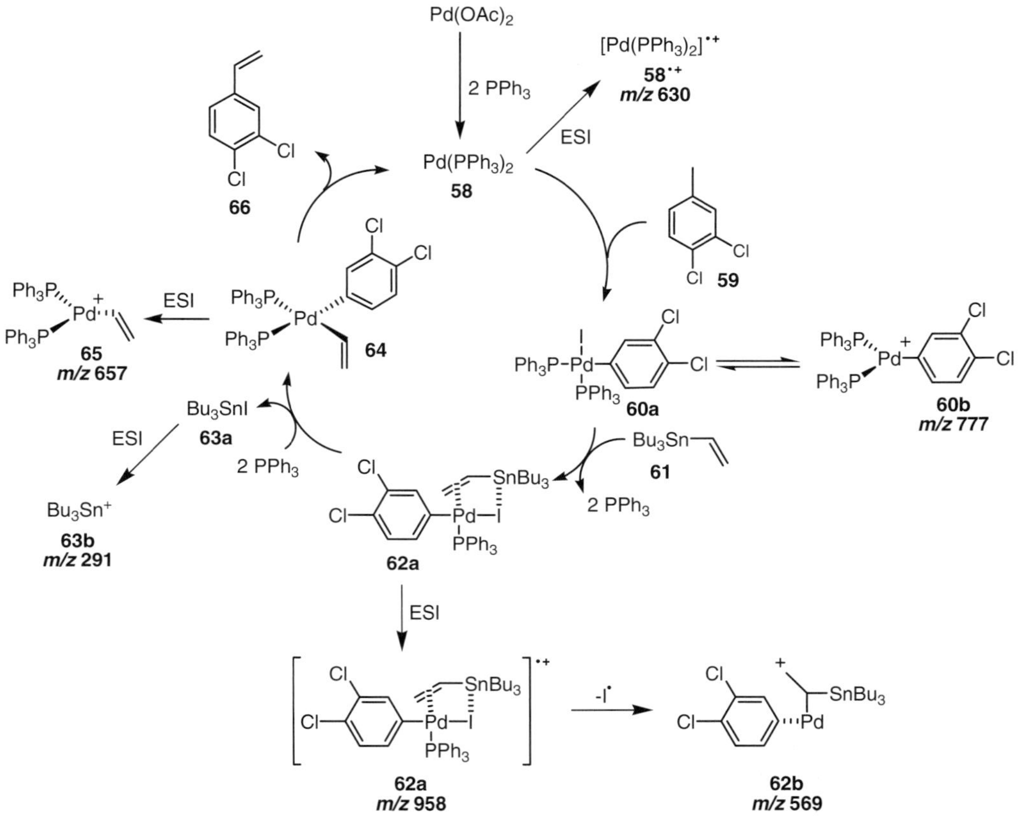

Scheme 3.20 Expanded mechanism based on ESI(+)-MS(/MS) data of the Pd(OAc)$_2$-mediated Stille reaction of 3,4-dichloroiodobenzene (**59**) and vinyltributyltin (**61**) in acetonitrile.

catalytically active Pd(0) species **58** are shown in association with their respective ionic species intercepted by ESI-MS and further characterized by ESI-MS(/MS).

For the Stille reaction, on-line ESI(+)-MS(/MS) monitoring allowed interception and characterization of (a) the actual catalytically active species Pd(Ph$_3$)$_2$, (b) the oxidative addition product **60a** as the corresponding ionic species **60b**, and (c) the transmetalation intermediate **62a** and two products of this process **63a** and **64**. Gas phase reductive elimination (for **65**$^+$) was observed. Therefore, for the first time, most of the major intermediates of a Stille reaction were intercepted, isolated, and characterized. Using ESI(-)-MS, the counteranion I$^-$ was the single species detected.

3.2.8
Three-Component Pd(0)-Catalyzed Tandem Double Addition-Cyclization Reaction

During an investigation on the chemistry of allenes [45], Guo and coworkers [46] developed the Pd(0)-catalyzed three-component tandem double addition-cyclization

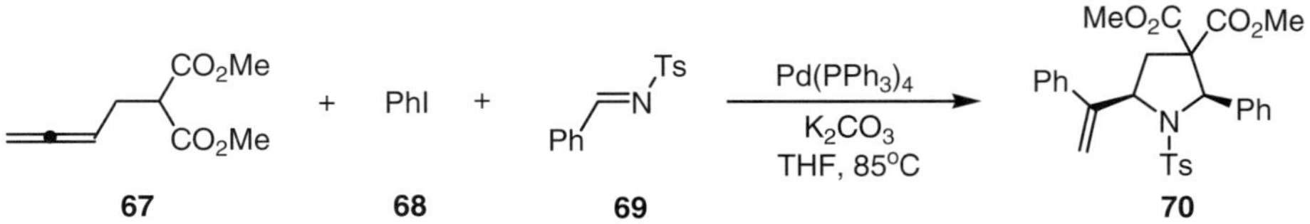

Scheme 3.21 Pd(0)-catalyzed three-component tandem double addition-cyclization reaction of 2-(2,3-allenyl)malonate (**67**), iodobenzene (**68**), and imine **69**.

reaction of 2-(2,3-allenyl)malonate **67**, iodobenzene **68**, and imine **69** for the stereoselective synthesis of *cis*-pyrrolidine derivative **70** (Scheme 3.21) [47]. In principle, this reaction could proceed through two different mechanisms: (i) carbopalladation forming π-allyl palladium species and (ii) azapalladation-reductive elimination mechanism [48].

Using ESI-FTMS experiments, the authors screened first for any possible intermediates in the reaction by gradual addition of the reactants according to the following protocols: (a) A solution of 2-(2,3-allenyl)malonate **67**, iodobenzene **68**, and Pd(PPh$_3$)$_4$ in THF was stirred at 85 °C in a nitrogen atmosphere. After the standard treatment, the ions **71** of *m/z* 707.1 and **73** of *m/z* 891.2 were detected. (b) A solution of 2-(2,3-allenyl)malonate **67**, iodobenzene **68**, imine **69**, and Pd(PPh$_3$)$_4$ in THF was stirred at 85 °C in a nitrogen atmosphere. After the same treatment for the samples taken between the reaction times of 30 min and 2 h, the ions **71** of *m/z* 707.1 and **73** of *m/z* 891.2 were detected. (c) A solution of 2-(2,3-allenyl)malonate **67**, iodobenzene **68**, imine **69**, Pd(PPh$_3$)$_4$, and K$_2$CO$_3$ in THF was stirred at 85 °C in a nitrogen atmosphere. In the first 5 min, **71** and **72** [**71** + CH$_3$CN]$^+$ were detected (Figure 3.7a); at 20 min, **71–73** (Figure 3.7b); at 40 min, **71–74**. Meanwhile, **75** [**67** + K]$^+$, **76** [**69** + K]$^+$, **77** [**70** + K]$^+$ and **78** [(**70**)$_2$ + K]$^+$ started to appear (Figure 3.7c). After 5 h, **71** became less abundant whereas **72** disappeared. After 15 h, **71**, **74**, and **75** also disappeared. After 24 h, **60** became very weak. After 36 h, **73** disappeared, and abundant ions **76–78** could be detected.

Based on the ESI-FTMS results, the authors proposed a reaction mechanism (Scheme 3.22) consisting of the oxidative addition of PhI with Pd(PPh$_3$)$_2$, formed through the disassociation of two molecules of PPh$_3$ from Pd(PPh$_3$)$_4$, yielding intermediate **71**, which can undergo carbopalladation with 2-(2,3-allenyl)malonate **67**, yielding the π-allyl palladium intermediate **73**. Deprotonation of the malonate moiety in **73** and the subsequent addition with imine **69** would yield intermediate **74**. Subsequent intramolecular allylic amination would yield **70** and regenerate the catalytically active Pd(PPh$_3$)$_2$.

3.2.9
Alkynilation of Tellurides Mediated by Pd(II)

Under palladium dichloride catalysis, vinylic tellurides couple efficiently with alkynes with retention of the double bond geometry. Looking for experimental

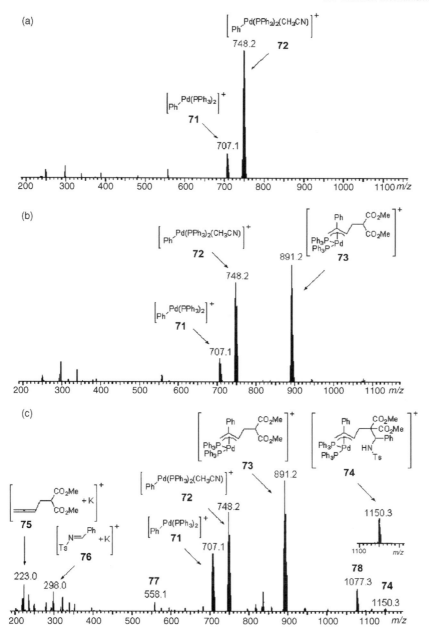

Figure 3.7 ESI(+)-MS for the reaction solution of a mixture of 2-(2,3-allenyl)malonate **67**, iodobenzene **68**, imine **69**, Pd(PPh₃)₄, and K₂CO₃ in THF stirred at 85 °C in a nitrogen atmosphere at the reaction times of (a) 5 min, (b) 20 min, and (c) 40 min.

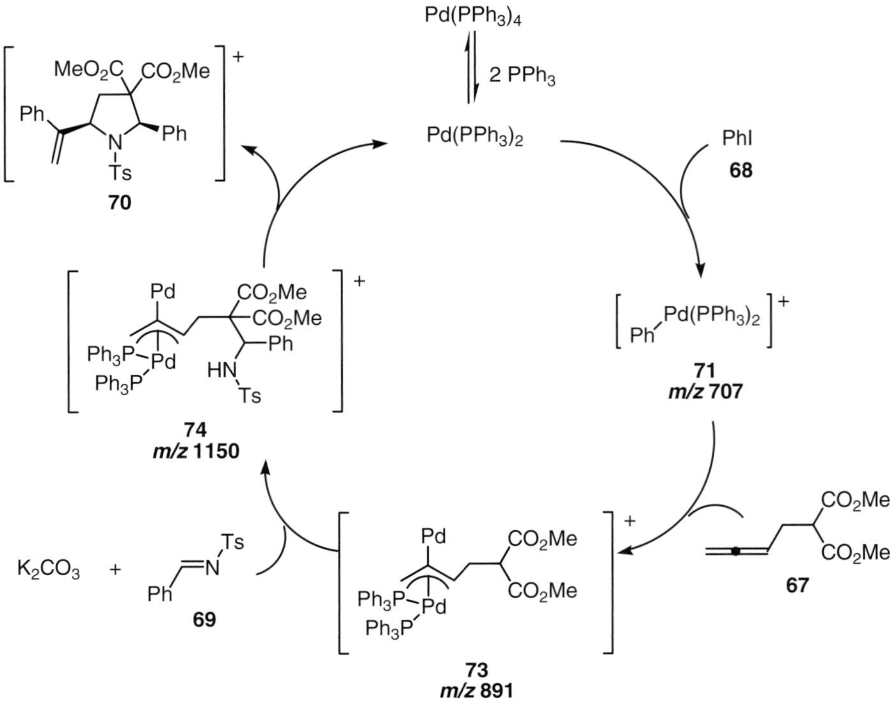

Scheme 3.22 The mechanism of Pd(0)-catalyzed three-component tandem double addition-cyclization reaction based on ESI(+)-FTMS data.

support to validate the catalytic cycle proposed for this reaction [49], Comasseto and coworkers [50] investigated the PdCl$_2$-promoted coupling reaction of vinylic tellurides with alkynes. Pd- and Te-containing cationic intermediates were fished directly from the reaction medium to the gas phase for ESI(+)-MS(/MS). Reaction of telluride **79** with alkyne **80** was performed according to Scheme 3.23 to give **81**. The reaction mixture was stirred for 1 h and the ESI-MS monitoring was able to detect a number of ions that were attributed to the species shown in Scheme 3.24. The most relevant data for the validation of the proposed catalytic cycle was the detection of three Te- and Pd-containing cationic complexes; **82** of m/z 717, **83** of m/z 499, and **84** of m/z 627. Ions **82** and **83** were likely formed by solution ionization of the neutral species L$_n$PdCl$_2$, that is, L$_n$PdCl$_2$ → L$_n$PdCl$^+$ + Cl$^-$. Cations analogous to **82** and **83**

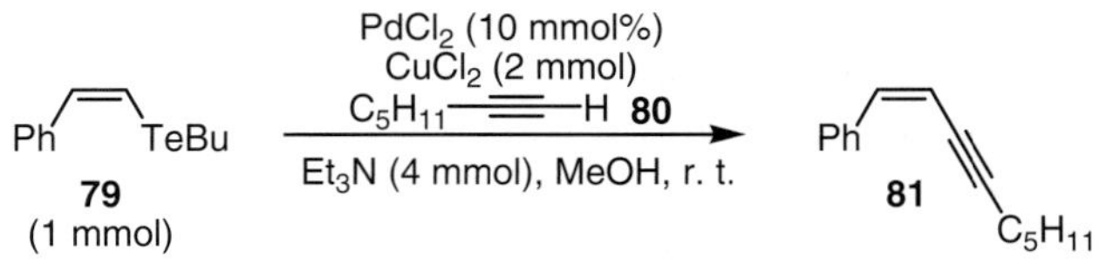

Scheme 3.23 Coupling of vinylic tellurides with alkynes catalyzed by PdCl$_2$ and CuCl$_2$.

3.2 Reaction Mechanisms

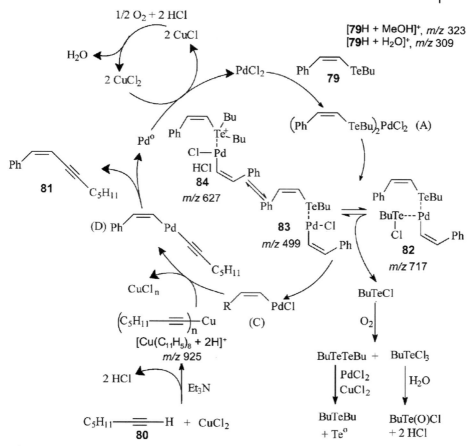

Scheme 3.24 Expanded mechanism based on data from ESI(+)-MS(/MS) monitoring of the reaction of **79** with alkyne **80** to give **81** using PdCl$_2$ and CuCl$_2$ in the absence of an inert atmosphere.

have been suggested [39, 51] using NMR and cyclic voltammetry in reactions where ArPd$^+$ complexes were postulated.

ESI-MS fishing of **82** and **83** in their cationic forms suggested that an equilibrium exists between these two species in solution in the palladium insertion process. It was also suggested that BuTeCl acts as a ligand to stabilize **82**, which may be formed by coordination of two styryl butyl tellurides to PdCl$_2$ followed by transmetalation. Since the relative intensity of **82** and **83** remained nearly constant for up to 36 h of reaction, as shown by continuous ESI-MS monitoring, the ligand exchange equilibrium between **82** and **83** is likely dynamic. The mechanism by which **84** is formed was not so straightforwardly rationalized, but a proposed route to **84** involves **83** in a ligand exchange process, that is, exchange of **79** by [79 + Bu]$^+$ and further association with neutral HCl present in the reaction medium. The species **82** and **83** are Pd-styryl complexes that can undergo β-hydrogen elimination to yield PhC≡CH. However, this acetylene was not observed as a byproduct in the coupling reaction. It is known that cations such as Li$^+$, Ag$^+$, and Tl$^+$ can inhibit such β-hydrogen eliminations [52]. It is also known that cation association (83 → 84) enhances the solubility of metal complexes [53]. Organocopper cluster intermediates were also identified and characterized as alkynylcopper species by characteristic Cu isotopic patterns and ESI-MS(/MS) dissociation. An expanded catalytic cycle for the coupling of vinylic tellurides with alkynes catalyzed by palladium dichloride was proposed (Scheme 3.24) [54].

3.2.10
TeCl$_4$ Addition to Propargyl Alcohols

TeCl$_4$ has been found to display reactivity toward alkynes similar to that of p-methoxyphenyltellurium trichloride, whereas it reacts with aromatic and 3-hydroxyalkynes by different mechanisms as shown by the characteristic stereochemistries of the products. The complete anti-stereospecificity of the additions of TeCl$_4$ to all propargyl alcohols studied is consistent with a cyclic chelated telluronium ion intermediate **87** in this reaction (Figure 3.8). Using ESI-MS and ESI-MS(/MS),

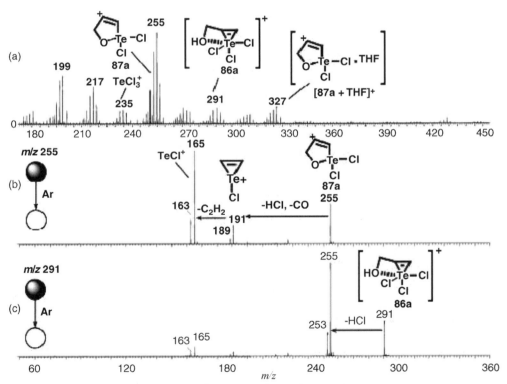

Figure 3.8 (a) ESI(+)-MS of a reaction solution of **85a**/TeCl$_4$ (1.0:0.9 equiv) in THF. ESI-MS(/MS) of the intermediate species (b) **87a** of m/z 255 and (c) **86a** of m/z 291.

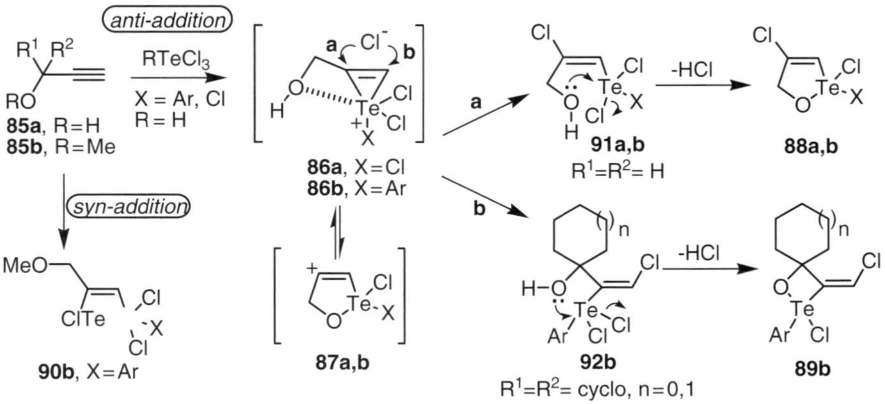

Scheme 3.25 Proposed mechanism of TeCl$_4$ addition to propargyl alcohols.

Santos et al. [55], were able to intercept and characterize the active electrophile TeCl$_3^+$ in the THF solution of TeCl$_4$, as well as its THF complex and several TeCl$_x$(OH)$_y^+$ derivatives. For the first time, therefore, on-line ESI-MS(/MS) monitoring permitted key Te(IV) cationic intermediates of the electrophilic addition of TeCl$_4$ to alkynes to be captured from the solution and to be gently and directly transferred to the gas phase for mass measurement, determination of isotopic patterns, and structural investigation via CID (Scheme 3.25). Two of the reaction products reported (the cyclic telluranes) have been found to act as cysteine protease inhibitors [56], and hence the mechanistic aspects of such reactions revealed in this study may assist in the design of new members of this class of potential antimetastatic agents.

3.2.11
S$_N$2 Reactions

A unique approach using ESI(−)-MS(/MS) monitoring was used to probe the mechanism of nucleophilic substitution reactions of methylated hydroxylamines, hydrazine and hydrogen peroxide with bis(2,4-dinitrophenyl) phosphate (BDNPP) [57]. Along with the ESI(−)-MS(/MS) results, kinetic and NMR spectroscopic data were collected, and the reaction of NH$_2$OH with BDNPP was also studied. It was concluded that there was both aromatic substitution, generating DNPP **95** and compound **96**, and initial phosphorylation of the OH group of NHMeOH by BDNPP (Scheme 3.26, Figure 3.9). The latter generates dinitrophenyl oxide (DNP), forming intermediate **94**, which breaks down slowly by two distinct pathways: (a) aromatic nucleophilic substitution described above, giving **96** and **98**, or (b) spontaneous rearrangement where the terminal NHMe group attacks the dinitrophenyl moiety to form a transient cyclic Meisenheimer complex **99**, as seen with the reaction with NH$_2$OH. The complex **99** rapidly ring opens, giving **97** as a long-lived product, perhaps through the mechanism proposed in Scheme 3.26. Ionic intermediates **94** and **97** (Schemes 3.26 and 3.27) are isomers. Their relative

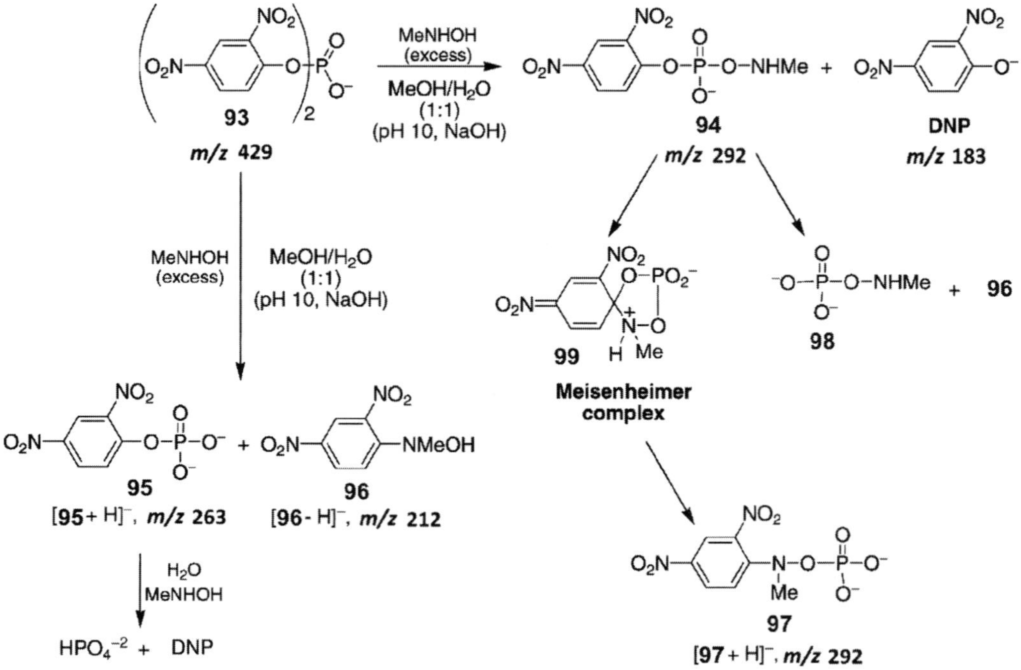

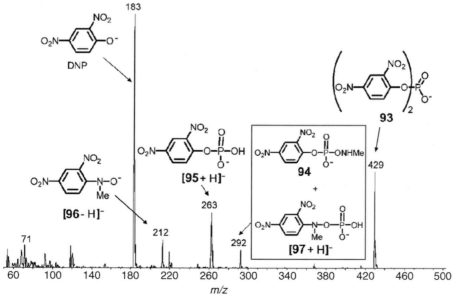

Scheme 3.26 Mechanism for the reaction of MeNHOH and **93** probed by ESI(−)-MS experiments, which corroborated kinetic studies.

Figure 3.9 ESI(-)-MS of the reaction mixture of 0.01 mol L^{-1} BDNPP with 0.1 mol L^{-1} NHMeOH in aqueous methanol (50% v/v) at pH 10.

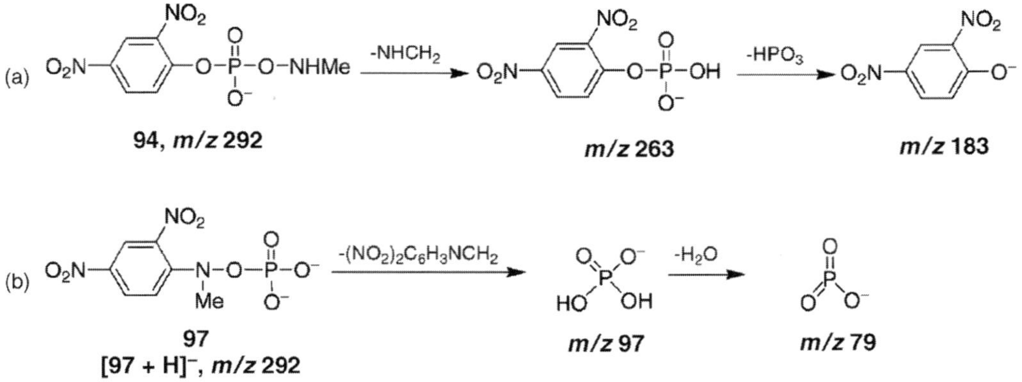

Scheme 3.27 Differentiation of the isomers **94** (a) and **97** (b) by contrasting dissociation pathways in ESI-MS(/MS) experiments. Isomer **94** (m/z 292) was proposed to form **97** (m/z 292) through a Meisenheimer complex intermediate.

amounts in the ionic population of ions of m/z 292 and therefore the ESI-MS(/MS) of such ions changed drastically with time. Samples taken after 10 min of reaction in solution showed that ions of m/z 292 dissociated mainly into four fragment ions of m/z 263, 183, 97, and 79. The fragment ions, of m/z 263 and 183, were formed in the gas phase from **94** (Scheme 3.27a), whereas dissociation to ions of m/z 97 and 79 was evidence for the presence of **97** (Scheme 3.27b). Therefore, both **94** and **97** were present in the reaction mixture at an early stage of the reaction. After 30 or 60 min of reaction, however, the ion of m/z 292 generated only the two fragment ions of m/z 97 (**100**) and 79. This temporal change in product ion distribution from the CID of the ion of m/z 292 confirmed that at a later stage of the reaction, intermediate **97**, rather than **94**, became a dominant species through rearrangement from **94** in solution.

3.2.12
Allylic Substitution Reaction

Based on the work developed by Chen [58], Markert and Pfaltz [59] used ESI-MS to perform the parallel screening of the Pd-catalyzed enantioselective allylation reaction of diethyl ethyl malonate with allylic esters. Enantiodiscrimination of a palladium catalyst with different chiral ligands was also studied by ESI-MS. In the catalytic cycle proposed (Scheme 3.28), the first step, the formation of Pd–allyl complexes **A** and **B** is fast, whereas the second step, nucleophilic addition to the allylic system to give products **102a** and **102b**, is slower. Mass-labeled enantiomers (pseudoenantiomers) were used, so the cationic intermediates **A** and **B** were observed separately in the ESI-MS with characteristic isotope distribution for palladium. The ratio **A:B** was inferred as the catalyst's ability to discriminate between two enantiomers with two different alkyl groups at the para position of the aryl group (Ar = 4-methylphenyl in **101a** and 4-ethylphenyl in **101b**). Five different catalyst precursors were tested with different

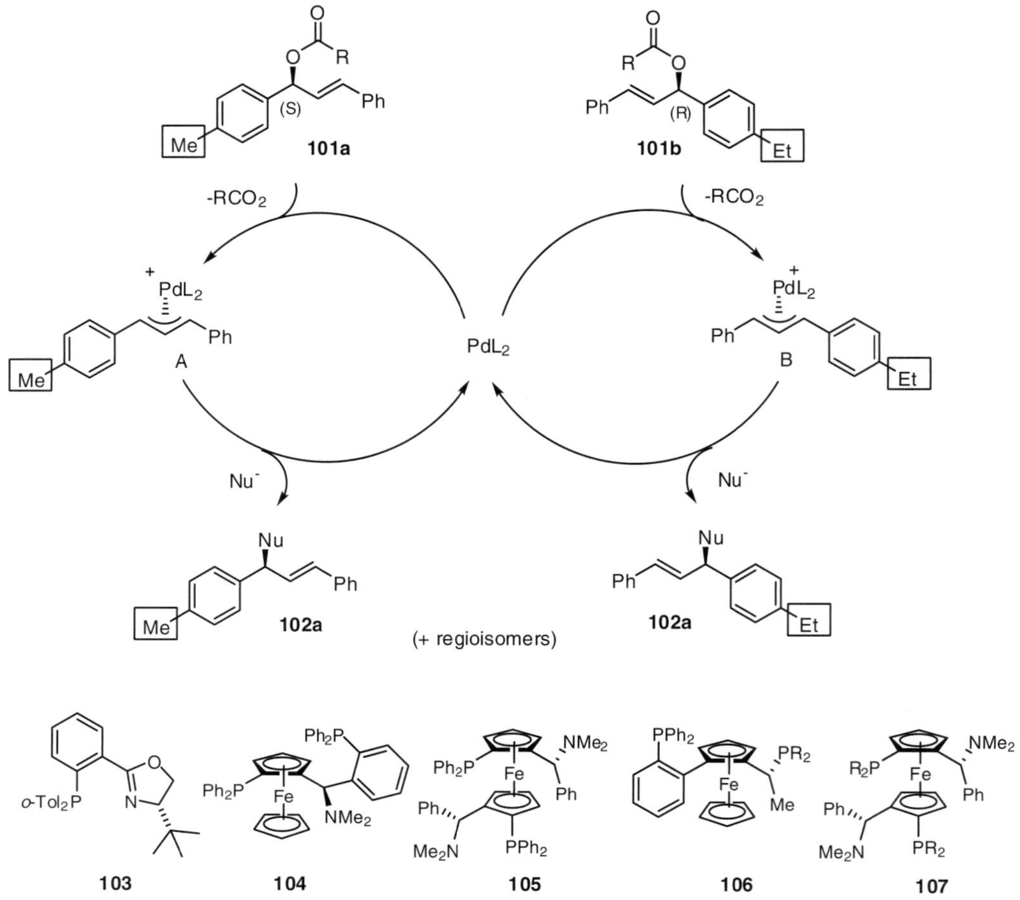

Scheme 3.28 Simultaneous screening of a mixture of five Pd catalysts (0.1 mmol L^{-1}) performed by Markert and Pfaltz.

stereoselectivities in homogeneous solutions containing pseudoracemate **101a,b** and the anion of diethyl ethyl malonate as nucleophiles. All 10 Pd–allyl intermediates were detected by ESI-MS after 2 min of reaction. A selectivity order was proposed from the ion abundance ratios based on the ligand as **106 > 105 > 104 ~ 103 ~ 107**, the complex with ligand **106** being the most selective catalyst. The authors proposed that ESI-MS monitoring may be used with an unlimited number of catalysts in simultaneous screening, as long as the ion signals do not overlap. Roglans and coworkers [60] also studied the substitution of allyl acetate by acetylacetone in presence and absence of Pd catalyst by ESI-MS.

Compounds **101a,b** were used as pseudoracemates in solution, as it was inferred by the authors that methyl and ethyl groups afforded no great differences in the reactivity of allyl groups toward Pd species, supposing almost the same reactivity behaviors for both.

3.2.13
Heterogeneous Fenton Reaction

The Fenton reaction [61] is one of the most efficient advanced oxidation processes for the destruction of organic contaminants in wastewaters. The classical Fenton system [62] uses a mixture of H_2O_2 and a soluble Fe(II) salt to generate *in situ* free hydroxyl radicals (HO•) according to the Haber-Weiss mechanism (Eq. (3.1)) [63]. The reaction of the strong oxidizing HO• mineralizes the organic compounds, thus yielding CO_2 and H_2O as the final harmless products.

$$Fe^{2+} + H_2O_2 \rightarrow Fe^{3+} + OH\bullet + OH^- \tag{3.1}$$

Moura *et al.* used ESI-MS(/MS) to monitor the oxidation of phenol by a novel heterogeneous Fenton system based on an FeO/Fe_3O_4 composite and H_2O_2 [64]. On-line ESI-MS(/MS) shows that this heterogeneous system promotes prompt oxidation of phenol to hydroquinone, which is subsequently oxidized to quinone, other cyclic poly-hydroxylated intermediates, and an acyclic carboxylic acid. A peroxide-type intermediate, probably formed via an electrophilic attack of HOO• on the phenol ring, was also intercepted and characterized. ESI-MS(/MS) monitoring of the oxidation of two other model aromatic compounds, benzene and chlorobenzene, indicates the participation of analogous intermediates. These results suggested that oxidation by the heterogeneous system is promoted by highly reactive HO• and HOO• radicals generated from H_2O_2 on the surface of the FeO/Fe_3O_4 composite via a classical Fenton-like mechanism.

For phenol (**108**), ESI-MS fished a single anion from the aqueous solution at near zero reaction time. As expected, this species corresponds to deprotonated phenol [**108** − H]$^-$ of m/z 93. Relatively intense anions of m/z 109, 113, 123, 125, and 157 were detected as the degradation proceeded for 20–40 min. These anionic intermediates most likely arise from deprotonation of their respective neutral forms in aqueous solutions. Despite the presumably low aqueous concentrations of these anions (because their neutral forms should predominate at pH ∼ 6), ESI(−)-MS detection was easily achieved. The ESI(−)-MS interception of the anionic species suggested successive OH radical attacks, likely at the activated *ortho* or *para* position of the phenol ring, or both. A reaction view (Scheme 3.29) via intermediates **108**–**113** was rationalized for the oxidation of phenol by the $FeO/Fe_3O_4/H_2O_2$ system.

3.2.14
Mimicking the Atmospheric Oxidation of Isoprene

Photooxidation of isoprene in the atmosphere was thought to yield only lighter and more volatile products such as formaldehyde, methacrolein, and methyl vinyl ketone. Evidence that the formation of hygroscopic polar products such as **115** can give rise to aerosols by gas-to-particle formation processes has opened a new panorama for the role of isoprene in the atmosphere. Under simulated atmospheric conditions, photooxidation of isoprene by ozone was shown to be relatively slow and to occur mainly by reaction with OH radicals [65]. Furthermore, when the OH•-initiated

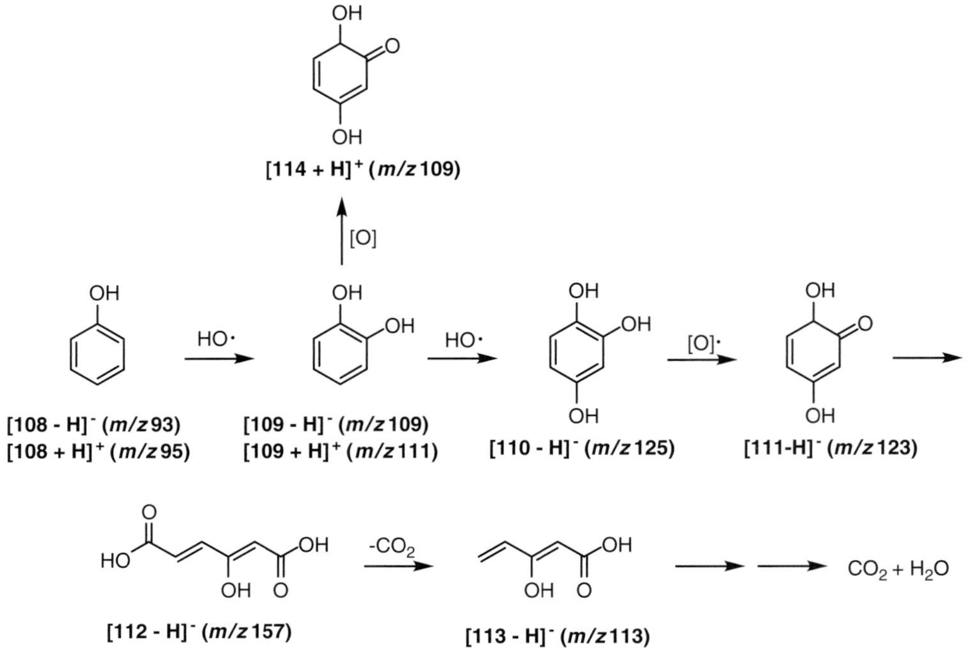

Scheme 3.29 A reaction view for phenol degradation by a heterogeneous Fenton system based on ESI-MS(/MS) data.

photooxidation of isoprene was performed in the absence of NO_x, the formation of 1,2-diol derivative **115** was observed (Scheme 3.30). These diasteromeric polyols have recently been identified in atmospheric aerosols of the Amazon forest [66], and their highly hygroscopic nature likely enhances the capability of aerosols to act as cloud condensation nuclei. Santos et al. tried to mimic the OH radical-mediated oxidation of isoprene in solution and to monitor the process on line and in real time by ESI(-)-MS(/MS) [67]. The OH radical-mediated oxidation of isoprene by H_2O_2 in solution initiated either photochemically or by AIBN seems to mimic adequately the atmospheric oxidation of isoprene, as revealed by ESI(−)-MS(/MS) monitoring. The *in situ* fishing, detection and structural characterization of **116** in its deprotonated form supports the 'C5 isoprene hypothesis' formulated [67] for the formation of the secondary organic aerosol (SOA) marker compounds, the diastereomeric 2-methyltetrols **116**. Other polyoxygenated compounds assigned as **115** (the diol precursors of **116**) and **117–120** have also been intercepted and structurally characterized. It is suggested that products **117–120** are formed by further oxidation of **115** and **116**, and they may therefore also be considered as plausible SOA components occasionally formed in normal or more extreme OH radical-mediated photooxidation of biogenic isoprene (Scheme 3.30). This study also revealed that ESI(−)-MS(/MS) monitoring has the potential to mimic and therefore to be used to study the OH radical-mediated oxidation of first-generation gas-phase oxidation products of isoprene and α-pinene, such as methacrolein and pinic acid.

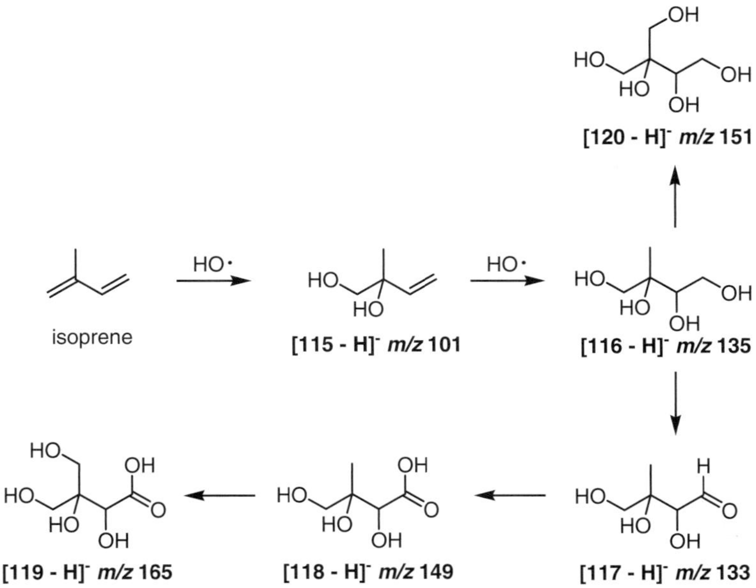

Scheme 3.30 Proposed route based on data from ESI-MS(/MS) monitoring of the formation of the oxygenated products **115–120** in the AIBN-initiated reaction of isoprene with hydroxyl radicals from hydrogen peroxide.

3.2.15
Advanced Oxidation Processes of Environmental Importance

Caffeine has been reported [68] to scavenge highly reactive free radicals, including hydroxyl radicals and excited states of oxygen, and to protect crucial biological molecules against these species. The antioxidant activity of caffeine is similar to that of the established biological antioxidant glutathione and significantly higher than that of ascorbic acid [69].

Using three different oxidation systems: UV/H_2O_2, TiO_2/UV, and Fenton, Eberlin and coworkers [70], have performed on-line and real-time ESI-MS(/MS) monitoring of advanced oxidation of caffeine in water to directly screen for intermediates and products of this environmentally important process. Together with high-accuracy MS measurements and GC-MS analysis, ESI-MS(/MS) showed that caffeine is first oxidized to N-dimethylparabanic acid, likely via initial OH insertion to the C4=C8 caffeine double bond. A second degradation intermediate, di(N-hydroxymethyl) parabanic acid, was identified by ESI-MS(/MS) and high-accuracy mass measurements. This polar and likely relatively unstable compound, which is not detected by offline GC-MS analysis, is likely formed via further oxidation of N-dimethylparabanic acid at both of its N-methyl groups and constitutes an unprecedented intermediate in the degradation of caffeine.

After UV irradiation for 90 min and then 150 min, ESI-MS was able to fish two increasingly abundant ions of m/z 143 (N,N-dimethylparabanic acid, **122**) and m/z

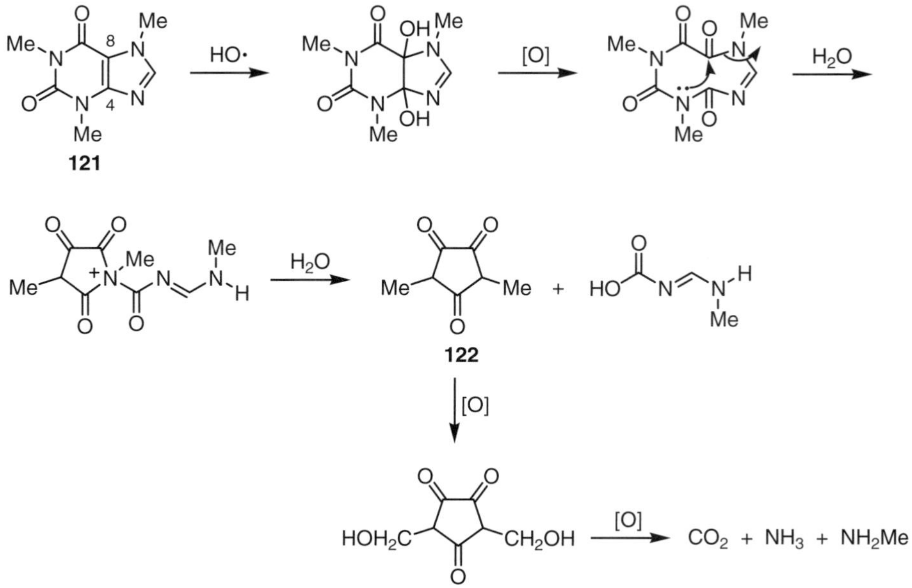

Scheme 3.31 Mechanism of caffeine degradation by H_2O_2/UV, TiO_2/UV or Fenton systems based on data from ESI-MS(/MS) monitoring.

175 [bis(*N*-hydroxymethyl)parabanic acid, **123**]. The conversion of **121** to **122** and then to **123** is further supported by the ESI-MS(/MS) of their protonated molecules, which display a series of structurally diagnostic fragment ions. The mechanism to convert caffeine (**121**) to dimethylparabanic acid (**122**) involves initially the (fast) attack of the hydroxyl radicals to the C4=C8 double bond of caffeine (Scheme 3.31). After successive hydroxylations and oxidations, **122** and **123** are formed, whereas **123** and its co-product are slowly mineralized to CO_2, NH_3, and NH_2Me.

3.2.16
Tröger's Bases

Tröger's bases constitute a relatively simple but geometrically rich, V-shaped class of bicyclic molecules that are commonly formed under acid catalysis by the simple condensation of anilines and formaldehyde. They have found a variety of applications, such as their use as relatively rigid chiral frameworks for the construction of chelating and biomimetic systems [71].

In general, it is now accepted that the mechanism for the formation of Tröger's bases involves a series of *in-situ* Friedel-Crafts reactions of **125**, but the detailed course of the reaction has not yet been fully established. In trying to elucidate major mechanistic aspects, different methylene sources as well as different anilines have

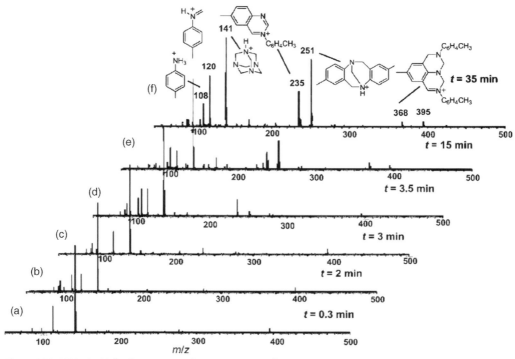

Figure 3.10 ESI(+)-MS for the reaction solution in neat TFA for the reaction of 4-toluidine and urotropine (**129**) after (a) 20 s; (b) 2 min; (c) 3 min; (d) 3.5 min; (e) 15 min; and (f) 35 min.

been used by Abella *et al.* [72] to form Tröger's bases. Direct infusion ESI-MS(/MS) was used to monitor Tröger's base formation from different anilines (4-toluidine and 4-aminoveratrol), using both formaldehyde and urotropine (**129**) as the methylene sources (Figure 3.10). To gain further evidence for the methylene transfer step, gas-phase ion/molecule reactions of protonated urotropine with two volatile amines were also performed.

Interestingly, protonated urotropine of m/z 141 was found to react with ethyl-amine and aniline in the gas phase to directly form the corresponding iminium ions of m/z 56 and 106 (Figure 3.11). Figure 3.11a shows that the iminion ion $C_2H_5-NH=CH_2^+$ of m/z 56 is formed in high yield in the ion/molecule reactions. The adduct of m/z 186, the first transient intermediate leading to methylene transfer, was also obtained from nucleophilic addition of the amine to protonated urotropine. Furthermore, in reactions with aniline (Figure 3.10b), the corresponding imine of m/z 106 (as for **125**, Scheme 3.32), that is, $Ph-NH=CH_2^+$, was formed as an abundant ion, and the transient adduct of m/z 234 was detected as a low abundance ion.

On the basis of the mechanistic data collected from on-line ESI-MS(/MS) monitoring and the gas-phase ion/molecule reactions just described, an experimentally probed mechanism for the formation of Tröger's bases using either formaldehyde or,

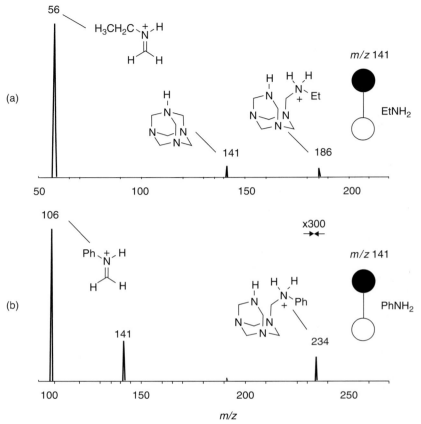

Figure 3.11 ESI(+)-MS(/MS) for gas-phase ion/molecule reactions of protonated urotropine of m/z 141 with (a) ethylamine and (b) aniline.

more particularly, urotropine as the source of methylene was presented (Scheme 3.32). This mechanism proposes the participation of all of the intermediates intercepted and characterized by ESI-MS(/MS).

3.2.17
The Three-Component Biginelli Reaction

In 1893, Pietro Biginelli published his pioneering findings on a three-component reaction that has become known as the Biginelli reaction. This three-component one-pot reaction leads to the synthesis of dihydropyrimidines (**130**), typically via the reaction of a benzaldehyde (**131**), an acetoacetate (**132**), and a (thio)urea (**133**) under acid catalysis [73]. The Biginelli reaction is quite versatile since it can be performed with variations in all three components, leading therefore to a myriad of dihydropyrimidines (Scheme 3.33).

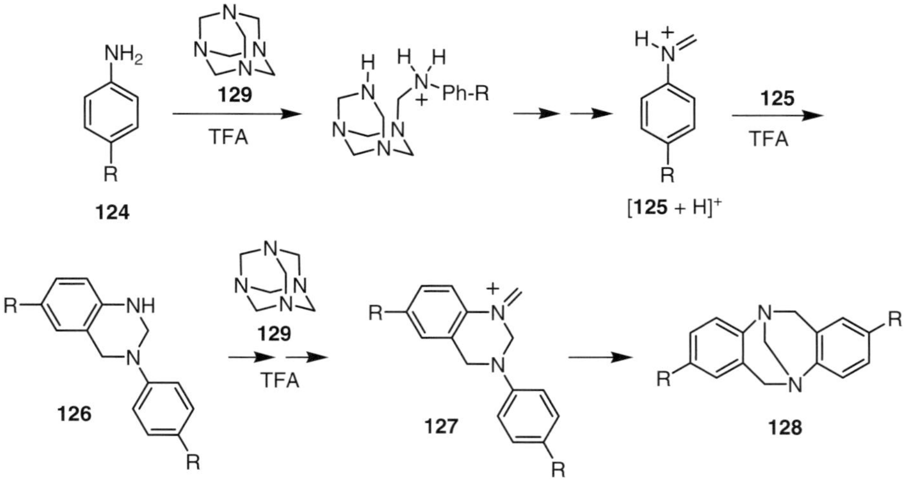

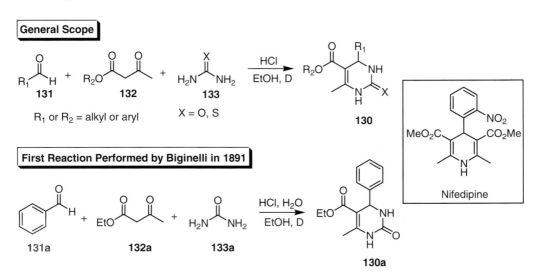

Scheme 3.32 Mechanism for the formation of Tröger's Bases based on data from ESI(+)-MS(/MS) monitoring and gas-phase ion/molecule reactions.

Biginelli reactions can be performed under a variety of conditions, and several improvements in its experimental procedure have been reported in recent years. A driving force in developing synthetic methodologies for Biginelli products **130** is their similar pharmacological profile with analogous drugs, such as nifedipine (Scheme 3.33), which act as calcium channel modulators [74].

Since its first report in 1893, a number of alternative mechanisms have been put forward for the Biginelli reaction: the iminium [75], the enamine [76], and the Knoevenagel [77] mechanisms, which Scheme 3.34 summarizes.

Scheme 3.33 The one-pot three-component Biginelli reaction.

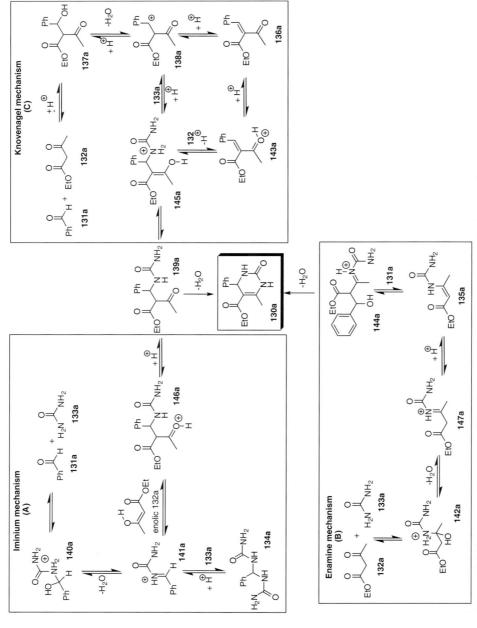

Scheme 3.34 The iminium (A), enamine (B) and Knoevenagel (C) mechanistic views for the Biginelli reaction.

ESI-MS and ESI-MS(/MS) were used to monitor this three-component reaction [78] and to try to intercept intermediates that would provide a detailed picture of the Biginelli mechanism in light of its three alternatives (Scheme 3.34). Additionally, a theoretical investigation using DFT calculations to evaluate the feasibility of the three competing mechanisms was also undertaken. The Biginelli intermediates were expected to be transferred directly from solution to the gas phase and detected by ESI-MS in the positive ion mode either in their natural cationic forms (such as **141a** and **138a**) or in protonated forms (such as **143a**). Based on the overall mechanistic view of Scheme 3.34, we first investigated the formation of the dormant bisureide **134a**. Benzaldehyde (**131a**, 1 mmol) and urea (**133a**, 2 mmol) were mixed in aqueous methanol (1 : 1 v/v, 5 mL) in the presence of a catalytic quantity of formic acid (0.1 mol%) and, most importantly, in the absence of ethyl acetoacetate (**132a**). After 5 min, a sample of the reaction solution was taken and its ESI(+)-MS recorded. Figure 3.12 shows that ESI(+)-MS was able to intercept key cationic species, namely protonated bisureide [**134a** + H$^+$] of m/z 209, the iminium ion **141a** of m/z 149, its hydrated precursor **140a** of m/z 167, and the reagents, protonated benzaldehyde [**131a** + H]$^+$ of m/z 107 and the protonated urea dimer of m/z 121. The intermediates intercepted by ESI-MS were then selected for further ESI(+)-MS(/MS) characterization, which, by showing predictable dissociation routes, provided evidence for the proposed structures (not shown).

We then performed the Biginelli reaction under the usual conditions with **131a**, **132a**, and **133a** (1 mmol each) in aqueous methanol (1 : 1 v/v, 5 mL) with a sub-stoichiometric amount of formic acid (0.1 mol%) as catalyst. Figure 3.13 shows the ESI(+)-MS collected after 5 min of reaction. Ions that were previously detected in the absence of **132a** (Figure 3.11) were again detected, but now three novel ions appeared which correspond to (a) intermediate **142a** of m/z 191, (b) the final Biginelli product in its protonated form [**130a** + H$^+$] of m/z 261, and (c) [**139a** + H$^+$] of m/z 279, that is most likely **146a** in Scheme 3.34 (perhaps in equilibrium with its isomeric form **145a**). These key ions were also characterized by ESI(+)-MS(/MS), and their dissociation chemistry was found to agree with the proposed structures.

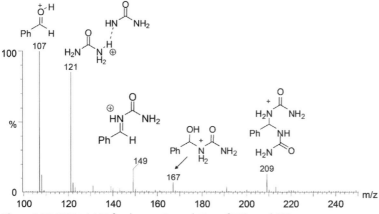

Figure 3.12 ESI(+)-MS for the reaction solution of **131a** and **133a**.

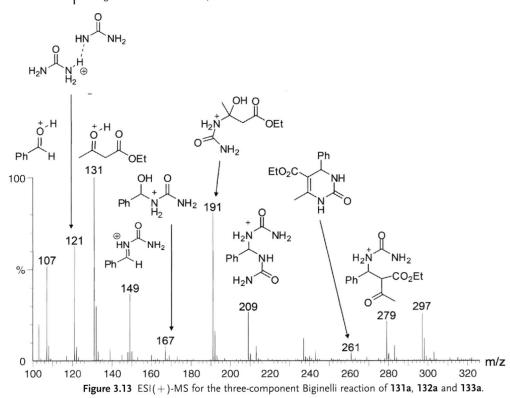

Figure 3.13 ESI(+)-MS for the three-component Biginelli reaction of **131a**, **132a** and **133a**.

Ions for the Knoevenagel mechanism such as the carbenium ion **138a** of m/z 219 or its isomer **143a** or the protonated form of the hydrated precursor [**137a** + H$^+$] of m/z 237 (Scheme 3.34) were not detected by ESI-MS (Figure 3.12), even after a reaction time of 2 h with continuous monitoring. The ESI(+)-MS interception of **140a**, **141a** and [**139a** + H$^+$] are therefore more consistent with the iminium mechanism (**A** in Scheme 3.34).

Kappe has questioned the participation of enamine **135a** (enamine mechanism **B**), arguing that the equilibrium for its formation lies far to the side of the reactants **132a** and **133a**. Via ESI(+)-MS monitoring we detected and characterized **142a** but failed to detect the enamine **135a** in its protonated form **147a** of m/z 173.

We also monitored, via ESI(+)-MS, the Knoevenagel condensation between **131a** (1 mmol) and **132a** (1 mmol) in aqueous methanol (1 : 1v/v, 5 mL) leading to **137a** in the presence of formic acid (0.1 mol%) and in the absence of **133a**. The reaction was monitored periodically, and after 6 h only the protonated reactants could be detected. The reaction solution was left stirring overnight and, after 24 h, the ESI(+)-MS of Figure 3.14 was collected. Analysis reveals that the protonated Knoevenagel adduct **143a** of m/z 219 and its protonated aldolic precursor [**137a** + H$^+$] of m/z 237 were finally detected. The Knoevenagel mechanism, evidenced by the detection of the intermediate **136a** (detected as its protonated form **143a**) after a 24 h reaction period, seems to be feasible but too slow

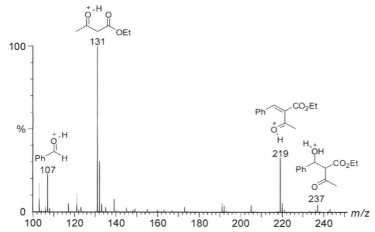

Figure 3.14 ESI(+) MS of the Knoevenagel condensation between **131a** and **132a** after 24 h.

in comparison to the much shorter time scale of the Biginelli reaction to contribute significantly to product formation under the conditions used in this study.

ESI(+)-MS monitoring of Biginelli reactions under three- and two-component conditions indicated therefore that the Knoevenagel pathway is too slow and should not significantly contribute to formation of the Bignelli product. In agreement with this finding, DFT calculations of the Knoevenagel pathway revealed a step with the greatest activation barrier for the proposed mechanisms. A single early intermediate (**142a**) associated with the enamine mechanism was intercepted, but it is postulated to be a dormant species that reverts to reagents during the course of reaction. Several intermediates associated with the iminium mechanism were intercepted and characterized by ESI(+)-MS(/MS). DFT calculations, including solvent effects, have also indicated that the iminium mechanism is the kinetically and thermodynamically favored route to the Biginelli product. Therefore, the combined experimental and theoretical results support the idea that the iminium mechanism (route **A**) is favored in Biginelli reactions. The most reasonable mechanistic interpretation would therefore assume three principal steps for the Biginelli reaction: (1) condensation of aldehyde **130** with the (thio)urea **133** in acidic media to form iminium ion **141** as the key intermediate; (2) addition of **141** to enolic acetoacetate **132** to form **139**, the immediate acyclic precursor of the final product, and (3) intramolecular condensation, via elimination of water from **139**, leading to formation of the final heterocyclic dihydropyrimidine product **130** via dehydration (Scheme 3.34).

3.2.18
Modeling the Ribonuclease Mechanism

The mechanism of ribonuclease A (RNase A) activity has been widely studied by evaluating different aspects such as roles of the catalytic amino acids, different substrates, thio effects, organic solvents, and pH and temperature dependence [79]. The usually accepted mechanism is the general acid/base pathway where a deprotonated imidazole group acts as a general base and another protonated imidazole group acts as a general acid interacting with the leaving group. This 'classical' mechanism has been questioned and a triester-like route has been considered with protonation of a nonbridging phosphoryl oxygen leading to a transition state resembling those typical of phosphate triester reactions [80]. Many bifunctional mimics have been studied, including oligonucleotides with imidazole residues and cyclodextrins with imidazole groups attached to the sugar residues. In the particular case of cyclodextrin mimics of RNase A, their catalytic efficiency is attributed not only to the imidazole groups but also to their hydrophobic binding to the substrate, although the importance of the hydrophobic effects has been questioned because catalytic effects are small for reactions of a variety of substrates with polymers having hydrophobic centers [81]. Although all these models resemble some aspects of RNase A reactions, they do not support fully the classical mechanism because no molecule truly models the action of RNase A, with imidazole and phosphate ester groups in the same molecule. The hydrolysis kinetics of a phosphate diester, (bis(2-(1-methyl-1H-imidazolyl)phenyl) phosphate), BMIPP, which bears two imidazole groups that may potentially act as the general acid/base catalysts conserved in the active site of RNase A (Scheme 3.35) have been performed [82].

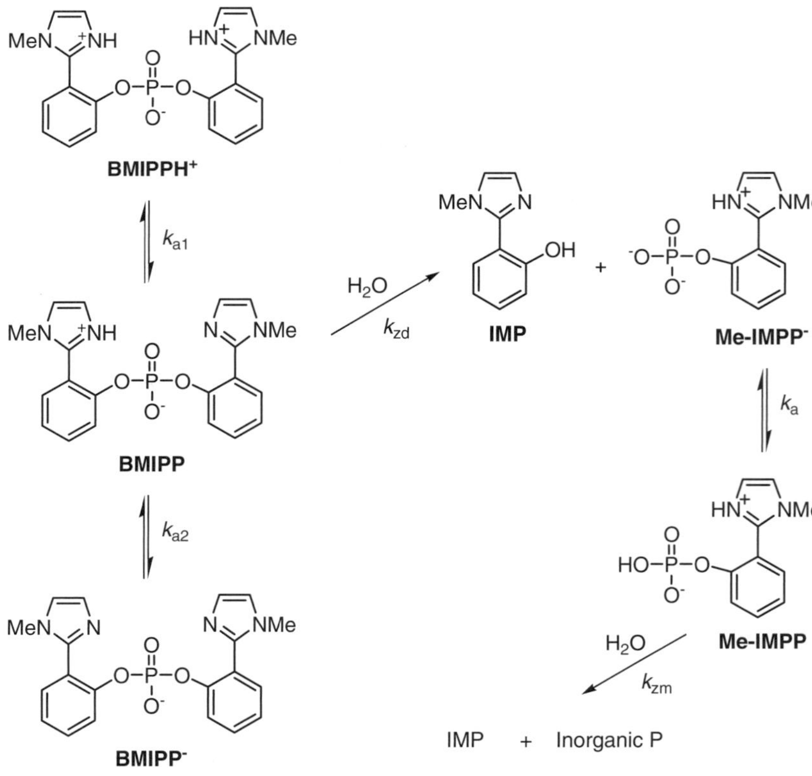

Scheme 3.35 Hydrolysis kinetics of bis(2-(1-methyl-1H-imidazolyl)phenyl) phosphate (BMIPP).

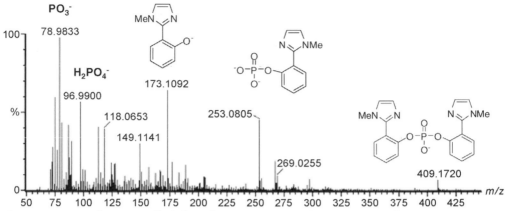

Figure 3.15 ESI-MS after 25 min of hydrolysis of BMIPP in aqueous solution at pH 6.5 and 60 °C.

To probe the mechanism of BMIPP hydrolysis, ESI-MS(/MS) was applied in the negative ion mode to monitor the course of this reaction [82]. Reagents, intermediates and products in anionic forms were transferred directly from the reaction solution to the gas phase, detected by ESI-MS, and then characterized by ESI-MS(/MS) via their unimolecular dissociation chemistry. After 25 min of hydrolysis of BMIPP in aqueous solution at pH 6.5 and 60 °C, a characteristic ESI-MS (Figure 3.15) was recorded. In this spectrum, a series of major anions are detected and identified as the mono-deprotonated forms of the reactant BMIPP of m/z 409, the monoester Me-IMPP of m/z 253, and the final phenolic product IMP of m/z 173, as well as inorganic P in the form of PO_3^- of m/z 79 and $H_2PO_4^-$ of m/z 97 (Scheme 3.35). ESI-MS(/MS) was then used to characterize these important species via CID. The resulting tandem mass spectra also supported the identifications: the Me-IMPP anion of m/z 253 is found to dissociate almost exclusively to PO_3^- of m/z 79 (Figure 3.16a), deprotonated IMP of m/z 173 dissociates mainly by two routes that lead either to the fragment ion of m/z 158 by the loss of a methyl radical or the fragment of m/z 118 by loss of 1-methyl-1H-azirine (C_3H_5N), thus forming the 2-cyanophenoxy anion (Figure 3.16b). Therefore, ESI-MS(/MS) data support the reaction pathway as depicted in Scheme 3.35.

3.2.19
Oxidative Cleavage of Terminal C=C bonds

Oxidative cleavage of terminal C=C bonds is a remarkable functional group transformation in synthetic organic chemistry. It is used to degrade large molecules or to introduce important oxygen functionalities [83]. Most of the available approaches for C=C oxidative cleavage suffer, however, from drawbacks such as accidents involving personal injury during ozonolysis, harsh reaction conditions, the need for expensive oxidants/catalysts, and/or tedious workup. Developing simple, fast and general methodologies for C=C oxidative cleavage to form ketones

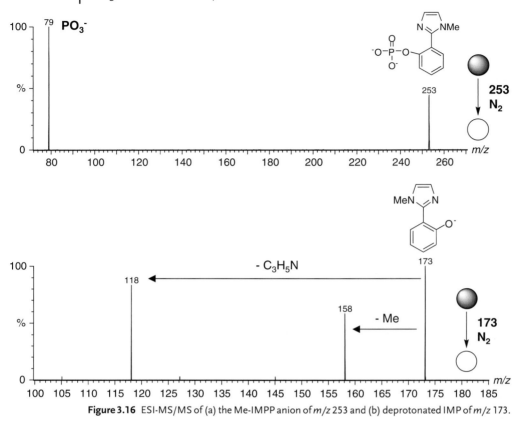

Figure 3.16 ESI-MS/MS of (a) the Me-IMPP anion of m/z 253 and (b) deprotonated IMP of m/z 173.

is therefore essential to further expand the scope of this synthetically relevant reaction. A reaction that uses m-chloroperbenzoic acid (m-CPBA) to promote oxidative cleavage of terminal C=C bonds, converting 1,1-diaryls into benzophenones under mild reaction conditions (Scheme 3.36) was recently monitored [84]. Oxidation was found to occur efficiently under soft conditions requiring no organometallic reagents or catalysts. In a representative spectrum (Figure 3.17), a series of key anions were detected and identified as the mono-deprotonated forms of the

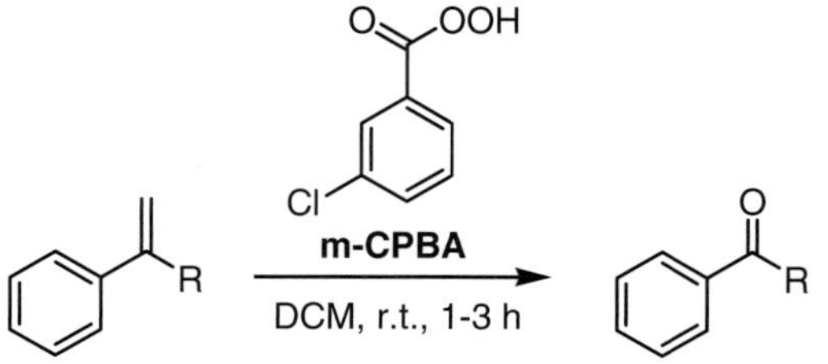

Scheme 3.36 Oxidative cleavage of terminal C=C bonds promoted by m-chloroperbenzoic acid (m-CPBA) converting 1,1-diaryls into benzophenones under mild reaction conditions.

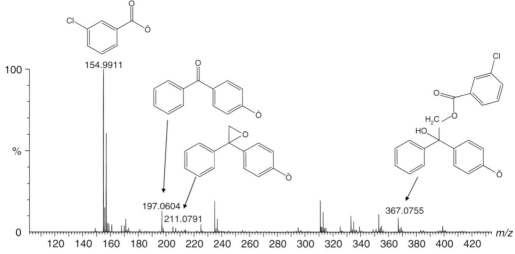

Figure 3.17 ESI(-)-MS for the reaction solution of a C=C oxidative cleavage reaction promoted by m-CPBA after 25 min.

reactant of m/z 197, the m-chloro benzoic acid (m-CBA) of m/z 155, and two intermediates of m/z 211 and m/z 367. These are important intermediates since they indicate that the reaction mechanism for the oxidation of double bonds involves epoxide formation (m/z 211) followed by epoxide opening promoted by m-CBA addition (m/z 367). ESI-MS(/MS) was then used to characterize these important intermediates (Figure 3.18). The resulting spectra supported the assignments: the anion of m/z 367 undergoes retro-addition either by losing a neutral m-CBA to yield the fragment ion of m/z 211 (the epoxide intermediate) or by losing the neutral epoxide to form deprotonated m-CBA of m/z 155. The epoxide

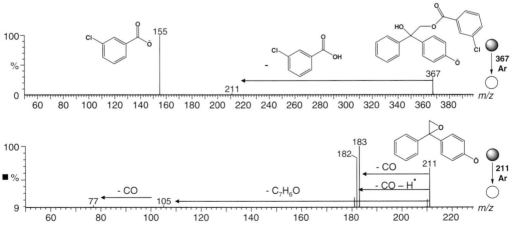

Figure 3.18 ESI(-)-MS(/MS) for the key anionic intermediates of m/z 367 and 211 after 25 min of C=C oxidative cleavage.

intermediate of m/z 211 loses mainly CO and then an H to yield the fragment ions of m/z 183 and m/z 182.

3.3
General Remarks

The representative studies discussed in this chapter exemplify the suitability of ESI-MS(/MS) monitoring to provide detailed mechanistic information for different types of organic reactions performed in solution under their most common conditions. The speed, selectivity and sensitivity of ESI-MS(/MS) monitoring has turned this technique into an indispensable tool to investigate mechanisms of organic reactions. Snapshots of ion compositions obtained via rapid and continuous ion fishing during the course of the reactions, provided by ESI-MS in the positive or negative ion modes, or both, and the characterization of the gaseous ionic species via on-line ESI-MS(/MS), have been shown to offer key information on the intermediates involved in major reaction steps. This information has probed established views or provided new or complementing mechanistic data that, in conjunction with data from other techniques, enable chemists to scrutinize such key processes in a fashion more deeply than ever before.

References

1 Fenn, J.B., Mann, M., Meng, C.K., Wong, S.F., and Whitehouse, C.M. (1989) *Science*, **246**, 64.
2 Santos, L.S., Kanaack, L., and Metzger, J.O. (2005) *Int. J. Mass Spectrom.*, **246**, 84; (b) Eberlin, M.N. (2007) *Eur. J. Mass Spectrom.*, **13**, 19; (c) Santos, L.S. (2008) *Eur. J. Org. Chem.*, 235.
3 (a) Nicolaou, K.C., Vourloumis, D., Winssinger, N., and Baran, P.S. (2000) *Angew. Chem. Int. Ed.*, **39**, 44; (b) Corey, E.J. (1991) *Angew. Chem. Int. Ed. Engl.*, **30**, 455.
4 Hill, J.S. and Isaacs, N.S. (1990) *J. Phys. Org. Chem.*, **3**, 285.
5 (a) Kaye, P.T. and Bode, M.L. (1991) *Tetrahedron Lett.*, **32**, 5611; (b) Fort, Y., Berthe, M.-C., and Caubère, P. (1992) *Tetrahedron*, **48**, 6371.
6 Santos, L.S., Pavam, C.H., Almeida, W.P., Coelho, F., and Eberlin, M.N. (2004) *Angew. Chem. Int. Ed.*, **43**, 4330.
7 Santos, L.S., DaSilveira Neto, B.A., Consorti, C.S., Pavam, C.H., Almeida, W.P., Coelho, F., Dupont, J., and Eberlin, M.N. (2006) *J. Phys. Org. Chem.*, **19**, 731.
8 Rosa, J.N., Afonso, C.A.M., and Santos, A.G. (2001) *Tetrahedron*, **57**, 4189.
9 Amarante, G.W., Milagre, H.M.S., Vaz, B.G., Ferreira, B.R.V., Eberlin, M.N., and Coelho, F. (2009) *J. Org. Chem.*, **74**, 3031.
10 Price, K.E., Broadwater, S.J., Jung, H.M., and McQuade, D.T. (2005) *Org. Lett.*, **7**, 147.
11 Aggarwal, V.K., Fulford, S.Y., and Lloyd-Jones, G.C. (2005) *Angew. Chem. Int. Ed.*, **44**, 1706.
12 Connon, S.J. (2006) *Chem. Eur. J.*, **12**, 5418.
13 Amarante, G.W., Benassi, M., Milagre, H.M.S., Braga, A.A.C., Maseras, F., Eberlin, M.N., and Coelho, F. (2009) *J. Am. Chem. Soc.*, submitted.

14 Tye, H. (2000) *J. Chem. Soc., Perkin Trans.*, **1**, 275.
15 Stiger, K.D., Mar-Tang, R., and Bartlet, P.A. (1999) *J. Org. Chem.*, **64**, 8409.
16 Lawrence, N.J., McGown, A.T., Nduka, J., Hadfield, J.A., and Pritchard, R.G. (2001) *Bioorg. Med. Chem. Lett.*, **11**, 429.
17 Milagre, C.D.F., Milagre, H.M.S., Santos, L.S., Lopes, M.L.A., Moran, P.J.S., Eberlin, M.N., and Rodrigues, J.A.R. (2007) *J. Mass Spectrom.*, **42**, 1287.
18 Benkovic, S.J., Benkovic, P.A., and Comfort, D.R. (1969) *J. Am. Chem. Soc.*, **91**, 1860.
19 Clayden, J., Greeves, N., Warren, S., and Wothers, P. (2001) *Organic Chemistry*, Oxford University Press Inc., New York.
20 Andersen, K.E., Sorensen, J.L., Lau, J., Lundt, B.F., Petersen, H., Huusfeldt, P.O., Suzdak, P.D., and Swedberg, M.D.B. (2001) *J. Med. Chem.*, **44**, 2152.
21 Ito, H., Omodera, K., Takigawa, Y., and Taguchi, T. (2002) *Org. Lett.*, **4**, 1499.
22 Ferraz, H.M.C., Pereira, F.L.C., Gonçalo, E.R.S., Santos, L.S., and Eberlin, M.N. (2005) *J. Org. Chem.*, **70**, 110.
23 Dounay, A.B. and Overman, L.E. (2003) *Chem. Rev.*, **103**, 2945.
24 (a) Shibasaki, M. and Vogl, E.M. (1999) *J. Organomet. Chem.*, **576**, 1; (b) Loiseleur, O., Hayashi, M., Keenan, M., Schmees, N., and Pfaltz, A. (1999) *J. Organomet. Chem.*, **576**, 16.
25 Beletskaya, I.P. and Cheprakov, A.V. (2000) *Chem. Rev.*, **100**, 3009.
26 Kikukawa, K., Nagira, K., Wada, F., and Matsuda, T. (1981) *Tetrahedron*, **37**, 31.
27 Severino, E.A. and Correia, C.R.D. (2000) *Org. Lett.*, **2**, 3039.
28 Severino, E.A., Costenaro, E.R., Garcia, A.L.L., and Correia, C.R.D. (2003) *Org. Lett.*, **5**, 305.
29 Sabino, A.S., Machado, A.H.L., Correia, C.R.D., and Eberlin, M.N. (2004) *Angew. Chem. Int. Ed.*, **43**, 2514.
30 Eberlin, M.N., Sabino, A.A., Machado, A.H.L., and Correia, C.R.D. submitted.
31 Horino, H. and Inoue, N. (1971) *Bull. Chem. Soc. Japn.*, **44**, 3210.
32 Horino, H. and Inoue, N. (1976) *J. Chem. Soc., Chem. Commun.*, **13**, 500–501.
33 Emrich, D.E. and Larock, R.C. (2004) *J. Organomet. Chem.*, **689**, 3756.
34 Kiss, L., Kurtan, T., Antus, S., and Brunner, H. (2003) *Arkivoc*, 69.
35 Buarque, C.D., Pinho, V.D., Vaz, B.G., da Silva, A.J.M., Eberlin, M.N., and Costa, P.R.R. (2007) *Org. Lett.*, submitted.
36 Fristrup, P., Le Quement, S., Tanner, D., and Norrby, P.O. (2004) *Organometallics*, **23**, 6160.
37 Widenhoefer, R., Zhong, H., and Buchwald, S. (1997) *J. Am. Chem. Soc.*, **119**, 6787.
38 Miyaura, N., Yamada, K., Suginome, H., and Suzuki, A. (1985) *J. Am. Chem. Soc.*, **107**, 972.
39 Aliprantis, A.O. and Canary, J.W. (1994) *J. Am. Chem. Soc.*, **116**, 6985.
40 (a) Moreno-Manãs, M., Pérez, M., and Pleixats, R. (1996) *J. Org. Chem.*, **61**, 2346; (b) Armendia, M.A., Lafont, F., Moreno-Manãs, M., Pleixats, R., and Roglans, A. (1999) *J. Org. Chem.*, **64**, 3592.
41 Milstein, D. and Stille, J.K. (1978) *J. Am. Chem. Soc.*, **100**, 3636.
42 Farina, V., Krishnamurthy, V., and Scott, W.J. (1998) *The Stille Reaction*, John Wiley & Sons Inc., New York.
43 Santos, L.S., Rosso, G.B., Pilli, R.A., and Eberlin, M.N. (2007) *J. Org. Chem.*, **72**, 5809.
44 Stille, J.R. (1986) *Angew. Chem. Int. Ed. Engl.*, **25**, 508.
45 Ma, S. (2003) *Acc. Chem. Res.*, **36**, 701.
46 Guo, H., Qian, R., Liao, Y., Ma, S., and Guo, Y. (2005) *J. Am. Chem. Soc.*, **127**, 13060.
47 Ma, S. and Jiao, N. (2002) *Angew. Chem., Int. Ed.*, **41**, 4737.
48 Ma, S. (2005) *Top. Organomet. Chem.*, **14**, 183.
49 Nishibayashi, S.Y., Cho, C.S., Ohe, K., and Uemura, S. (1996) *J. Organomet. Chem.*, **526**, 335.
50 Raminelli, C., Prechtl, M.H.G., Santos, L.S., Eberlin, M.N., and Comasseto, J.V. (2004) *Organometallics*, **23**, 3990.
51 Aramendía, M.A. and Lafont, A. (1999) *J. Org. Chem.*, **64**, 3592.

52 (a) Wang, Z., Zhang, Z., and Lu, X. (2000) *Organometallics*, **19**, 775; (b) Zhang, Z., Lu, X., Zang, Q., and Han, X. (2001) *Organometallics*, **20**, 3724; (c) Liu, G. and Lu, X. (2002) *Tetrahedron Lett.*, **43**, 6791; (d) de Meijere, A. and Meyer, F.E. (1994) *Angew. Chem., Int. Ed. Engl.*, **33**, 2379 and references therein.

53 Alexakis, A., Berlan, J., and Besace, Y. (1986) *Tetrahedron Lett.*, **27**, 1047.

54 Sonogashira, K., Tohda, Y., and Hagihara, N. (1975) *Tetrahedron Lett.*, 4467.

55 Santos, L.S., Cunha, R.L.O.R., Comasseto, J.V., and Eberlin, M.N. (2007) *Rapid Commun. Mass Spectrom.*, **21**, 1479.

56 Cunha, R.L.O.R., Urano, M.E., Chagas, J.R., Almeida, P.C., Bincoletto, C., Tersariol, I.L.S., and Comasseto, J.V. (2005) *Bioorg. Med. Chem. Lett.*, **15**, 755.

57 (a) Domingos, J.B., Longhinotti, E., Brandao, T.A.S., Bunton, C.A., Santos, L.S., Eberlin, M.N., and Nome, F. (2004) *J. Org. Chem.*, **69**, 6024; (b) Domingos, J.B., Longhinotti, E., Brandao, T.A.S., Santos, L.S., Eberlin, M.N., Bunton, C.A., and Nome, F. (2004) *J. Org. Chem.*, **69**, 7898.

58 Chen, P. (2003) *Angew. Chem. Int. Ed.*, **42**, 2832.

59 Market, C. and Pfaltz, A. (2004) *Angew. Chem. Int. Ed.*, **43**, 2498.

60 Chevrin, C., Le Bras, J., Hénin, F., Muzart, J., Pla-Quintana, A., Roglans, A., and Pleixats, R. (2004) *Organometallics*, **23**, 4796.

61 Pera-Titus, M., Garcia-Molina, V., Banos, M.A., Gimenez, J., and Esplugas, S. (2004) *Appl. Catal. B*, **47**, 219.

62 Zazo, J.A., Casas, J.A., Mohedano, A.F., Gilarranz, M.A., and Rodriguez, J. (2005) *J. Environ. Sci. Technol.*, **39**, 9295.

63 Petigara, B.R., Blough, N.V., and Mignerey, A.C. (2002) *Environ. Sci. Technol.*, **36**, 639.

64 Moura, F.C.C., Araujo, M.H., Dalmázio, I., Alves, T.M.A., Santos, L.S., Eberlin, M.N., Augusti, R., and Lago, R.M. (2006) *Rapid Commun. Mass Spectrom.*, **20**, 1859.

65 Atkinson, R. and Carter, W.P.L. (1984) *Chem. Rev.*, **84**, 437.

66 Claeys, M., Graham, B., Vas, G., Wang, W., Vermeylen, R., Pashynska, V., Cafmeyer, J., Guyon, P., Andreae, M.O., Artaxo, P., and Maenhaut, W. (2004) *Science*, **303**, 1173.

67 Santos, L.S., Dalmázio, I., Eberlin, M.N., Claeys, M., and Augusti, R. (2006) *Rapid Commun. Mass Spectrom.*, **20**, 2104.

68 (a) Devasagayam, T.P.A. and Kesavan, P.C. (1996) *Indian J. Exp. Biol.*, **34**, 291; (b) Devasagayam, T.P.A., Kamat, J.P., Mohan, H., and Kesavan, P.C. (1996) *Biochim. Biophys. Acta*, **1282**, 63; (c) Kesavan, P.C. and Powers, E.L. (1985) *Int. J. Radiat. Biol.*, **48**, 223; (d) Kesavan, P.C. (1992) *Curr. Sci. India*, **62**, 791.

69 Kesavan, P.C. and Sarma, L. (1995) *Subcellular Biochemistry*, Plenum, New York.

70 Dalmázio, I., Santos, L.S., Lopes, R.P., Eberlin, M.N., and Augusti, R. (2005) *Environ. Sci. Technol.*, **39**, 5982.

71 Webb, T.H. and Wilcox, C.S. (1993) *Chem. Soc. Rev.*, 383.

72 Abella, C.A.M., Benassi, M., Santos, L.S., Eberlin, M.N., and Coelho, F. (2007) *J. Org. Chem.*, **72**, 4048.

73 Biginelli, P. (1893) *Chim. Ital.*, **23**, 360. For reviews see: (a) Kappe, C.O. (1993) *Tetrahedron*, **49**, 6937; (b) Kappe, C.O. (2000) *Acc. Chem. Res.*, **33**, 879.

74 Dondoni, A., Massi, A., Minghini, E., and Bertolasi, V. (2002) *Helv. Chim. Acta*, **85**, 3331.

75 Kappe, C.O. (1997) *J. Org. Chem.*, **62**, 7201.

76 Folkers, K. and Johnson, T.B. (1933) *J. Am. Chem. Soc.*, **55**, 3781.

77 Sweet, F.S. and Fissekis, J.D. (1973) *J. Am. Chem. Soc.*, **95**, 8741.

78 De Souza, R.O.M.A., da Penha, E.T., Milagre, H.M.S., Garden, S.J., Esteves, P.M., Eberlin, M.N., and Antunes, O.A.C. (2009) *J. Org. Chem.*, in press.

79 Raines, R.T. (1998) *Chem. Rev.*, **98**, 1045.

80 Herschlag, D. (1994) *J. Am. Chem. Soc.*, **116**, 11631.

81 Liu, L., Rozenman, M., and Breslow, R. (2002) *J. Am. Chem. Soc.*, **124**, 12660.
82 Orth, E.S., Brandão, T.A.S., Milagre, H.M.S., Eberlin, M.N., and Nome, F. (2008) *J. Am. Chem. Soc.*, **130**, 2436.
83 Haines, H.A. (1985) *Methods for the Oxidation of Organic compounds*, Academic Press, New York.
84 Singh, F.V., Milagre, H.M.S., Eberlin, M.N., and Stefani, H.A. (2007) *Organic Letters*, in press.

4
Studies of Reaction Mechanism Intermediates by ESI-MS
Rong Qian, Jing Zhou, Shengjun Yao, Haoyang Wang, and Yinlong Guo

4.1
Introduction

Applications of atmospheric pressure ionization mass spectrometry (API-MS) to the study of reaction intermediates and mechanisms are reviewed. API-MS, especially ESI-MS, has provided many opportunities to intercept and characterize the key intermediates from the reaction mixtures. Combined with tandem mass spectrometric (MS/MS) methods, this technique has been extensively used for structural characterization of organic compounds and mechanism deduction of some organic reactions. Furthermore, API-MS affords a straightforward approach to trapping and identifying short-lived intermediates.

4.2
Studies on the Intermediates and Mechanisms of Pd-Catalyzed Reactions

In 1994, Aliprantis and coworkers studied the catalytic intermediates in the Suzuki reaction by ESI-MS [1]. The currently accepted catalytic cycle involves oxidative addition, transmetalation, isomerization, and reductive elimination (Scheme 4.1). In order to form the protonated intermediates detectable by ESI-MS, pyridyl bromide and three phenylboronic acids were chosen for the reaction. Intermediate ions of $[(pyrH)Pd(PPh_3)_2Br]^+$, diaryl Pd(II) species, and some other derivative palladium species were detected in the reaction mixture.

Although this study did not provide quantitative or detailed structural information about the intermediates, it was able to analyze the reaction mixture and observe mass spectra that correspond to the two key intermediates in the proposed mechanism for the Suzuki reaction. These results demonstrate that ESI-MS can be a valuable mechanistic tool when used in conjunction with other techniques.

Recently, we have used electron spray ionization Fourier transform ion cyclotron mass spectrometry (ESI-FTICRMS) to intercept and characterize several organometallic species in Pd-catalyzed reactions of 2,3-allenoate with two organoboronic acids

Reactive Intermediates: MS Investigations in Solution. Edited by Leonardo S. Santos.
Copyright © 2010 WILEY-VCH Verlag GmbH & Co. KGaA, Weinheim
ISBN: 978-3-527-32351-7

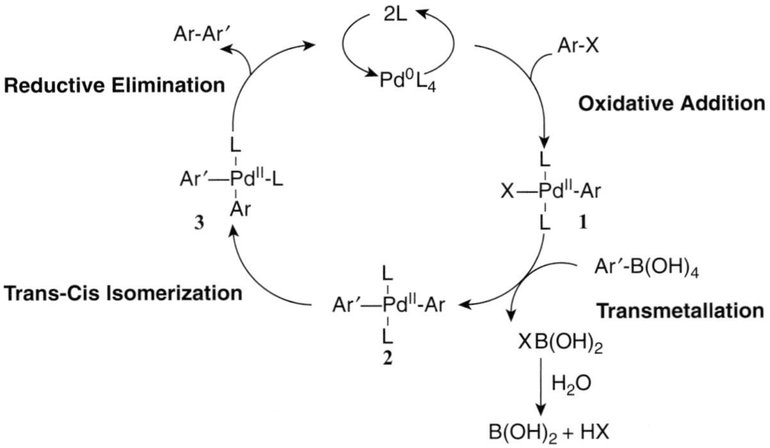

Scheme 4.1 Proposed mechanism for the Suzuki reaction.

in the presence of AcOH [2]. The palladium-catalyzed highly regio- and stereoselective hydroarylation or hydroalkenylation of allenes to form tri- or tetrasubstituted alkenes has become an important area of study in synthetic organic chemistry. Despite the importance of its synthetic utility, this reaction still has a number of unaccountable features, including the mechanistic details.

After some preliminary studies, we found that ESI-FTICRMS monitoring of this reaction yielded some interesting information about the reaction intermediates. The palladium intermediates involved in the catalytic cycle were intercepted, detected, and characterized by ESI-FTICRMS in the positive-ion mode. The first key palladium-containing intermediate ion of m/z 689.1 was detected after 3 min of the reaction with organoboronic acid **4a**. Continuous monitoring of these intermediates by ESI-FTICRMS also enabled us to detect three other important ions: the palladium cation of m/z 833.2, which usually disappeared five minutes after initiation of the reaction, the palladium cation of m/z 834.2, which might result from the oxidation of the neutral intermediate HPd(PPh$_3$)$_2$–vinyl species during the ESI process, and the cationic π-allyl intermediate ion of m/z 909.2, which may be formed by sequential oxidation during the ESI process and elimination of H• from the (η^3-allyl)palladium complex in the gas phase (Figure 4.1).

Through the ESI-FTICRMS studies, some key intermediates of the Pd(0)-catalyzed addition reaction of allenes with boronic acids have been successfully intercepted and characterized, and this has led to the mechanism shown in Scheme 4.2. The result indicates that this ionization techneique in combination with MS is a very useful tool for structural elucidation of metal complexes and organometallic systems, and opens up a straightforward approach to trapping and identifying reactive intermediates.

Another major reaction we studied by ESI-FTICRMS was the Pd(0)-catalyzed three-component tandem double addition cyclization reaction [3]. This reaction provides an efficient route to polysubstituted *cis*-pyrrolidine derivatives with matched relay and excellent regio- and stereoselectivity. In principle, there are two types of

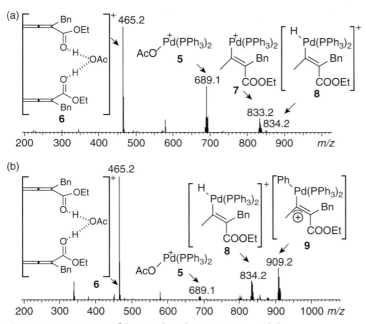

Figure 4.1 ESI-(+)-MS of the samples taken at (a) 3 min and (b) 2 h after mixing allenoate and organoboronic acid **4a**.

mechanisms: (1) carbopalladation forming π-allyl palladium species and (2) an azapalladation reductive elimination mechanism.

All the key intermediates were detected by ESI-FTICRMS over time, such as those ions of m/z 707.1, 891.2 and 1150.3 (Figure 4.2). After these intermediate ions were further characterized by sustained off-resonance irradiation collision-induced dissociation (SORI-CID), one mechanism was proposed as shown in Scheme 4.3. The mechanism involving the carbopalladation with 2-(2,3-allenyl)malonate yielding the π-allyl palladium intermediate (Scheme 4.3) was confirmed.

ESI-MS was also employed to identify reactive intermediates of some reactions mediated by platinum-alumina, titanium alkoxide, and so on [3–11]. This method has proved to be a good choice for an assay of multiple, competitive, and simultaneously occurring catalytic reactions.

4.3
Studies on Some Reactive Intermediates and Mechanisms of Radical Reactions

Metzger and coworkers have studied reactive intermediates of chemical reactions in solution by using a microreactor coupled to an ESI mass spectrometer. The highly stereo- and regioselective dimerization of *trans*-anethole to give the head-to-head *trans*, *-anti*, *trans*-cyclobutane initiated by aminium salt proceeds by a radical cation chain mechanism (Scheme 4.4) and this method was further used to study the transient radical cations intermediates in electron transfer-initiated D-A reactions [12–14].

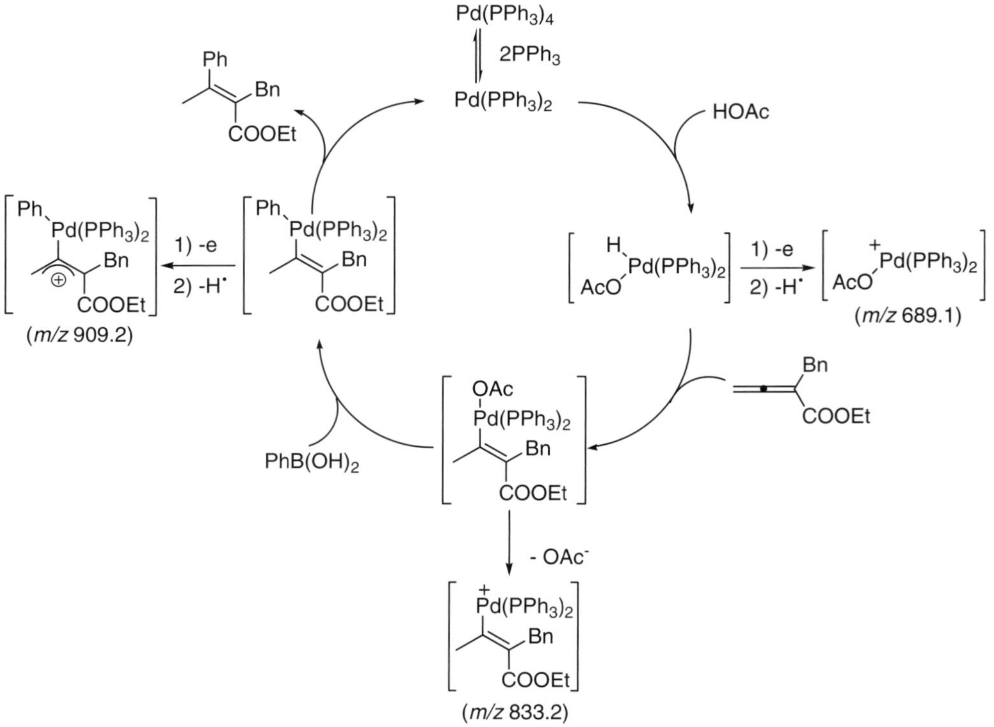

Scheme 4.2 Proposed catalytic cycle for the palladium-catalyzed addition of an organoboronic acid to an allene in the presence of AcOH based on the MS study.

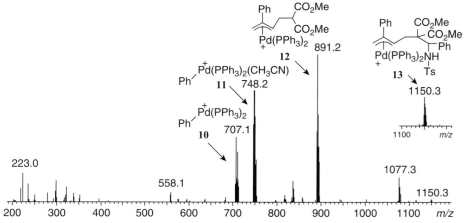

Figure 4.2 ESI(+)-MS for a solution of the sample taken from the reaction mixture of 2-(2,3-allenyl)malonate, iodobenzene, imine, Pd(PPh$_3$)$_4$, and K$_2$CO$_3$ in THF stirred at 85 °C in a nitrogen atmosphere at the reaction time of 40 min.

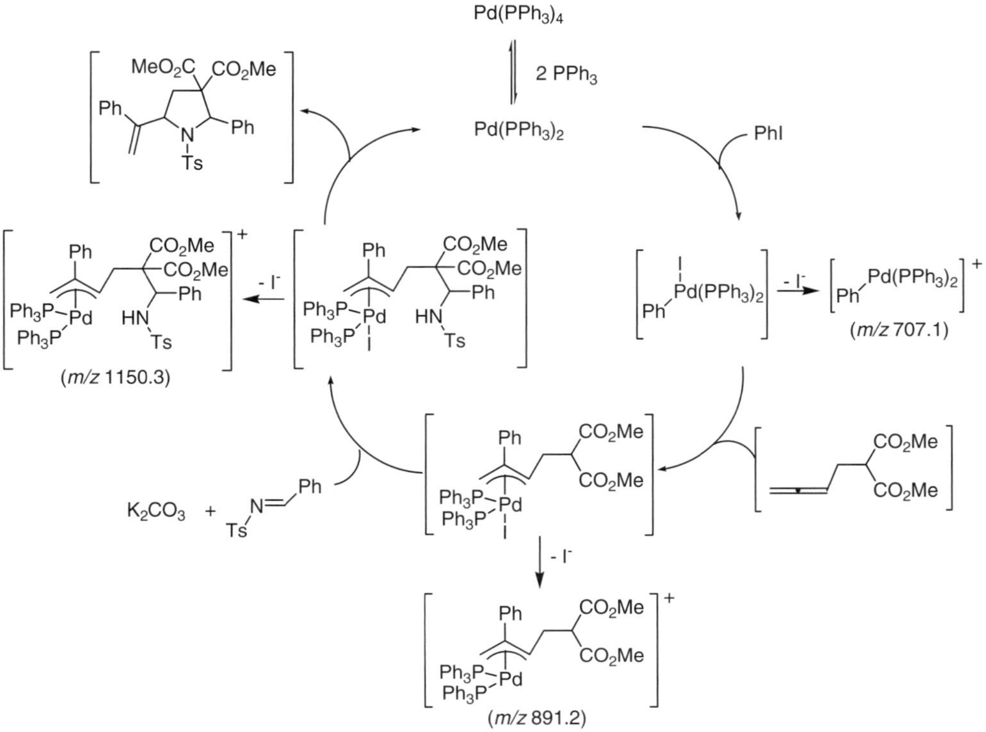

Scheme 4.3 The mechanism of Pd(0)-catalyzed three-component tandem double addition-cyclization reaction based on the study of ESI-FTICRMS.

The reacting solution was first examined by atmospheric pressure chemical ionization mass spectrometry (APCI-MS). The solutions were injected with a dual syringe pump into an effective micromixer, which was coupled directly to the ion source of the mass spectrometer. For comparison, two compounds were examined separately by APCI-MS, and the respective quasimolecular ions were observed. The radical cations **14** and **15** were characterized unambiguously by APCI-MS/MS (Figure 4.3).

The ESI-MS result accordingly shows an intense signal for radical cation *tris(p-bromophenyl)aminium*, and no intense signals for **14** (m/z 148) and **15** (m/z 296) are observed. However, very weak ion currents appear at these m/z values. By using the MS/MS technique, both radical cations **14** and **15** could be detected and identified clearly by comparison with the MS/MS spectra of the species **14** (Figure 4.3a) and **15** (Figure 4.3b) produced by APCI. The fragmentation of the ions at m/z 148 and m/z 296, which were examined under ESI-MS/MS conditions, agrees very well with that of the authentic ions **14** and **15**, respectively, which were characterized by APCI-MS/MS.

The reactions were studied by on-line coupling of a microreactor system to APCI-MS and ESI-MS. The transient radical cations were detected and characterized by ESI-MS/MS.

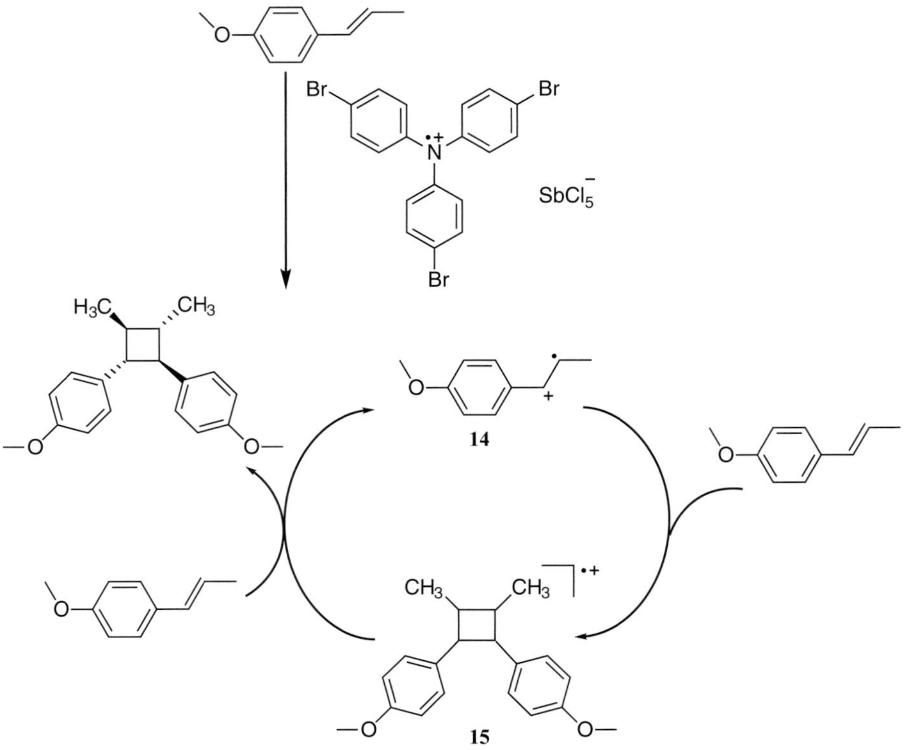

Scheme 4.4 Radical chain reaction of *trans*-anethole mediated by tris(*p*-bromophenyl)aminium hexachloroantimonate to give the dimer product 1,2-bis(4-methoxyphenyl)-3,4-dimethylcyclobutane via two reaction intermediates.

In the present work, we have reported the use of ESI-MS and ESI-MS/MS to monitor reaction solutions of Selectfluor (a fluorine donor reagent) with triphenylethylene and **16b** by the ESI/TOF-MS under the same conditions [15–18]. The aim is to intercept the radical cationic intermediates m/z 256 and m/z 332 resulting from the SET pathway, though the radical intermediates are too unstable to detect even using the radical traps such as TEMPO.

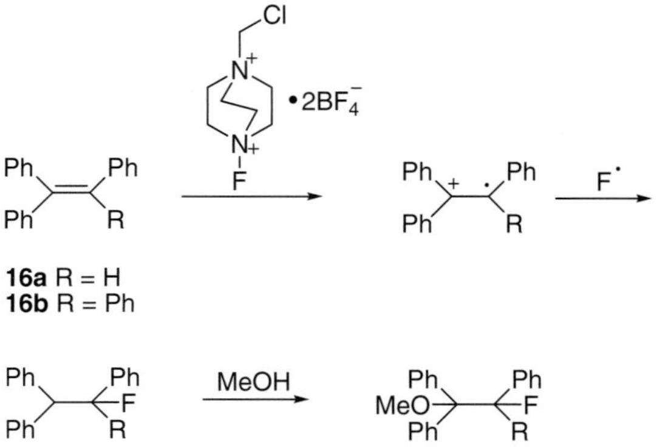

16a R = H
16b R = Ph

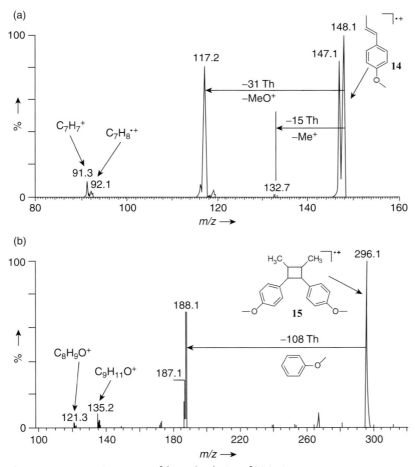

Figure 4.3 (a) MS/MS spectrum of the molecular ion of **14** (m/z 148); (b) MS/MS spectrum of the molecular ion of **15** (m/z 296).

Shortly after 1–3 min of the reaction of olefin **16a**, some species were detected as major ions: m/z 161, m/z 267, m/z 179, m/z 197, m/z 256 and m/z 275 (Figure 4.4). While **16b** was used as substrate, the ion of m/z 256 was replaced by m/z 332, and m/z 275 was replaced by m/z 351. The results of ESI-TFICRMS and ESI-TOFMS indicate that the proposed ion structures correspond to the only chemically reasonable elemental compositions.

Through this analysis, reliable characterization of the reaction intermediates was accomplished by ESI-MS and ESI-MS/MS experiments, confirming the assumption that the fluorination of multiply phenyl-substituted olefins follows a SET mechanism.

The fundamental understanding of the ES ionization mechanism and the CID fragmentation behavior of ions with open and closed shells sets the stage for the analysis of solution-phase reaction mixtures. The presented examples clearly dem-

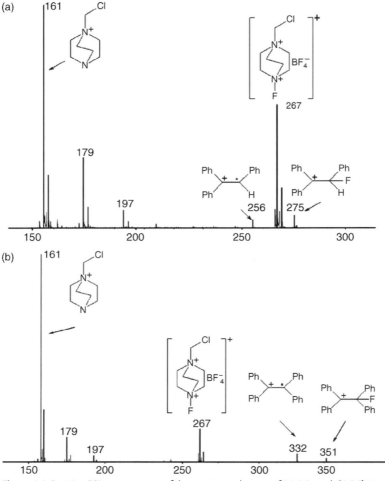

Figure 4.4 Positive ESI mass spectra of the reaction solutions of (a) **16a** and (b) **16b** in methanol.

onstrate that the successful elucidation of complex reaction mechanisms demands the sensitive transfer of transient intermediates from the solution to the gas phase, which can be accomplished by ESI-MS. Subsequent selection of relevant precursor ions for MS^n experiments offers extensive sets of data for structure elucidation. It must be noted that structure assignments based solely on fragmentation patterns need further confirmation by exact mass measurements of the product ions to assure reliable assumptions.

However, reaction mechanism studies based on ESI-MS and MS^n require appropriate control experiments to clarify the origins and identities of detected species unambiguously as real transient intermediates of a reaction path under consideration. It is of special importance that these control experiments be carefully chosen and properly conducted to reliably exclude the formation of artefact ions by ESI, which can dramatically mislead the interpretations.

4.4
Studies on the Intermediates and Mechanism of Organocatalysis Reactions

Probing the mechanisms of nonmetal-catalyzed reactions by ESI-MS/MS have also been carried out by chemists. Eberlin and coworkers have described the use of ESI-MS for the interception and characterization of the organocatalytic intermediates in the DABCO-catalyzed Baylis-Hillman reaction (Scheme 4.5) [19]. Through the ESI-MS study, most proposed intermediates were successfully intercepted and characterized as protonated species, which provided strong evidence for the currently accepted catalytic cycle. In addition, the Baylis-Hillman reaction catalyzed by Lewis acids [20] and dissolved in ionic liquids [21] has also been investigated by Eberlin and coworkers.

ESI-MS was also employed by Metzger and coworkers in conjunction with the microreactor to identify crucial intermediates in the aldol reaction catalyzed by L-proline [22]. They have successfully intercepted and characterized all the intermediates assumed for the catalytic cycle (Scheme 4.6), thus confirming the mechanism currently accepted for this catalytic cycle. Moreover, they also described the use of ESI-MS for the characterization of catalytic intermediates in the L-prolinamide-catalyzed α-halogenation (Cl, Br, I) of aldehydes [23]. Both the intermediates assumed to participate in the catalytic cycle and the transients were intercepted and charac-

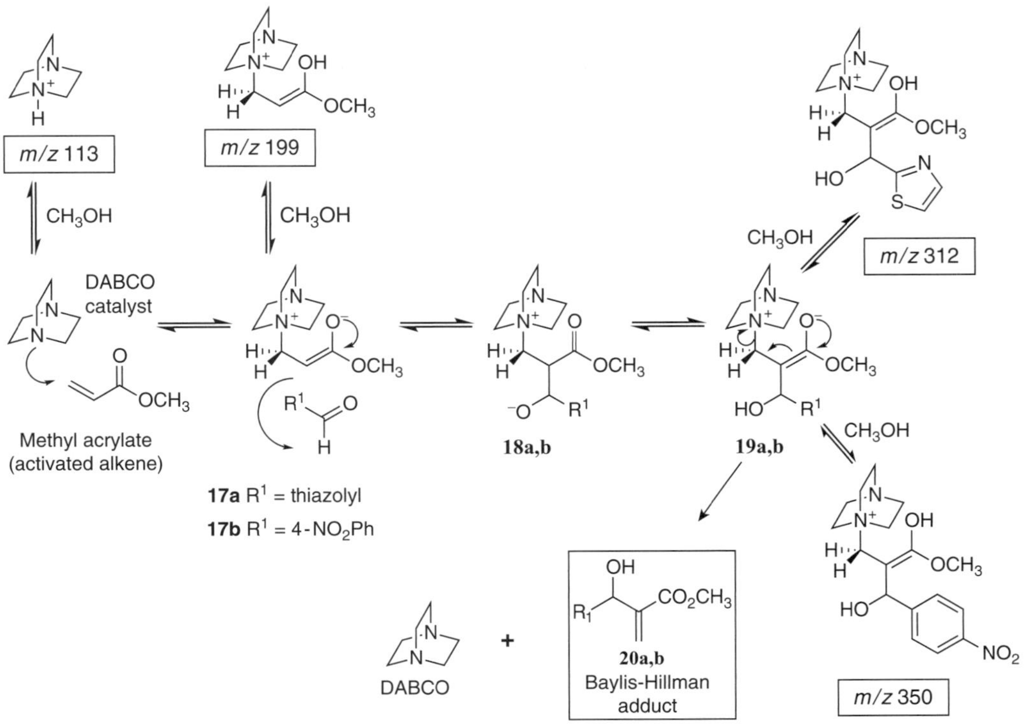

Scheme 4.5 ESI-MS study of the mechanism of Baylis-Hillman reaction catalyzed by DABCO.

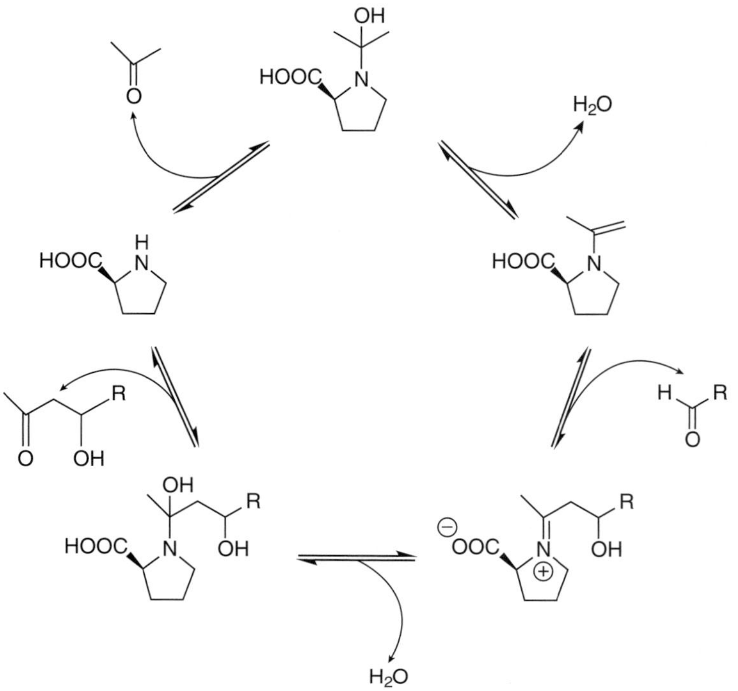

Scheme 4.6 Proposed mechanism for the L-proline-catalyzed aldol reaction.

terized, which led to a more detailed understanding of the reaction process (Scheme 4.7).

Furthermore, the ESI-MS study of NQO1-catalyzed reduction of estrogen *ortho*-quinones has also provided original an contribution by circumventing the problem of nonenzymatic reduction of estrogen quinone by NAD(P)H [24]. This has spilled over into the study of biological metabolism, and several groups are making major headway in using ESI-MS to probe the mechanism of the physiological metabolic reaction.

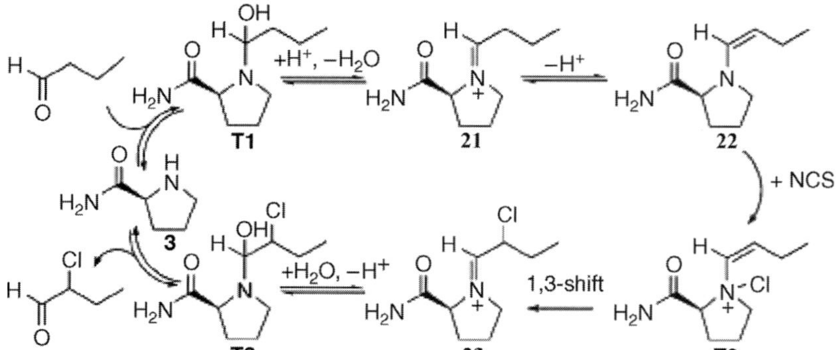

Scheme 4.7 ESI-MS study of the mechanism of L-prolinamide-catalyzed α-chlorination.

4.5
Studies on the Intermediates and Mechanism of Transition Metal-Catalyzed Polymerization Reactions

Investigations into the mechanism of the polymerization process have been a major research focus of polymer scientists. The two most common mass spectrometry techniques for the analysis of polymers are MALDI-TOF-MS and ESI-MS [25]. Although ESI-MS is not very well suited to mapping out entire molecular-weight distributions because of the limited mass range accessible (mainly limited to m/z 4000), it has its own advantages for the study of polymerization process [26, 27]. Firstly, ESI-MS is a softer ionization process, which has been shown to limit the fragmentation of polymer end groups during the ionization process, secondly, ESI-MS can show the transient ionic species in gas phase, which are the same as those originally present in solution, and thirdly, ESI-MS/MS can further demonstrate the observed catalytic activity of the species by the ion/molecule reaction in the gas phase. ESI-MS applied to the in-depth study of polymerization processes has been an integral tool for elucidating reaction mechanisms since the late 1990s. In this section we feature representative examples of the application of electrospray ionization mass spectrometry techniques to polymerization.

For the observation of neutral species in the solution phase, the first problem is how to make the species become ions. As a first example, Peter Chen *et al.* reported an electrospray ionization mass spectrometric method for the identification of uncharged active species in a solution phase ROMP reaction based on derivatization of the active species with a charged substrate [28].

As depicted in Scheme 4.8, norbornene with a pendant prosthetic ammonium or phosphonium cation is added, together with ordinary norbornene, to the otherwise neutral [(Cy$_3$P)$_2$(Cl)$_2$Ru=CHPh] (Cy=cyclohexyl) catalyst in solution.

After a short time, the electrospray mass spectrum (Figure 4.5) showed the detection of an electrically neutral catalyst by trapping with cationized substrate. The intramolecular P complex, rather than the bisphosphane complex, was investigated as the resting state in the catalytic cycle. In the hypothetical case, in which there were several metathesis-active complexes at work, each would yield its own series, with the mass of the catalytic 'head group' distinguishing one series from the others.

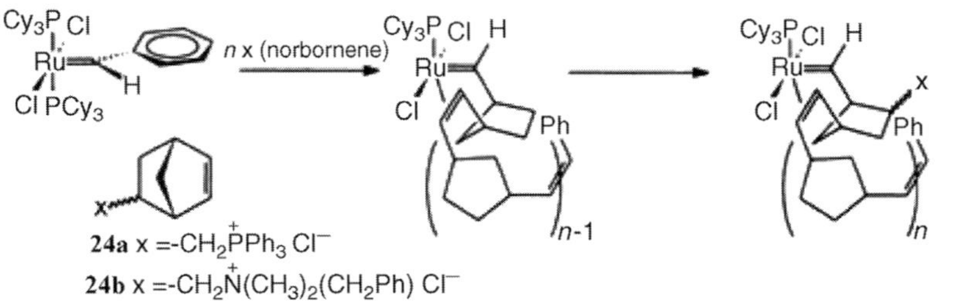

Scheme 4.8 Derivatization of ruthenium catalysts with cationic norbornene species enables detection by ESI-MS.

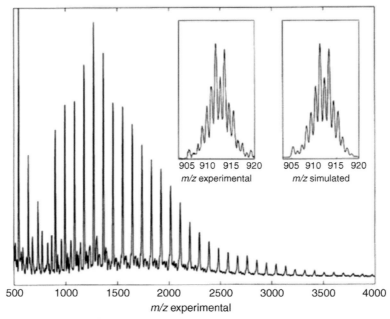

Figure 4.5 ESI-MS of the catalyst-bound ROMP oligomers obtained by reaction of [(Cy$_3$P)$_2$Cl$_2$Ru=CHPh] with norbornene and a covalently functionalized norbornene derivative. Complexes that incorporated the functionalized norbornene are visible in the ESI-MS. The inset compares the isotope pattern of one oligomer against a computed distribution.

The Ziegler–Natta polymerization of olefins is a technical process of the utmost importance for the synthesis of high-molecular-weight polymers. The technical process is based on the activation of a suitable zirconocene precursor, typically Zp$_2$ZrMe$_2$ or Zp$_2$ZrCl$_2$, by a strong Lewis acid as co-catalyst (e.g., methyl aluminoxane, MAO).

Mechanistic studies of Ziegler–Natta-like polymerization has mainly focused on the role of cationic alkylzirconocene species in the catalysis. In the 1980s, Jordan *et al.* first isolated and characterized an alkylzirconocene cation in the form of the salt [Cp$_2$ZrMe(THF)]$^+$[BPh$_4$]$^-$ [29, 30]. However, by the later 1990s, the reactivity of isolated alkylzirconocene cations in the gas phase was confirmed by Plattner and Chen (Scheme 4.9) [31]. Figure 4.6 shown the reaction products of [Cp$_2$Zr-(CH$_2$CHEt)CH$_3$]$^+$ with 1-butene in octopole under near-zero energy, multiple-

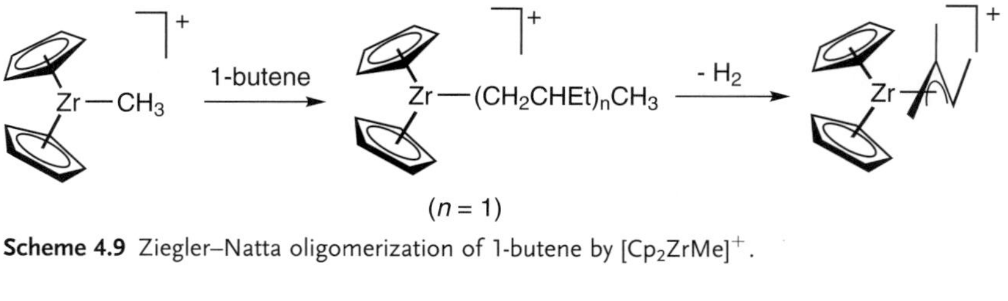

Scheme 4.9 Ziegler–Natta oligomerization of 1-butene by [Cp$_2$ZrMe]$^+$.

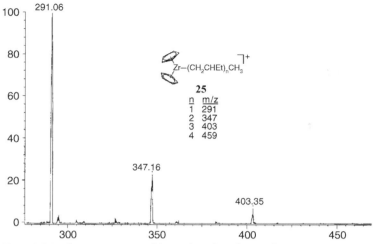

Figure 4.6 Product-ion spectrum of the products from the gas-phase reaction of [Cp$_2$Zr-(CH$_2$CHEt)CH$_3$]$^+$ (m/z 291) with 1-butene.

collision conditions. The peaks at m/z 347, m/z 403 and m/z 459, which represent the addition of up to four units of 1-butene, can be observed. These results showed that the Ziegler–Natta polymerization reaction proceeds in the mass spectrometer with completely unsolvated complexes.

The Ziegler–Natta polymerization of ethene was also studied by Santos and Metzger [32]. The cation [Cp$_2$Zr(CH$_2$CH$_2$)$_n$CH$_3$]$^+$ (n = 1–31) was easily detected in the reaction solution of [Cp$_2$ZrCl$_2$] and MAO by on-line microreactor coupled with ESI-MS (Figure 4.7). The spectrum also shows the odd- and even-numbered-chain

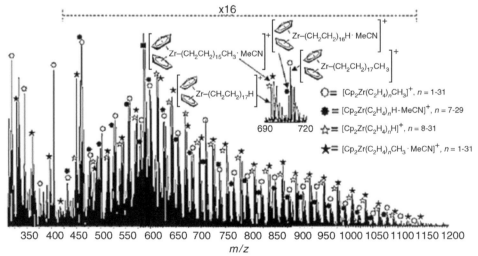

Figure 4.7 Ziegler–Natta polymerization of ethene: Positive-mode ESI mass spectrum of the MeCN-quenched reaction solution of [Cp$_2$ZrCl$_2$]/MAO (1 : 1.2 equiv) and C$_2$H$_4$.

{Cp$_2$Zr}–alkyl species, as well as the MeCN adducts. The catalytic activity of these cations was directly demonstrated by the ion/molecule reaction of [Cp$_2$Zr(CH$_2$CH$_2$)$_n$CH$_3$]$^+$ and ethene in the gas phase.

Chen *et al.* reported their screening methodology for investigating highly active, cationic carbene catalysts. The system of Hofmann ruthenium-catalyzed olefin metathesis (Scheme 4.10) was successfully studied through systematic variation of structural features of the catalyst and ESI-MS/MS by probing metathesis activity by reaction with vinyl ethers (Scheme 4.11) [33].

As shown in Figure 4.8, the metathesis activity could be probed by reaction with vinyl ethers (from m/z 649 to m/z 603 in the daughter-ion spectrum). The extent of

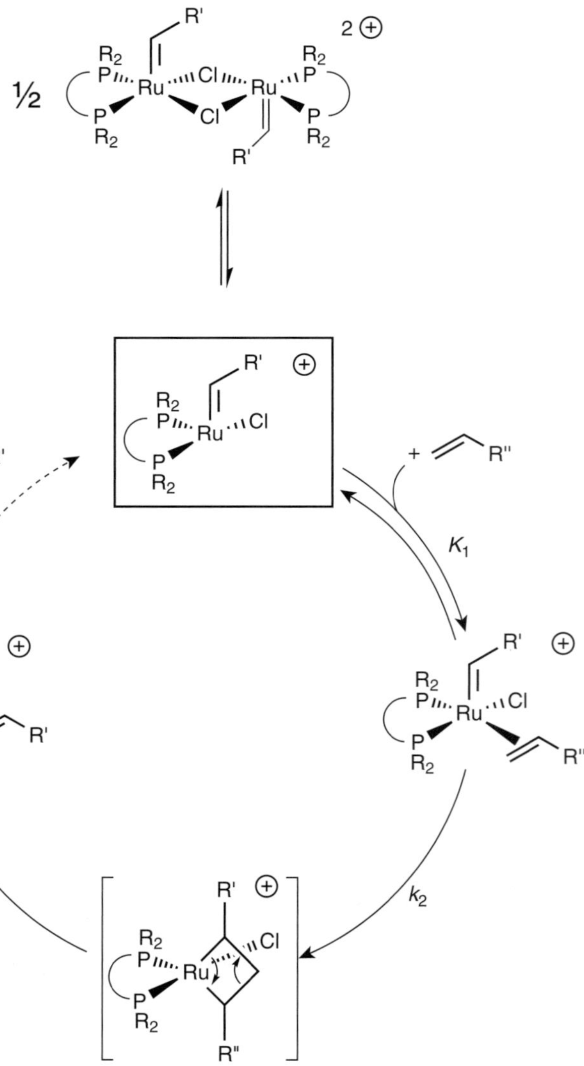

Scheme 4.10 Chauvin mechanism for catalytic ROMP.

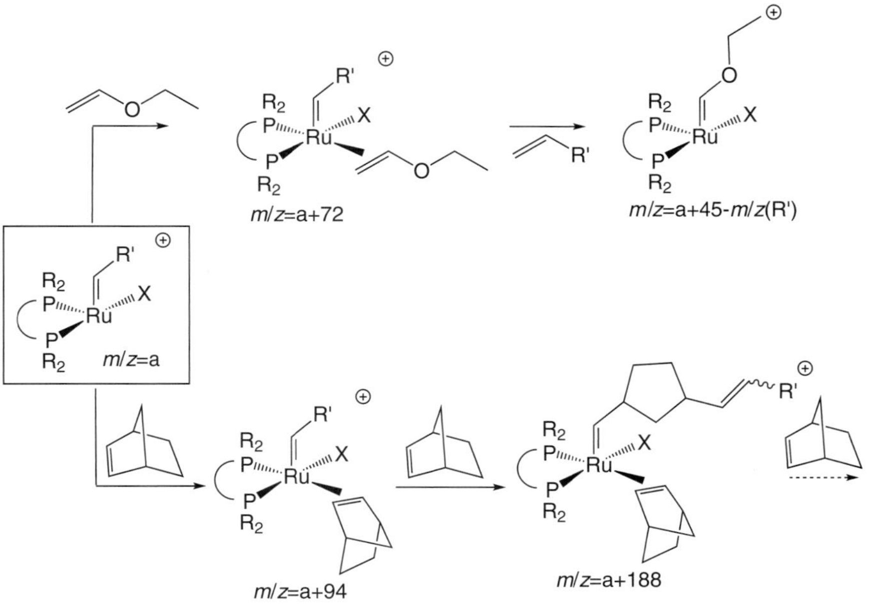

4.5 Studies of Transition Metal-Catalyzed Polymerizations

Scheme 4.11 Examples of olefin metathesis with acyclic or cyclic substrates in the gas phase.

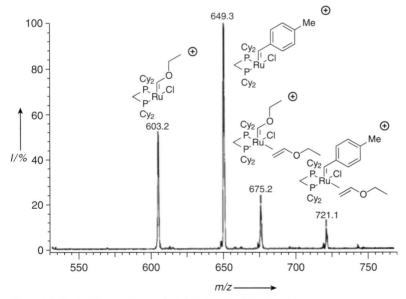

Figure 4.8 Product-ion spectrum of catalytic ion at m/z 649, and collision with ethyl vinyl ether in the collision octopole at low energy (nominal 1 eV).

reaction during the finite residence time in the octopole is a measure of relative rate. Reaction with norbornene instead of vinyl ethers gave direct access to ROMP efficiency because the multiple additions of norbornene units could be directly observed.

The crudest interpretation of the norbornene addition data would be simply to take the catalyst which makes the longest chains in the gas-phase as the best catalyst. However, a more detailed examination of the relative intensities, modeled with a simple kinetic model, allows the determination of the relative rates of addition of the first norbornene unit versus all subsequent norbornene units (Scheme 4.12).

Based on these, they established that the reactivity trends determined in the gas phase parallel solution-phase reactivity. The overall rate for the monocations in the gas phase depends on the P-complex pre-equilibrium and metalacyclobutane formation, which was found to be the rate-determining step.

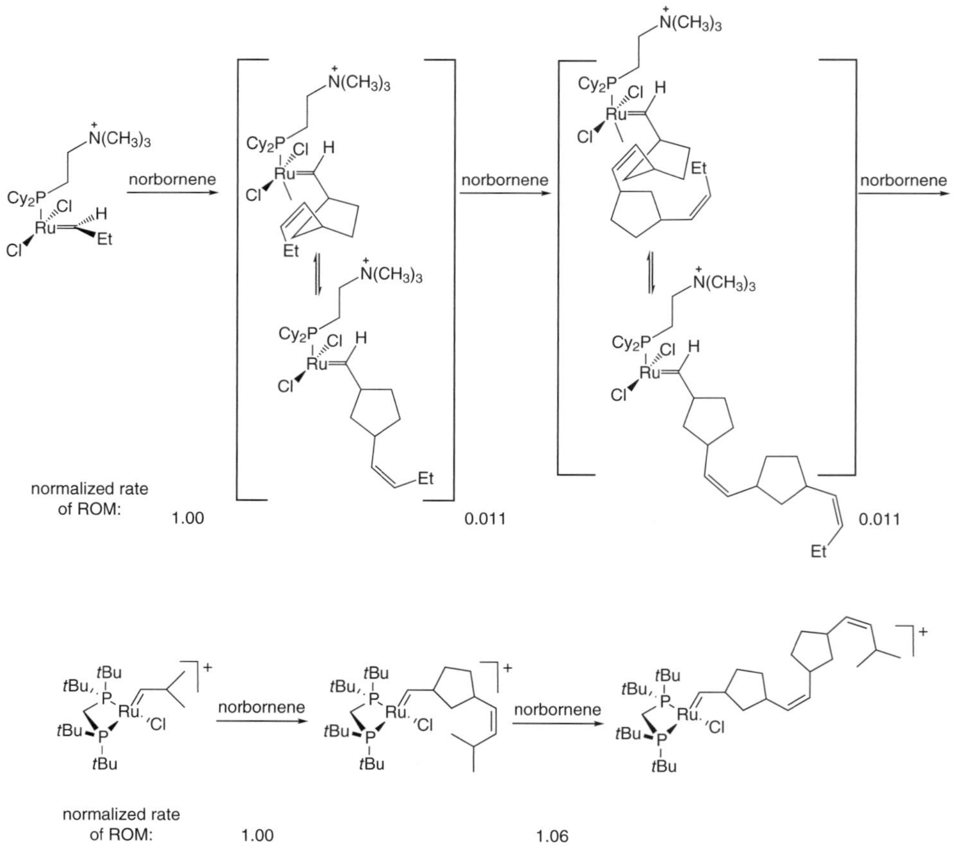

Scheme 4.12 Kinetic models for calculating the relative rate of addition of the first norbornene unit and the subsequent addition steps.

The rapid assay of a variety of structural effects on metathesis rate, combined with mechanistic analysis, leads to predictions for optimized catalysts for given applications. For typical organometallic complexes in homogeneous catalysis, the synthesis consumes the greatest fraction of expended time. The methodology of screening before synthesis described here offers great savings in time and effort.

Electrospray ionization mass spectrometry applied to polymerization process is a relatively young field and the technology underpinning the instrumentation is under constant development. The shortcomings mentioned above are thus being addressed, making the technology even more attractive for polymer scientists interested in reaction mechanisms.

References

1 Aliprantis, A.O. and Canary, J.W. (1994) Observation of catalytic intermediates in the Suzuki reaction by electrospray mass spectrometry. *J. Am. Chem. Soc.*, **116**, 6985–6986.

2 Qian, R., Guo, H., Liao, Y., Guo, Y., and Ma, S. (2005) Probing the mechanism of palladium-catalyzed addition of organoboronic acids to allenes in the presence of AcOH with ESI-FTMS. *Angew. Chem. Int. Ed.*, **44**, 4771–4774.

3 Guo, H., Qian, R., Liao, Y., Ma, S., and Guo, Y. (2005) ESI-MS studies on the mechanism of Pd(0)-catalyzed three-component tandem double addition-cyclization. *J. Am. Chem. Soc.*, **127**, 13060–13064.

4 Griep-Raming, J., Meyer, S., Bruhn, T., and Metzger, J.O. (2002) Investigation of reactive intermediates of chemical reactions in solution by electrospray ionization mass spectrometry: radical chain reactions. *Angew. Chem. Int. Ed.*, **41**, 2738–2742.

5 Wilson, S.R., Perez, J., and Pasternak, A. (1993) ESI-MS detection of ionic intermediates in phosphine-mediated reactions. *J. Am. Chem. Soc.*, **115**, 1994–1997.

6 Sabino, A.A., Machado, A.H.L., Correia, C.R.D., and Eberlin, M.N. (2004) Probing the mechanism of the Heck reaction with arene diazonium salts by electrospray mass and tandem mass spectrometry. *Angew. Chem. Int. Ed.*, **43**, 2514–2518.

7 Masllorens, J., Moreno-Mañas, M., Pla-Quintana, A., and Roglans, A. (2003) First Heck reaction with arenediazonium cations with recovery of Pd-triolefinic macrocyclic catalyst. *Org. Lett.*, **5**, 1559–1561.

8 Moreno-Mañas, M., Pleixats, R., Sebastian, R.M., Vallribera, A., and Roglans, A. (2004) Organometallic chemistry of 15-membered tri-olefinic macrocycles: catalysis by palladium(0) complexes in carbon-carbon bond-forming reactions. *J. Organomet. Chem.*, **689**, 3669–3684.

9 Bonchio, M., Licini, G., Modena, G., Moro, S., Bortolini, O., Traldi, P., and Nugent, W.A. (1997) Use of electrospray ionization mass spectrometry to characterize chiral reactive intermediates in a titanium alkoxide mediated sulfoxidation reaction. *Chem. Commun.*, 869–870.

10 Bonchio, M., Licini, G., Modena, G., Moro, S., Bortolini, O., Traldi, P., and Nugent, W.A. (1999) Enantioselective Ti(IV) sulfoxidation catalysts bearing C_3-symmetric trialkanolamine ligands: solution speciation by ^{1}H NMR and ESI-MS analysis. *J. Am. Chem. Soc.*, **121**, 6258–6268.

11 Santos, L.S., Knaack, L., and Metzger, J.O. (2005) Investigation of chemical reactions in solution using API-MS. *Int. J. Mass. Spectrom.*, **246**, 84–104.

12 Fürmeier, S. and Metzger, J.O. (2004) Detection of transient radical cations in electron transfer-initiated Diels–Alder reactions by electrospray ionization mass spectrometry. *J. Am. Chem. Soc.*, **126**, 14485–14492.

13 Meyer, S., Koch, R., and Metzger, J.O. (2003) Investigation of reactive intermediates of chemical reactions in solution by electrospray ionization mass spectrometry: radical cation chain reactions. *Angew. Chem. Int. Ed.*, **42**, 4700–4703.

14 Fürmeier, S., Griep-Raming, J., Hayen, A., and Metzger, J.O. (2005) Chelation-controlled radical chain reactions studied by electrospray ionization mass spectrometry. *Chem. Eur. J.*, **11**, 5545–5554.

15 Zhang, X., Liao, Y., Qian, R., Wang, H., and Guo, Y. (2005) Investigation of radical cation in electrophilic fluorination by ESI-MS. *Org. Lett.*, **7**, 3877–3880.

16 Zhang, X., Wang, H., and Guo, Y. (2006) Interception of the radicals produced in electrophilic fluorination with radical traps (TEMPO, DMPO) studied by electrospray ionization mass spectrometry. *Rapid Commun. Mass Spectrom.*, **20**, 1877–1882.

17 Zhang, X. and Guo, Y. (2006) Electrospray ionization mass spectrometric study on the alpha-fluorination of aldehydes. *Rapid Commun. Mass Spectrom.*, **20**, 3477–3480.

18 Zhang, X., Wang, H., Liao, Y., Ji, H., and Guo, Y. (2007) Study of methylation of nitrogen-containing compounds in the gas phase. *J. Mass Spectrom.*, **42**, 218–224.

19 Santos, L.S. *et al.* (2004) Probing the mechanism of the Baylis-Hillman reaction by electrospray ionization mass and tandem mass spectrometry. *Angew. Chem. Int. Ed.*, **43**, 4330–4333.

20 Amarante, G.W. *et al.* (2006) Formation of substituted N-oxide hydroxyquinolines from O-nitrophenyl Baylis-Hillman adduct: a new key intermediate intercepted by ESI-(+)-MS(/MS) monitoring. *Tetrahedron Lett.*, **47**, 8427–8431.

21 Santos, L.S. *et al.* (2006) The role of ionic liquids in co-catalysis of Baylis-Hillman reaction: interception of supramolecular species via electrospray ionization mass spectrometry. *J. Phy. Org. Chem.*, **19**, 731–736.

22 Marquez, C. and Metzger, J.O. (2006) ESI-MS study on the aldol reaction catalyzed by L-proline. *Chem. Commun.*, **14**, 1539–1541.

23 Marquez, C.A., Fabbretti, F., and Metzger, J.O. (2007) Electrospray ionization mass spectrometric study on the direct organocatalytic alpha-halogenation of aldehydes. *Angew. Chem. Int. Ed.*, **46**, 6915–6917.

24 Gaikwad, N.W., Rogan, E.G., and Cavalieri, E.L. (2007) Evidence from ESI-MS for NQO1-catalyzed reduction of estrogen ortho-quinones. *Free Radical Biology and Medicine*, **43**, 1289–1298.

25 Barner-Kowollik, C., Davis, T.P., and Stenzel, M.H. (2004) Probing mechanistic features of conventional, catalytic and living free radical polymerizations using soft ionization mass spectrometric techniques. *Polymer*, **45**, 7791–7805.

26 Chen, P. (2003) Electrospray ionization tandem mass spectrometry in high-throughput screening of homogeneous catalysts. *Angew. Chem. Int. Ed.*, **42**, 2832–2847.

27 Murgasova, R. and Hercules, D.M. (2002) Polymer characterization by combining liquid chromatography with MALDI and ESI mass spectrometry. *Anal. Bioanal. Chem.*, **373**, 481–489.

28 Adlhart, C. and Chen, P. (2000) Fishing for catalysts: mechanism-based probes for active species in solution. *Helv. Chim. Acta*, **83**, 2192–2196.

29 Jordan, R.F., Dasher, W.E., and Echols, S.F. (1986) Reactive cationic

dicyclopentadienylzirconium(IV) complexes. *J. Am. Chem. Soc.*, **108**, 1718–1719.

30 Jordan, R.F., Bajgur, C.S., Willett, R., and Scott, B. (1986) Ethylene polymerization by a cationic dicyclopentadienylzirconium(IV) alkyl complex. *J. Am. Chem. Soc.*, **108**, 7410–7411.

31 Feichtinger, D., Plattner, D.A., and Chen, P. (1998) Ziegler–Natta-like olefin oligomerization by alkylzirconocene cations in an electrospray ionization tandem mass spectrometer. *J. Am. Chem. Soc.*, **120**, 7125–7126.

32 Santos, L.S. and Metzger, J.O. (2006) Study of homogeneously catalyzed Ziegler–Natta polymerization of ethene by ESI-MS. *Angew. Chem. Int. Ed.*, **45**, 977–981.

33 Volland, M.A.O., Adlhart, C., Kiener, C.A., Chen, P., and Hofmann, P. (2001) Catalyst screening by electrospray ionization tandem mass spectrometry: Hofmann carbens for olefin metathesis. *Chem. Eur. J.*, **7**, 4621–4632.

5
On-line Monitoring Reactions by Electrospray Ionization Mass Spectrometry

Leonardo S. Santos

5.1
Introduction

In the last decade, mass spectrometry has developed at a tremendous rate. This expansion has been driven by the growing knowledge of ionization methods at atmospheric pressure (API), mainly electrospray ionization (ESI) [1], which makes the investigation of liquid solutions possible by mass spectrometry. ESI is used for ionic species in solution, and this 'ionization' method opened up the access to the direct investigation of chemical reactions in solution via mass spectrometry. In principle, ESI make possible the detection and study not only of reaction substrates and products, but even short-lived reaction intermediates as they are present in solution, providing new insights into the mechanism of several studied reactions.

ESI-MS and its tandem version ESI-MS/MS are rapidly becoming the techniques of choice for solution mechanistic studies in chemistry. Depending on the conditions set in the equipment, the ESI process can be used to transfer analyte species generally ionized in the condensed phase into the gas phase as isolated entities. An interesting method of studying intermediates using ESI is the 'ion fishing' technique applied by Chen [2]. The suspected catalytic ionic species is 'fished' from solution and transferred to the collision cell of the mass spectrometer, where the gas-phase catalytic reaction can be further studied by ion/molecule reactions [3].

This chapter deals with applicability of on-line mechanistic investigations of catalyzed reactions through ESI-MS in the condensed phase. It is focused on the interception of intermediates in organic reactions previously proposed based on experimental evidence, and isolation of the different products, validating (or not) the empiric proposals. More detailed information about the technique can be found in some excellent reviews [4] that summarize current thinking on the various stages of the ESI process [5, 6]. However, a brief description of the ESI mechanism is appropriate at his point.

Reactive Intermediates: MS Investigations in Solution. Edited by Leonardo S. Santos.
Copyright © 2010 WILEY-VCH Verlag GmbH & Co. KGaA, Weinheim
ISBN: 978-3-527-32351-7

5.2
Preservation of the Charge in the Transit of Ions from Solution to the Gas Phase Using the ESI Technique

The electrospray process can be described with relative simplicity. A solution of the analyte is passed through a capillary held at high potential. The high voltage generates a mist of highly charged droplets, which passes through potential and pressure gradients toward the analyzer portion of the mass spectrometer. During this transition, the droplets decrease in size by evaporation of the solvent and by droplet subdivision resulting from the coulombic repulsions caused by the high charge density achieved in the shrinkage. The final result is that the ions become completely desolvated [7].

The charge state of the isolated ions is assumed to closely reflect the charge state in solution (multiply charged species due to ion/molecule reactions in the interface are sometimes observed), since the transfer of ions to the gas phase is not an energetic process – the desolvation is indeed a process that effectively cools the ion [8–10]. Therefore, it can be assumed that ESI involves only the stepwise disruption of noncovalent interactions, principally the removal of molecules of solvation, and interception of this process may allow the preservation of relatively strong noncovalent interactions of analytical significance [9]. For example, in a detailed study by Kebarle and Ho, the transfer to the gas phase of different ions dissolved in a wide variety of solvents was possible by ESI. It included singly and multiply charged inorganic ions, for example, alkali metals, alkaline earths, transition metals, organometallic species, and singly and multiply protonated or deprotonated organic compounds, for example, amines, peptides, proteins, carboxylic acids, and nucleic acids. The experimental information available so far on the preservation of the charge is ambiguous [10]. Single-electron transfer can occur during the electrospray process, because the ESI capillary may act as an electrolytic half-cell when extreme conditions are employed [11]. It is quite common to find reports concerning redox reactions in ferrocene derivatives [12], metalloporphyrins [13], metal complexes of amino acids [14], and some organic molecules. Some neutral organic molecules that possess no sites of protonation/deprotonation have also been analyzed using unconventional conditions for ESI technique [15]. In all these examples, high ion-source voltages or modifications in the configuration of equipment were carried out, producing the electrolytic cell-like behavior of ESI-MS.

There are some documented cases where the charge state is changed after the desolvation process, such as multiply charged inorganic and organic ions which are sometimes not seen in their expected charge state. The change occurs mainly for ions with the charge localized on one single atom or small group of them, for example, M^{z+} with $z > 1$, SO_4^{2-}, and PO_4^{3-} [16]. When the multiple charge is confined to such a small volume, the high coulombic repulsions present force the ion to undergo a charge reduction by intra-cluster proton transfer reaction in the case of protic solvents, or by charge separation between the metal center and the bonded solvent molecules when aprotic solvents are used. Such reduction processes are promoted by collisions with residual gas molecules in the source interface region [17].

The interception of multiply charged naked ions can be achieved when the charge is stabilized by solvent molecules or ligands.

The ESI technique is not only qualitative: the intensity of the detected gas phase ions and the corresponding concentrations of these ions in the electrosprayed solution are related. However, in cases where the solution contains compounds able to react with these ions, the intensities may change drastically, and much more complex relationships may prevail. The coordination properties of supramolecular compounds may also differ between the solution and the gas phase. It is well known that polydentate ligands capable of forming stable solution complexes with transition metal ions favor the transfer of the metal into the gas phase by stabilizing it [18]. Using salts of doubly charged ions ($M^{II}X_2$) as an example, ESI of [$M^{II}L + X_2$] complexes may afford singly [$M^{II}L + X$]$^+$ or doubly [$M^{II}L$]$^{2+}$ charged ions by losing one or two counter-ions (X^-) respectively. In some cases, it is possible to observe ions in the ESI mass spectrum corresponding to [$M^{II}L - H$]$^+$ resulting from the elimination of one counter-ion plus a subsequent expulsion of the counter-anion acid (HX). The proton is therefore provided by the ligand. Meanwhile in all these cases the oxidation state of the metal center does not change [18]. The coordination number can indeed change, as Vachet and coworkers suggested. For a number of Ni^{II} and Cu^{II} complexes, the coordination numbers differ on going from solution to gas phase [19].

All these cases suggest the necessity to perform a rigorous analysis to probe unambiguously that the species detected by ESI are the ones prevailing in solution, and more importantly to confirm that they are indeed reactive intermediates on the reaction path. An outstanding methodology is to isolate in the gas phase the species assumed to participate in the reaction mechanism and perform ion/molecule reactions with the substrate of the reaction solution. This methodology is a very powerful way to reject side-products and to assure the reliability of the analysis. Another important method is to study well-known reactions and compare the data obtained by ESI-MS with other spectroscopic techniques.

5.3
Developing Methods to Study Reaction Mechanisms

5.3.1
Monitoring Methods

There are two different possibilities for studying a reaction using API methods, namely off-line and on-line screenings.

5.3.1.1 **Off-Line Monitoring**
Early investigations using API techniques were usually performed off line [20, 21]. A sequence of events in off-line study of a reaction in solution by mass spectrometry could be first to investigate the specific reaction conditions by mixing the reagents for the detection of the different intermediates, and then to determine the solution

composition over time when the reactants are progressively transformed into products [22, 23]. This latest operation can be accomplished by direct screening by MS of the reaction intermediates at pre-defined intervals and characterization by MS/MS, always if there is a reasonable concentration of them in solution and these species are not degraded within a few seconds/minutes. The overall time resolution has to match the rate of the process to yield the desired information. It is determined by the interval elapsed between consecutive sampling operations and is a direct function of the time required to secure and quench each aliquot in off-line methods. However, there are some inherent limitations to this approach. Transient species cannot be analyzed by this method because of their short lifetime in solution.

Hinderling and Chen [24, 25] used off-line monitoring in developing a mass spectrometric assay of polymerization catalysts for combinational screening. A homogeneous Brookhart polymerization catalyzed by a Pd(II) diimine complex using various ligands was quenched and then studied by ESI-MS. The mass spectrometric method required milligram quantities of catalyst, took place in only few minutes, and was suitable for both pooled and parallel screens of catalyst libraries. This intriguing method of high-throughput screening of homogeneous catalyst by electrospray ionization tandem mass spectrometry was recently reviewed [4c].

5.3.1.2 On-Line Monitoring
In a second scenario, the kinetic and mechanistic information about the reactions in solution can be studied using reactors coupled to the ion-source. The first on-line mass spectrometric investigation in electrochemical reactions using thermospray ionization was reported in 1986 [26]. It proved the potential-dependent formation of dimers and trimers in the electrooxidation of N,N-dimethylaniline in an aqueous solution. Another example of on-line investigations is the reaction of ferric bleomycin and iodosylbenzene that was reported by Sam and coworkers [27] in 1995, where a low dead volume mixing tee directly attached to the spray source was used. This feature provides mass-specific characterization of stable products and reactive intermediates with lifetimes down to the millisecond time regime, and subtle changes in the reactional medium could also be observed. The simplest reactor coupled on-line to the mass spectrometer is its own syringe. It allows screening of the reaction in real time and trapping of transient species. Several groups developed different devices to study the mechanism in solution of radical-initiated, photochemical, electrochemical and organometallic reactions, reducing the transit time of the species during the experiments.

5.3.2
Microreactors

5.3.2.1 PEEK Mixing Tee as Microreactor
An interesting commercially available PEEK mixing tee (Alltech, Figure 5.1) can be seen as a useful microreactor [28, 29]. It can be directly connected to the ESI spray capillary, allowing reaction times from 0.7 to 28 s in a continuous-flow mode to be

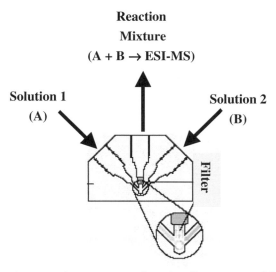

Figure 5.1 The microreactor allows the effective mixing of the reactants in solution and is coupled directly to the electrospray ion source.

investigated. Longer reaction times can easily be covered by introducing a fused silica transfer capillary of variable length between the microreactor and the spray capillary. The reaction time is controlled by flow rates from syringes and by capillary diameters. The chemical reaction in this system takes place by mixing two liquid flows containing the substrate and the reagent in the close proximity of the ionization source. At the moment of mixing the two solutions, the reaction is initiated, and the mass spectra of the reacting solution under steady state conditions can be acquired (Figure 5.1).

5.3.2.2 Capillary Mixer Adjustable Reaction Chamber

The Wilson's on-line continuous-flow apparatus [30] consists of two concentric capillaries, each of which is connected to a syringe to allow for infusion of two reactant solutions. The two solutions are mixed at the end of the inner (fused silica) capillary, and the reaction is allowed to proceed until pneumatically assisted ESI occurs at the outlet of the outer (stainless steel) capillary. The reaction time is controlled by the flow rates from the syringes and by the volume between the mixing point and the outer capillary outlet. Mass spectra can be recorded for selected reaction times by controlling the syringe's flow rate and the volume available between the capillaries ends. To acquire kinetic data (i.e., intensity time profiles), the inner capillary is withdrawn continuously such that time-dependent changes in signal intensity can be observed for all ions in the mass spectrum. A continuous-flow setup for 'time-resolved' ESI-MS is depicted in Figure 5.2. This system represents a concentric capillary mixer with adjustable reaction chamber volume. This instrument operates under laminar flow conditions, a fact that has to be taken into account

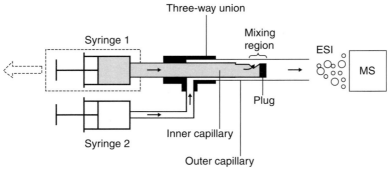

Figure 5.2 Schematic representation of a continuous-flow mixing setup for kinetic measurements by time-resolved ESI-MS. Syringes 1 and 2 deliver a continuous flow of reactants; mixing of the two solutions initiates the reaction of interest. The plug at the end of the inner capillary ensures that all the liquid from syringe 1 is expelled through a notch at the capillary end into the narrow inter-capillary space. The inner capillary can be automatically pulled back, together with syringe 1 (as indicated by the dashed arrow). This provides a means to control the 'age' (reaction time) of the mixture at the capillary outlet. Arrows indicate the direction of liquid flow.

for the data analysis. Reactant solutions are continuously expelled from two syringes. The first of these syringes is connected to the inner capillary, whereas the solution delivered by the second syringe flows through the outer capillary. The reaction of interest is initiated by mixing the two solutions at the end of the inner capillary. The reaction then proceeds, while the mixture flows toward the outlet of the apparatus, where ESI takes place. The reaction time is determined by the solution flow rate, the diameter of the outer tube and, most importantly, by the distance between the mixing point and the outlet. This distance can be modified by adjusting the position of the inner capillary, as indicated in Figure 5.2. Kinetic experiments are typically performed by initially suppressing the volume between capillaries, which corresponds to a reaction time of zero. The mass spectrometer then continuously monitors the ions emitted from the capillary outlet; meanwhile the volume – the reaction time – is increased by slowly pulling back the inner capillary. These experiments provide three different sets of data, that is, the reaction time determined by the mixer position, the information on the identity of the species in the reaction mixture provided by the m/z values, and the concentration of each of these species, which is related to the intensity recorded for the ions in the mass spectra. The time resolution of this system is suitable for measuring rate constants in the range from $1.0\,s^{-1}$ up to at least $100\,s^{-1}$.

5.3.2.3 Photolysis Cell

Arakawa on-line apparatus operates by directly irradiating a sample solution passing through a quartz photolysis cell located in the middle of the ESI spray tip to detect intermediates with lifetimes of more than few minutes (Figure 5.3). A light shutter equipped with a UV cutoff filter is mounted at the exit of the lamp to control photoirradiation. This set allows two different modes of irradiation. The first one, defined as cell mode irradiation, consists of the direct irradiation of the sample

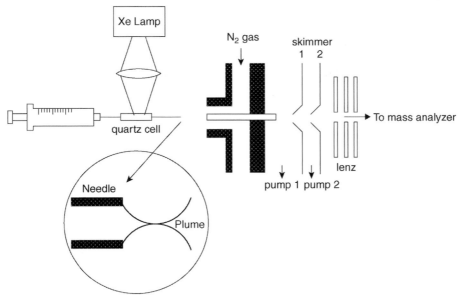

Figure 5.3 Schematics of electrospray mass spectrometry for on-line analysis of photochemical reactions. Two modes of photoirradiation ($\lambda > 420$ nm; a UV cutoff filter) were used: the cell mode, involving irradiation of the sample solution in a quartz cell, and the spray mode, involving irradiation of charged droplets distributed in the plume at the tip of needle.

solution as it passes through the cell (Figure 5.3). It takes about 2 min for the flowing sample to pass across the irradiated area in the cell and about 40 s to arrive at the tip of the needle for spraying, which allow the mass analysis of photoproducts with lifetimes of minutes to be performed [31]. The second mode, defined as spray mode, involves the irradiation of the charged droplets distributed in a plume fashion at the tip of the needle, making possible the detection of intermediates with lifetime of milliseconds.

5.3.2.4 Photochemical Reactor

Similar time resolution was described by Brum and Dell'Orco [32] monitoring the photolysis of idoxifene in a jacketed reactor. In this particular study, the signal of the relevant $[M + H]^+$ ions was directly used without previous isolation, demonstrating the potential utility of such MS interfaces for *in situ* probing.

The experimental approach consists of a jacketed reaction flask equipped with a looping stainless steel capillary (Figure 5.4), where the solution is irradiated. A pump povides circulation of the solution through the capillary at a high flow rate (5–10 mL min^{-1}); meanwhile another pump separates part of the circulating solution into PEEK tubing. Finally, a third pump allows the addition of buffer solution or make-up solvent appropriate for the ionization method through a static mixing tee. The accuracy of the fluid flow is ensured by the use of a metering valve in the connection

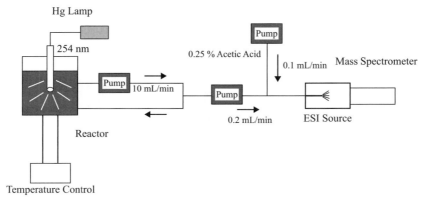

Figure 5.4 Basic schematic of the experimental apparatus developed by Brum and Dell'Orco.

between the looping capillary and the PEEK tubing. This apparatus allows the introduction of samples to the mass spectrometer with an approximate dead time of few minutes, making it unsuitable for extremely rapid reactions. However, it can be used to study a very great variety of not only photochemical, but also organic reactions in real time [33] and in a processing scale model of the reactions.

5.3.2.5 Nanospray Photochemical Apparatus

Amster's apparatus is an on-line ESI-MS technique for the study of photochemical reactions that greatly reduces the transit time of photogenerated species [34]. Figure 5.5 shows sample solutions that are irradiated directly in the optically transparent nanospray tip of the ESI source. Subsequent thermal reactions of the primary photoproducts take place in the region between the photolysis zone and the tip end. The transit time of a photoproduct depends on the volumetric flow rate of the sample, the inner diameter of the tip, and D, the distance between the midpoint of the irradiated zone and the tip end. For example, with $D = 0.84$ mm, a tip diameter of 40 μm, and a flow rate of 40 μL h^{-1}, products require 95 ms to arrive at the tip end for spraying. All chemical reactions are quickly quenched (ms time scale) once the

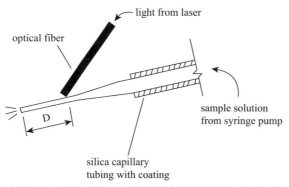

Figure 5.5 Photochemical apparatus for nanospray tip developed by Amster.

sample solution leaves the tip owing to rapid desolvation of the solute that occurs during the electrospraying process. Consequently, ionic species with solution lifetimes in the millisecond range or longer can be detected by this technique.

5.3.2.6 Electrochemical Cell

Lev and coworkers [35] developed a radial-flow electrochemical cell coupled to an electrospray interface to study electrochemical reactions involving parallel and consecutive chemical reactions (Figure 5.6). The electrochemical flow cell consists of two cylindrical Delrin blocks (1 and 2) separated by a 50 μm Teflon spacer (3). A 1.6 mm diameter Pt disk electrode (WE) is pressed into the lower block. The electrolyte flows in a radial, inward direction. A central aperture in the Teflon spacer

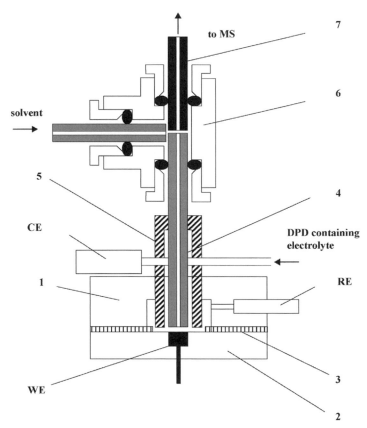

Figure 5.6 Schematic Lev's radial thin-layer flow electrochemical cell on-line with electrospray emitter. (1) Upper Delrin block with tubings; (2) bottom Delrin block with Pt working disk electrode; (3) Teflon spacer; (4) reactor-PEEK tube OD 1.57 mm; (5) PEEK tube OD 3.2 mm, ID 1.6 mm; (6) T-junction; (7) grounded stainless steel tube approx. 5 cm long, ID 0.13 mm, OD 1.6 mm.

(3) allows passage of the electrolyte toward the center of the disk electrode where it is collected into the axial inner tube (4). The volume of the compartment of the working electrode, which is the volume of electrolyte located between the working electrode and the walls of the central tube (4), is only 0.1 µL. The working disk electrode and the inner tube (4) are coaxial. The channel in the central tube (4) forms a reactor in which subsequent chemical reactions can occur, and its volume can be adjusted to 0.5, 2.0, 2.8, and 8.0 µL through different inner PEEK tubes, covering reaction times from 4.3 to 480 s. A detailed description of on-line coupling of electrochemical apparatus to mass spectrometers can be found in the review by Diehl and Karst [36]. Despite the many applications of electrochemistry–mass spectrometry [37] it must be stated that there are limitations, because of the restriction to electrochemically active analytes or electroactive labels. Electrochemistry–mass spectrometry is not a technique that has been generally used, but it has good potential to solve analytical problems when the electrochemical properties of the (derivatized) analytes can be employed for analytical purposes.

5.4
Probing Reactivity of Intermediates

We and others have introduced an important methodology based on the isolation of species in the gas phase assumed to participate in the reaction mechanism and perform ion/molecule reactions with neutral substrates (in the collision cell) of the reaction solution. This methodology seems to be very useful for discarding side-products and to assure the reliability of the analyses.

Initially, the traditional methods used for gas-phase ion/molecule reactions were limited to neutral compounds that were sufficiently volatile and thermostable to be transferred to the gas phase by heating. Some years ago it was shown that using API conditions (ESI or APCI), the range of neutral molecules participating in ion/molecule reactions could be extended toward those of lower volatility and thermal stability.

Today, this new approach is widely used by mass spectrometrists, exploiting the outstanding speed, sensitivity, and selectivity of the diverse environments that only mass spectrometry offers. The rational relationship between structure and chemical reactivity is used to investigate the structure of gaseous neutral molecules and ions through an arsenal of selective ion/molecule reactions. These investigations, which have been based mainly on EI (electron ionization) [38], reveal the intrinsic mechanistic details of reactions that occur in the unique solvent- and counter-ion-free environments of mass spectrometers. Now, the effectiveness of ion/molecule reactions is used to probe complex problems in organometallic chemistry. However, there are few examples of relevance using this approach, with a focus on those reactions performed under the well-defined conditions of sequential mass spectrometry using mass-selected ions. Key mechanistic details and potential applications of ion/molecule reactions have been shown to occur efficiently under API conditions. These innovative strategies have greatly widened the scope of application of such

5.4 Probing Reactivity of Intermediates

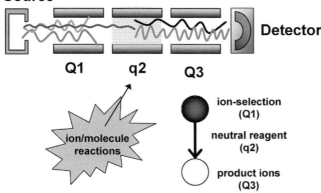

Figure 5.7 Representative 'triple-quadrupole' equipment that is used to study reaction mechanism and further ion/molecule reactions to probe reactivities of trapped species from solutions.

reactions to charged molecules in solutions, which are today easily obtained using API techniques (ESI, APCI, APPI).

From a synthetic point of view, mass spectrometrists use 'triple-quadrupoles' as a useful tool for further refined mechanistic studies (Figure 5.7). Triple-quadrupoles, Q-traps, and other devices certainly afford one of the most complete laboratories for the studies of gas-phase ion/molecule reactions. The remarkable success of this concept has inspired many scientists to devote more attention to this potentially advantageous strategy. Thus, using triple-quadrupoles, ions can be obtained from the ion-source (ESI, APCI), purified via mass-selection in Q1 and reacted in q2 (collision cell) under controlled conditions, and further mass analysis of the reaction products can be performed in Q3. The main advantage is that all these steps are carried out 'on line' in very short time intervals under conditions that can be maintained for long periods and can be easily described and reproduced. Another important method in the study of well-known reactions is to compare the data obtained by ESI-MS with other spectroscopic techniques.

5.4.1
Reaction Mechanism Studies

5.4.1.1 Radical Fenton Reaction

Some synthetically important radical reactions have been studied by API-MS methods. The transient radicals were unambiguously detected and characterized by MS/MS methods. Hess and coworkers [39] reported the degradation products resulting from modified Fenton reactions with the nitroaromatic compounds trinitrotoluene (TNT) and trinitrobenzene (TNB) through electrospray ionization tandem mass spectrometry (off-line ESI-MS^n). Several hydroperoxide adducts were tentatively identified as initial, one-electron reduction products of TNT, and their structure was confirmed by tandem mass spectrometry (Scheme 5.1).

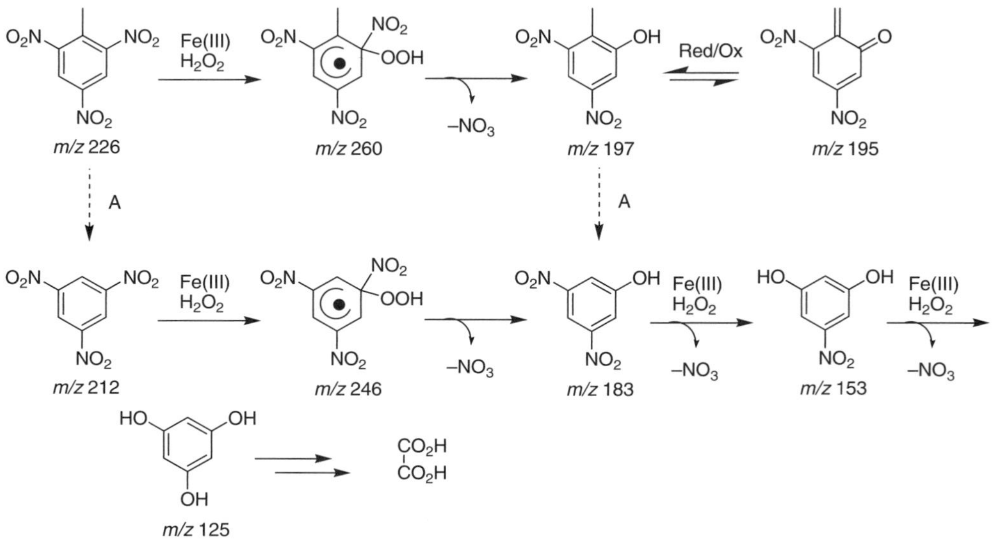

Scheme 5.1 Proposed degradation pathway of TNT and TNB by modified Fenton reaction by Hess [39]. Initial TNT- and TNB-hydroperoxyl radical formation led to step denitration and hydroxylation of the aromatic ring. The oxidation of phloroglucinol (1,3,5-trihydroxybenzene) produced oxalic acid. Nitrate and oxalate were two major end-products of this Fenton system. A dashed arrow represents oxidation of the methyl group and subsequent oxidative decarboxylation of 2,4,6-trinitrobenzene carboxylic acid leading to 1,3,5-trinitrobenzene.

5.4.1.2 Heterogeneous Fenton System

Relatively few examples of radical-monitoring reactions have appeared in the literature [29, 40]. Structural investigation of a novel heterogeneous Fenton system based on an FeO/Fe_3O_4 composite and H_2O_2 was reported [41]. On-line ESI-MS and ESI-MS/MS show that model aromatic compounds (phenol, benzene, and chlorobenzene) are successively oxidized to phenolic, benzoquinonic and ring-opened carboxylic acid-type intermediates. These results suggest that highly reactive hydroxyl radicals are generated from H_2O_2 on the surface of the FeO/Fe_3O_4 heterogeneous catalyst by a classical Fenton-like mechanism. The oxidation reactions were performed with the system $FeO/Fe_3O_4/H_2O_2$ for phenol, benzene, and chlorobenzene used as model organic contaminants with on-line sequential monitoring via both ESI(−)-MS and ESI(+)-MS. ESI(−)-MS was used for the $FeO/Fe_3O_4/H_2O_2$ oxidation of phenol. As expected, at the very beginning (zero reaction time), ESI-MS 'fishes' a single anion from the aqueous solution, that is, deprotonated phenol of m/z 93. Relatively intense anions of m/z 109, 113, 123, 125, and 157 were detected as the degradation proceeds for 20–40 min. These anionic intermediates most likely arise from deprotonation of their respective neutral forms in aqueous solution. Despite the disfavored equilibria with their neutral forms expected at a pH of about 6 and the presumably low aqueous concentrations of these anions, clear ESI(−)-MS detection was achieved. The ESI(−)-MS interception of the anionic species suggested successive hydroxyl radical attacks, likely at either the *ortho* or *para*-activated positions

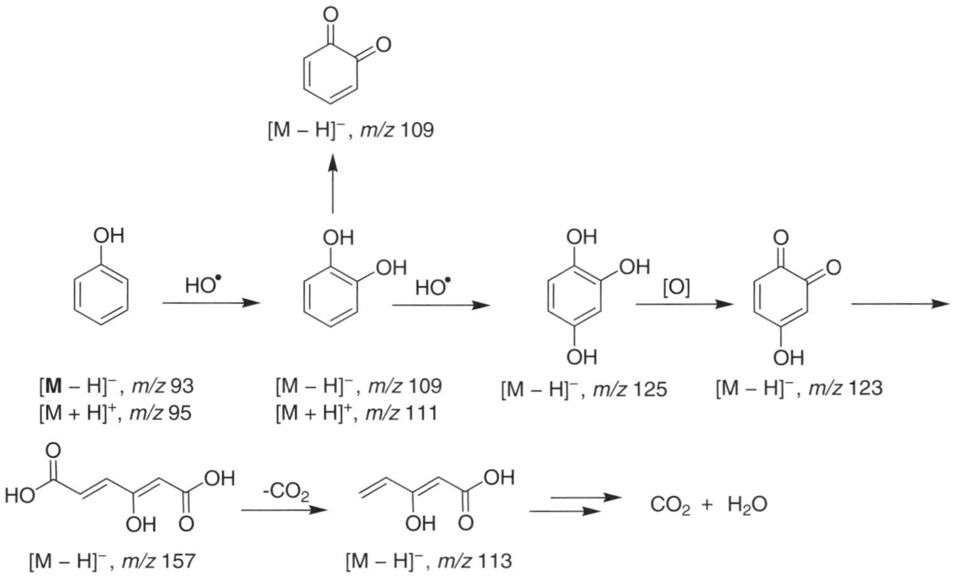

Scheme 5.2 ESI-MS monitoring of phenol degradation through heterogeneous Fenton system [41].

of the phenol ring, or both. A basic reaction scheme with intermediates for the oxidation of phenol by the $FeO/Fe_3O_4/H_2O_2$ system was rationalized as depicted in Scheme 5.2. This additional example of on-line ESI-MS and ESI-MS/MS monitoring of the degradation of model organic molecules in aqueous solution using this heterogeneous Fenton system shows the consecutive formation of phenolic, benzoquinonic, and carboxylic intermediates. This series of polyhydroxylated intermediates indicates degradation via the successive attack by $HO^\bullet$ radicals formed likely from reactions of H_2O_2 with Fe^{2+} surface species. The $FeO/Fe_3O_4/H_2O_2$ system is suggested to act via a classical Fenton-like mechanism.

5.4.1.3 Radical Cation Chain Reactions

Griep-Raming and Metzger [42] studied the thermal dissociation of the triphenylmethyl dimer and of tetra-(*p*-anisyl) hydrazine, operating the ESI source as an electrolytic cell to ionize neutral species, for example, the triphenylmethyl radical. In this study, an electrospray ionization source able to heat the spray capillary was used.

5.4.1.4 [2 + 2]-Cycloaddition of Trans-Anethole

Meyer and Metzger [29, 40a] studied the *tris*(*p*-bromophenyl)aminium hexachloroantimonate ($1^{+\bullet}SbCl_6$) mediated [2 + 2]-cycloaddition of *trans*-anethole (**2**) to give 1,2-*bis*-(4–methoxyphenyl)-3,4-dimethyl cyclobutane (**3**) (Scheme 5.3). The reaction proceeds as a radical cation chain reaction via transients $2^{+\bullet}$ and $3^{+\bullet}$ that were unambiguously detected and characterized by ESI-MS/MS directly in the reacting solution. At first, the reaction was studied by APCI-MS, because substrate **2** and product **3** are not ionized by ESI. A solution of $1^{+\bullet}SbCl_6$ and a solution of **2**, both in

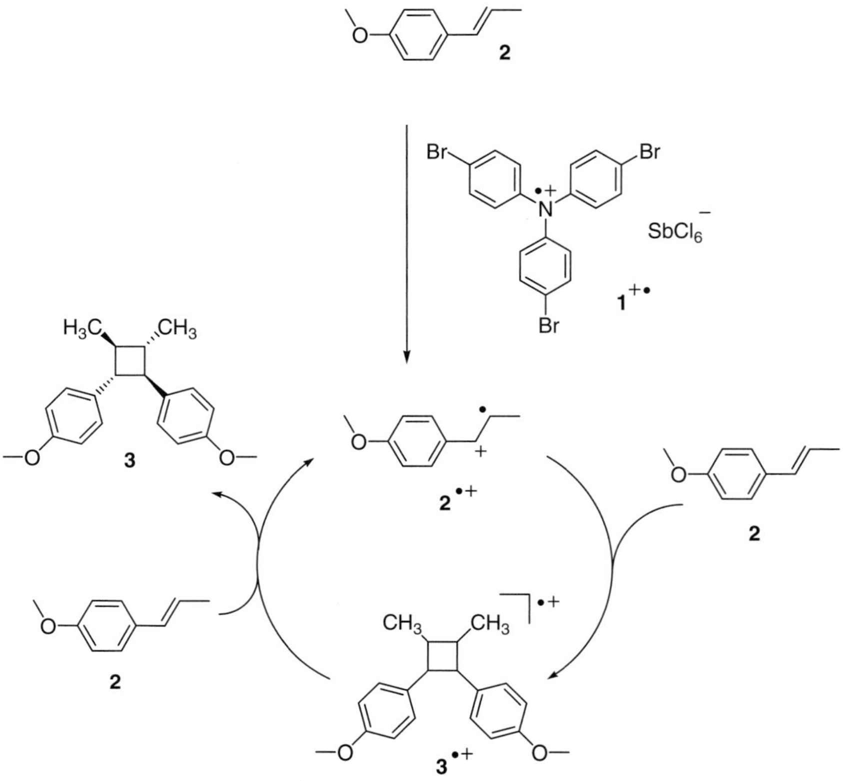

Scheme 5.3 Tris(p-bromophenyl)aminium hexachloroantimonate (**1**$^{+\bullet}$SbCl$_6^-$)-mediated radical cation chain reaction of trans-anethole (**2**) to give head-to-head trans,anti,trans-1,2-bis-(4–methoxyphenyl)-3,4-dimethyl cyclobutane (**3**) via the reactive intermediates **2**$^{+\bullet}$ and **3**$^{+\bullet}$.

dichloromethane, were mixed in a micro reactor system, continuously feeding the APCI-MS. Thus, the reaction could be followed by APCI, making sure that a solution containing an ongoing radical cation chain reaction was introduced in the ion source. This is of central importance because an ongoing reaction process is necessary for a successful detection by ESI-MS of intermediates under steady-state conditions.

When the reacting solution of **1**$^{+\bullet}$SbCl$_6^-$ and **2** was examined by ESI-MS, the spectrum depicted in Figure 5.8a was obtained. The ESI-MS spectrum only displays an intense signal of radical cation **1**$^{+\bullet}$. Substrate **2** and product **3** are not ionized during the ESI process and cannot be observed. The transient radical cations **2**$^{+\bullet}$ of m/z 148 and **3**$^{+\bullet}$ of m/z 296 were not recognized in the spectrum directly, since the quasi stationary concentration of the intermediates in the radical cation chain reaction is estimated to be approximately 10^{-7} M, three orders of magnitude lower than the concentration of radical cation **1**$^{+\bullet}$. The signals of the transient radical cations are expected to disappear in the chemical noise. However, zooming into the chemical noise, it is possible to recognize the very weak signals at the expected m/z of

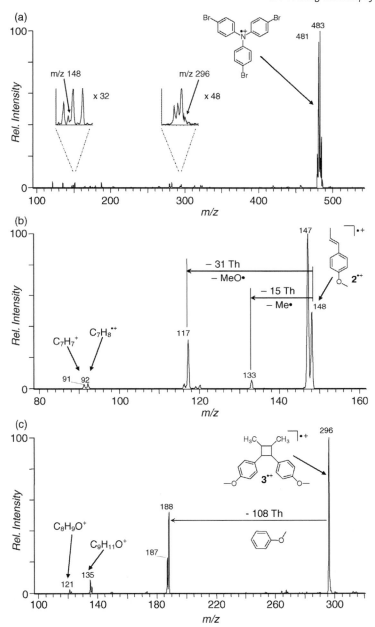

Figure 5.8 (a) ESI mass spectrum in the positive ion mode of the reacting solution of *trans*-anethole (**2**) (0.5 mmol L^{-1}) and *tris*(*p*-bromophenyl)aminium hexachloroantimonate (**1**) (5.0 mmol L^{-1}) in dichloromethane (reaction time approx. 7 seconds). (b) ESI-MS/MS of the ion of *m/z* 148 of the same reacting solution showing identical fragmentations compared to the APCI-MS/MS spectrum of the authentic radical cation **2**$^{+\bullet}$. (c) ESI-MS/MS of the ion of *m/z* 296 of the same reacting solution showing identical fragmentations compared to the APCI-MS/MS spectrum of the authentic radical cation **3**$^{+\bullet}$.

the radical cations. Using the MS/MS technique, radical cations $2^{+\bullet}$ (Figure 5.8b) and $3^{+\bullet}$ (Figure 5.8c) were clearly detected. The ESI-MS/MS spectrum was in accordance with those of the respective authentic radical cations characterized by APCI-MS/MS.

5.4.1.5 Electron Transfer Initiated Diels–Alder Reactions

Using a microreactor-coupled API-MS (Figure 5.1), Fürmeier and Metzger investigated the *tris(p*-bromophenyl)aminium hexachloroantimonate ($4^{+\bullet}SbCl_6^-$, 5.0 mmol L^{-1}) initiated reaction of phenylvinylsulfide (**5**, 0.5 mmol L^{-1}) and cyclopentadiene (**6**, 2.5 mmol L^{-1}) in dichloromethane, which gives the respective Diels–Alder product **7** [40b]. This radical cation chain reaction proceeds via the transient radical cations $5^{+\bullet}$ of the dienophile and $7^{+\bullet}$ of the respective Diels–Alder addition product (Scheme 5.4). These radical cations could be detected directly and characterized unambiguously in the reacting solution by ESI-MS/MS. Their identity was confirmed by comparison with MS/MS spectra of authentic radical cations $5^{+\bullet}$ and $7^{+\bullet}$ obtained by APCI-MS and by CID experiments of the corresponding molecular ions generated by EI-MS. In addition, substrates and products could be monitored easily in the reacting solution by APCI-MS.

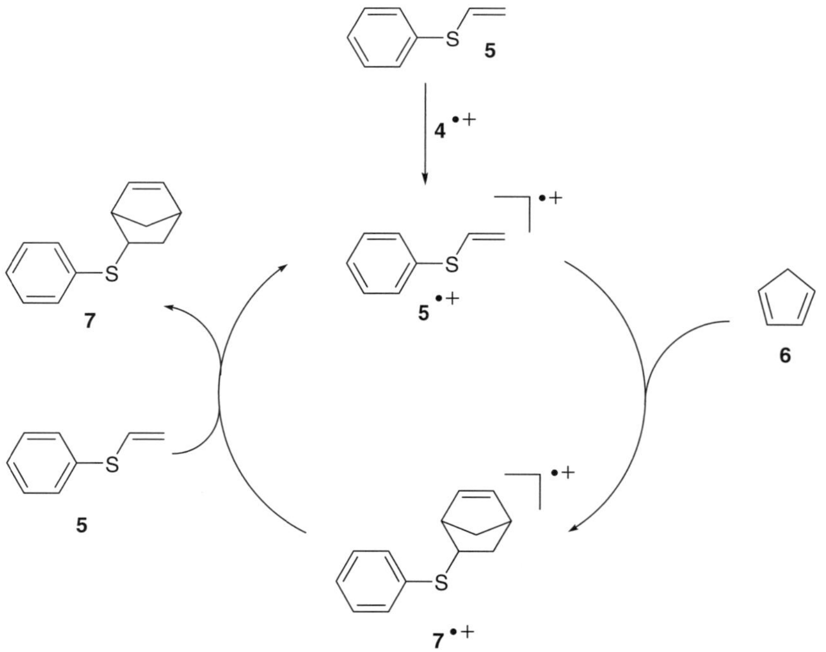

Scheme 5.4 *Tris(p*-bromophenyl)aminium hexachloroantimonate ($4^{+\bullet}SbCl_6^-$) initiated radical cation chain reaction of phenylvinylsulfide (**5**) and cyclopentadiene (**6**) to give the Diels–Alder product 5-(phenylthio)norbornene (**7**) via the reactive intermediates $5^{+\bullet}$ and $7^{+\bullet}$.

Furthermore, Fürmeier and Metzger used the same procedure to study the electron transfer-initiated Diels–Alder reaction of isoprene and anethole, as well as the Diels–Alder dimerization of 1,3-cyclohexadiene. They unambiguously detected and characterized the respective transient radical cations of the dienophiles and of the Diels–Alder addition products.

5.4.1.6 Radical Chain Reactions

Free radicals are neutral species and therefore cannot normally be detected by ESI-MS. However, it is well known that radical reactions can be mediated by Lewis acids if the substrate acts as a Lewis base by chelating the metal atom of a Lewis acid in solution. Fürmeier and coworkers studied tin hydride-mediated radical additions to dialkyl 2-alkyl-4-methyleneglutarates in the presence of Lewis acids (Scheme 5.5) [28, 43]. Diesters such as dialkyl glutarates were able to chelate the Lewis acid and form the dialkyl glutarate–Lewis acid complex, thus allowing the detection of Lewis acid–ester complexes by ESI-MS. For example, the complex with $Sc(OTf)_3$ dissociates to form a chelate complex cation, and a triflate anion was intercepted. In addition to monomeric complex ions, dimeric complex ions $[8_2 \cdot Sc_2(OTf)_5]^+$ were also observed, giving evidence of the respective dimeric complexes in solution [44].

The reaction was carried out by mixing a solution of glutarate **9** with **8** and $Sc(OTf)_3$ in diethyl ether saturated with air with a solution of tributyltin hydride containing triethylborane under argon in the microreactor on-line coupled to the ESI ion source (Figure 5.9). The reacting solution was fed continuously into the mass spectrometer. In Figure 5.9a, the mass spectrum of the reacting solution after a reaction time of

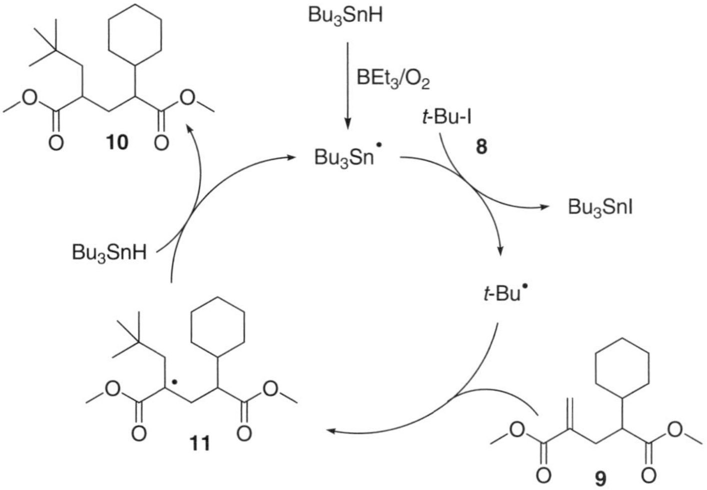

Scheme 5.5 Tributyltin hydride-mediated addition of *tert*-butyl iodide (**8**) to dimethyl 2-cyclohexyl-4–methylene glutarate (**9**), stereoselectively giving *syn*-dimethyl 2-cyclohexyl-4-neopentyl glutarate (**10**) via transient adduct radical 11 in the presence of $Sc(OTf)_3$.

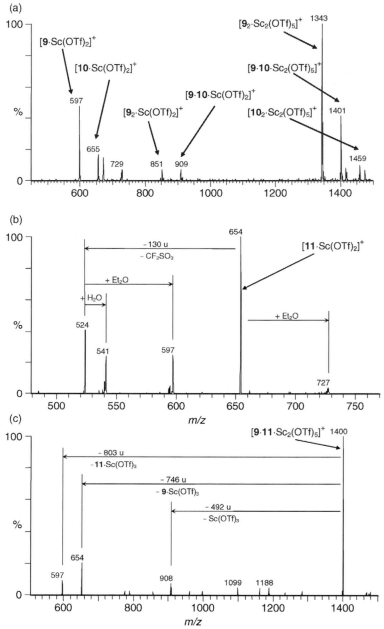

Figure 5.9 (a) ESI mass spectrum in the positive ion mode of the reacting solution of the tributyltin hydride-mediated addition (1.25 mmol L^{-1}) of tert-butyl iodide (**8**) (2.0 mmol L^{-1}) to dimethyl 2-cyclohexyl-4-methyleneglutarate (**9**) (0.5 mmol L^{-1}) in the presence of scandium triflate (0.6 mmol L^{-1}) in diethyl ether resulting in addition product **10** after a reaction time of approximately 30 s. (b) ESI-MS/MS of the radical complex ion [**11**·Sc(OTf)$_2$]$^+$ (m/z 654). (c) ESI-MS/MS of the substrate-radical complex ion [**9·11**·Sc$_2$(OTf)$_5$]$^+$ of m/z 1400 from the same reacting solution.

approximately 30 s is depicted. Monomeric and dimeric complex ions of substrate **9** and product **10** can be observed.

Furthermore, heterodimeric complex ions of substrate and product [9·10·Sc(OTf)$_2$]$^+$ and [9·10·Sc$_2$(OTf)$_5$]$^+$, respectively, were observed. An intermediate radical complex ion [11·Sc(OTf)$_2$]$^+$ with an expected m/z ratio of 654 could not be unambiguously detected in the mass spectrum (Figure 5.9a) because of the steady-state concentration of radical **11** in the radical chain reaction, which is estimated to be approximately 10^{-7} M, four orders of magnitude lower than the concentration of substrate **9** and of product **10**.

Using MS/MS it was also possible to detect and characterize the intermediate radical **11** as a monomeric complex ion [11·Sc(OTf)$_2$]$^+$ (m/z 654, Figure 5.8b) as well as a heterodimeric complex ion with substrate **9** and product **10**, respectively. The MS/MS spectrum of the heterodimeric complex ion of substrate and radical [9·11·Sc$_2$(OTf)$_5$]$^+$ (m/z 1400) is depicted in Figure 5.9c.

Two main and characteristic fragmentations of this ion are the dissociation by the loss of neutral radical complex **11**·Sc(OTf)$_3$ (−803 u) that gives the substrate complex ion [9·Sc(OTf)$_2$]$^+$ (m/z 597), and the loss of neutral substrate complex **9**·Sc(OTf)$_3$ (−746 u) results in the radical complex ion [11·Sc(OTf)$_2$]$^+$ (m/z 654). Additionally, a fragmentation to complex ion [9·11·Sc(OTf)$_2$]$^+$ (m/z 908) by loss of Sc(OTf)$_3$ (−492 u) was observed. The elemental composition of the ions [11·Sc(OTf)$_2$]$^+$, [9·11·Sc$_2$(OTf)$_5$]$^+$, and [10·11·Sc$_2$(OTf)$_5$]$^+$ were confirmed via their accurate masses determined by Q-TOF measurements.

The radical chain allylation of diethyl 2-iododiadipate with allyltributyltin in the presence of Sc(OTf)$_3$ to give diethyl 2-allyladipate via a radical intermediate was studied quite analogously. The transient diester radical was detected and characterized by MS/MS [28].

5.4.1.7 Photochemical Reactions

Arakawa and coworkers [45] developed the on-line photoreaction cell depicted in Figure 5.3 and performed a series of studies on the detection of reaction intermediates in photosubstitution and photooxidation of Ru(II) complexes. The photosubstitution of Ru(bpy)$_2$B^{2+} [bpy = 2,2′-bipyridine; B = 3,3′-dimethyl-2,2′-bipyridine (dmbpy) or 2-(aminomethyl)pyridine (ampy)] was studied in acetonitrile and pyridine. Irradiation of Ru(bpy)$_2$B^{2+} and related complexes yields a charge-transfer excited species with an oxidized Ru center and an electron localization on the bpy moiety. The excited-state complex underwent ligand substitution via a stepwise mechanism that includes an η^1 bidentate ligand (Scheme 5.6). Photoproducts such as Ru(bpy)$_2$S$_2^{2+}$ (S = solvent molecule) and intermediates with a monodentate (mono-*hapto*-coordination) B ligand, Ru(bpy)$_2$BS^{2+}, and Ru(bpy)$_2$BSX$^+$ (X = ClO$_4^-$, PF$_6^-$) were detected. Other studies also identified photo-oxidized products of several mixed-valence Ru(II) complexes upon irradiation ($\lambda >$ 420 nm) [31b, 46].

5.4.1.8 Photochemical Switching Reaction

An interesting photochemical switching reaction utilizing ESI-MS was reported by Arakawa and coworkers [47] (Scheme 5.7). This process involved a cation (e.g., K$^+$,

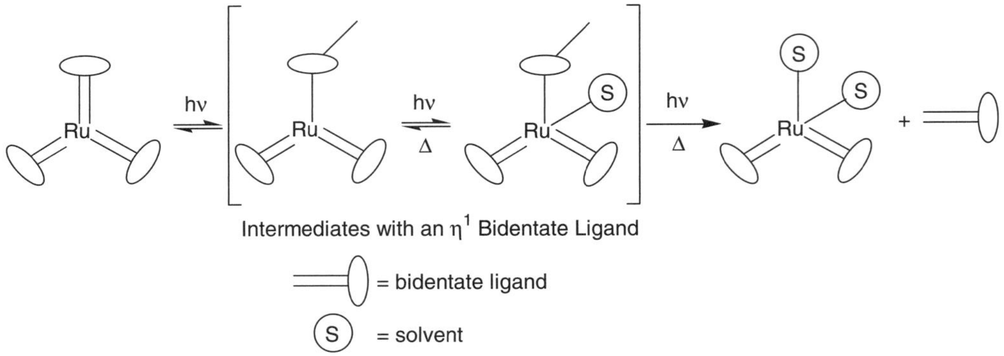

Scheme 5.6 Photooxidation of ruthenium complexes.

Na$^+$) binding complex that consisted of a malachite green derivative incorporating a bis(monoazacrown ether), **12**. The irradiation (240–400 nm) of a solution of crowned malachite green containing an equimolar amount of potassium or sodium perchlorate resulted in an ESI-MS signal which was assigned to the corresponding quinoid cation and subsequent loss of the metal ion, **13**. Presumably, the metal ion was released upon charge-charge repulsion, followed by the intramolecular ionization of the leuconitrile moiety (Scheme 5.7). The 'off' or 'on' mode of a cation-binding host could be manipulated by UV irradiation and confirmed by ESI-MS analysis.

The same on-line photochemical ESI-MS methodology was employed in the analysis of photoallylation reactions of dicyanobenzene (DCB) by allylic silanes via photoinduced electron transfer [48]. Brum and Dell'Orco [32] reported on investigations of the photolysis of a substituted triphenylethylene derivative using the reactor depicted in Figure 5.4. The approach allowed the monitoring of the parent species and five relevant reaction products simultaneously. The rate constants for photoconversion could be derived from the time-ion current data. Herein, Schuster, and coworkers used ESI-MS to study off line some photoadditions to C$_{60}$ [49].

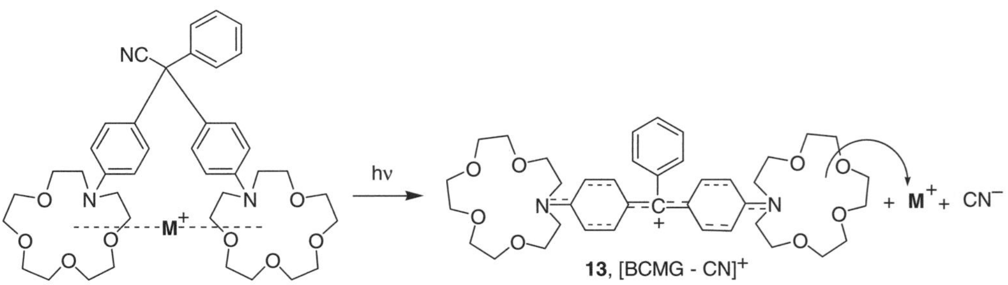

Scheme 5.7 Photochemical switching reaction with Na$^+$ and K$^+$ as an 'off' or 'on' mode of a cation binding host. The screening was performed with substrate concentrations around 0.2 mmol L^{-1} in acetonitrile as solvent.

5.4.1.9 Photoinitiated Polymerization Reaction

Amster and coworkers [34] reported the use of ESI-MS to probe the solution photochemical behavior of [CpFebz]PF$_6$, where Cp is η^5-cyclopentadienyl and bz is η^6-benzene, applying the apparatus proposed by the group shown in Figure 5.5. The interest in the [CpFe(η^6-arene)]$^+$ family arises from their use as visible-light-sensitive photoinitiators for the polymerization of epoxides and other monomers. Mechanistic studies by several groups have established key features of the solution photochemistry of these mixed-ring sandwich complexes [50]. The proposed mechanism, corroborated by Amster, proceeds by ligand field photoexcitation inducing loss of arene to produce [CpFe(L)$_3$]$^+$. The irradiation of an acetonitrile solution of [CpFebz]$^+$ in the nanospray tip of the ESI source yields two major series of ionic products: [CpFe(MeCN)$_{2-3}$]$^+$ and [Fe(MeCN)$_{3-6}$]$^{2+}$. The first series results from the photoinduced release of benzene from the parent complex (Eq. (2.1)), while the second series reflects the subsequent thermal disproportionation of the half-sandwich product (Scheme 5.8). From the experimental data, a lifetime of 95 ms for [CpFe(MeCN)$_3$]$^+$ in room temperature acetonitrile was derived. Irradiating [CpFebz]$^+$ in an MeCN solution containing up to 400 mM cyclohexene oxide (CHO) yielded the same products as those observed in the pure solvent. In addition, species containing coordinated CHO, such as [Fe(MeCN)$_3$(CHO)]$^{2+}$ and [Fe(MeCN)$_4$(CHO)]$^{2+}$, become increasingly abundant at the higher epoxide concentrations. However, no products containing more than one epoxide molecule are observed, even at the highest CHO concentrations used in the experiments. This result indicates that CHO cannot compete effectively with acetonitrile for coordination sites on the metal center. Photolyzing [CpFebz]$^+$ and CHO in the poorly coordinating solvent, 1,2-dichloroethane, resulted in a much richer assortment of products. Half-sandwich complexes of general formula [CpFe(H$_2$O)(CHO)$_{0-5}$]$^+$, as well as fully ring-deligated complexes

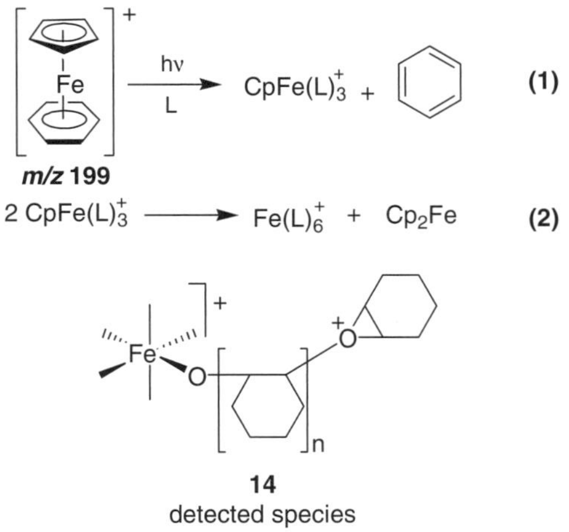

$$[CpFebz]^+ \xrightarrow[L]{h\nu} CpFe(L)_3^+ + C_6H_6 \quad (1)$$

m/z 199

$$2\ CpFe(L)_3^+ \longrightarrow Fe(L)_6^+ + Cp_2Fe \quad (2)$$

14
detected species

Scheme 5.8 Iron-containing product **14** detected in photolyzed solutions of [CpFebz]$^+$ (0.041 mmol L^{-1}) and cyclohexene oxide (CHO, 40 mmol L^{-1}) in 1,2-dichloroethane employing the microreactor depicted in Figure 5.5.

$$[Zn^{II}Pc^{-2}][Cl^-]^- + N_2H_4 \cdot H_2O + h\nu \longrightarrow [Zn^{II}Pc^{-3}][Cl^-]^{-2} + \text{hydrazine radical products} \quad (a)$$

$$[Zn^{II}Pc^{-3}][Cl^-]^{-2} + CBr_4 + h\nu \longrightarrow [Zn^{II}Pc^{-2}][Cl^-]^- + Br^- + \text{other radical products} \quad (b)$$

Scheme 5.9 Proposed mechanism of photoreduction (a), and photooxidation (b) of zinc phthalocyanine.

of general formula $[(H_2O)Fe(CHO)_{2-12}]^{2+}$ (with transit time of 50 ms) were detected. A chemically reasonable description of the structures of ions such as $[(H_2O)Fe(CHO)_{10}]^{2+}$, $[Fe(CHO)_8]^{2+}$, and $[CpFe(CHO)_5]^+$ involving a growing polymer chain bound directly to the metal center (detected as **14**, Scheme 5.8) was suggested. However, no further studies have been made to confirm this suggestion.

Stillman and coworkers [51] reported the direct measurements of the products following the photooxidation and photoreduction of a metallophthalocyanine π-ring. ESI-MS was used to detect the anion zinc(II) (1,4,8,11,15,18,22,25-octafluoro, 2,3,9,10,16,17,23,24-octaperfluoroisopropylphthalocyaninechloride, $[ZnperF_{64}Pc(-2)(Cl)]^-$ and its π-ring anion radical species, $[ZnperF_{64}Pc(-3)(Cl)]^{2-}$. The anion radical species was then photooxidized as a sacrificial photoinduced oxidizing agent using CBr_4 to produce $[ZnperF_{64}Pc(-2)(Cl)]^{2-}$ (Scheme 5.9). The complete reaction cycle was detected directly by ESI-MS, and reaction times of some minutes are provided.

5.4.2
Electrochemical Reactions

Lev and coworkers studied the electrochemical oxidation of N,N'-dimethyl-p-phenylenediamine (DPD) in aqueous electrolyte using the reactor depicted in Figure 5.6 [35]. The competitive paths of the reaction mechanism include the parallel coupling reactions of unreacted DPD with quinonediimine (the first path) and quinonemonoimine (the second path). The latter is formed by hydrolysis of quinonediimine. The basic transformations of DPD (compound **15**) are depicted in Scheme 5.10. The authors mentioned that products were detected after passing through the reactor, where just chemical processes occurred due to the relative multistage chemical coupling reactions. If the chemical reactions did not take place in the flow cell, two electrons could be removed from each DPD molecule entering the cell, which would result in a complete transformation of DPD into quinonediimine.

5.4.3
Heck Reaction

The Heck arylation with arenediazonium salts was studied by Eberlin and coworkers through ESI-MS tandem mass spectrometry experiments. The group detected and structurally characterized the main cationic intermediates of this catalytic cycle directly from solution to the gas phase [52]. They proposed a detailed catalytic cycle of

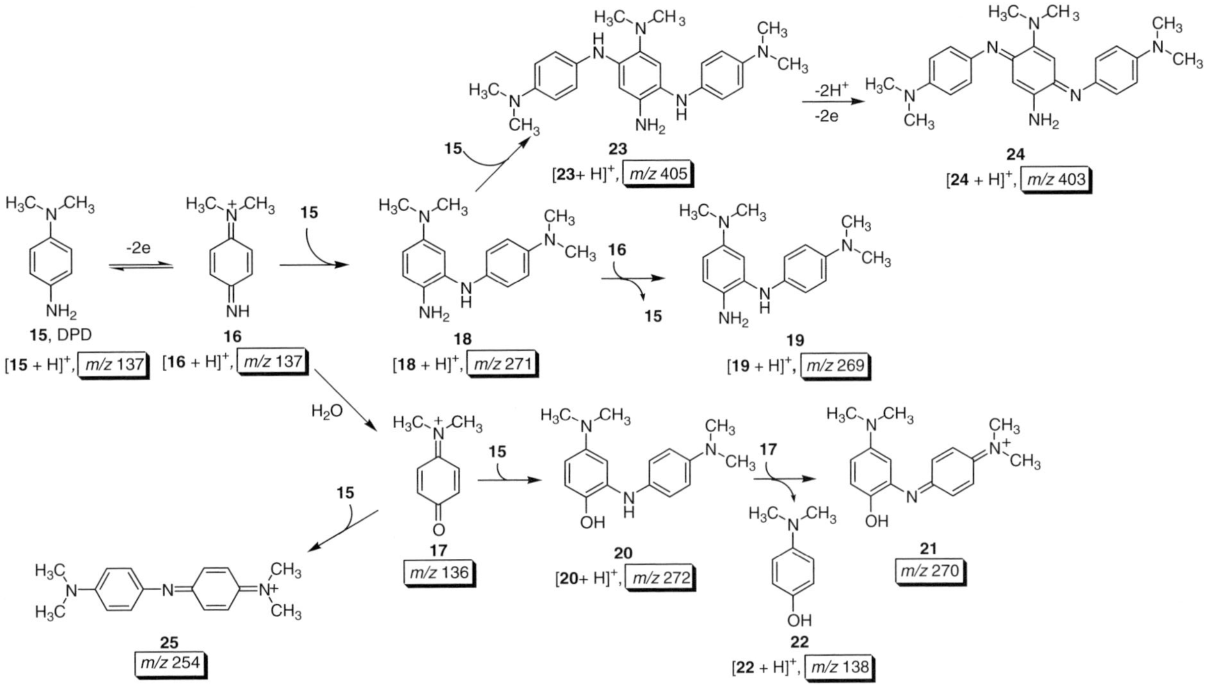

Scheme 5.10 Lev's on-line electrochemical-MS studies of the mechanism of oxidation of N,N′-dimethyl-p-phenylenediamine (DPD, 5.0 mmol L^{-1}) in aqueous electrolytes (0.1 mol L^{-1}).

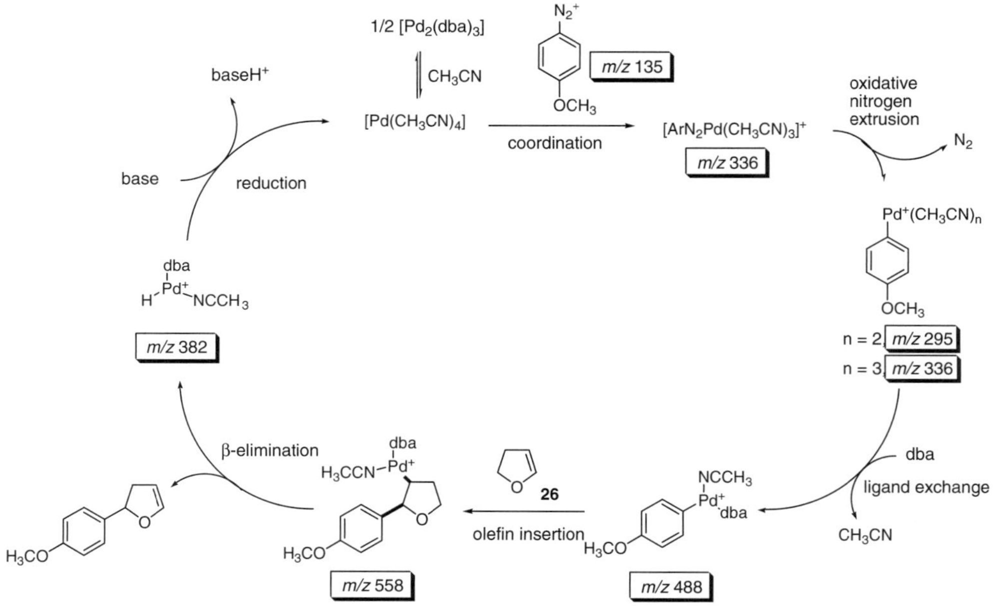

Scheme 5.11 Proposed catalytic cycle of the Heck reaction with arene diazonium salts.

the Heck reaction with arene diazonium salts on the basis of the results obtained from ESI experiments using alkene **26** as example, as depicted in Scheme 5.11.

Roglans and coworkers [53, 54] also studied the Heck reaction of arene diazonium salts using a palladium(0) complexes of 15-membered macrocyclic triolefines.

5.4.4
Suzuki Reaction

The Suzuki coupling reaction of pyridyl halide **27** with phenylboronic acids **29a–c** was studied by Aliprantis and Canary (Scheme 5.12) [22]. They performed off-line ESI-MS investigation, identifying pyridylpalladium(II) complexes as the cation species $[(pyrH)Pd(PPh_3)_2Br]^+$ (**28a**) and $[(pyr)Pd(PPh_3)_2]^+$ (**28b**). Additionally, diarylPd(II) $[(pyrH)(R^1R^2C_6H_3)Pd(PPh_3)_2]^+$ (**30**) and arylPd(II) $[(R^1R^2C_6H_3)Pd(PPh_3)_2]^+$ were observed in the ESI mass spectra. A few years later, some new results were published by Roglans and coworkers on ESI-MS investigation of Pd-catalyzed self-coupling reaction of areneboronic acids [23, 55].

5.4.5
Pd-Catalyzed Enantioselective Allylation Reaction

Pfaltz and coworkers [56] applied the ESI-MS technique for parallel screening of Pd-catalyzed enantioselective allylation reaction of diethyl ethyl malonate with allylic esters. Enantiodiscrimination of a palladium catalyst with different chiral ligands was

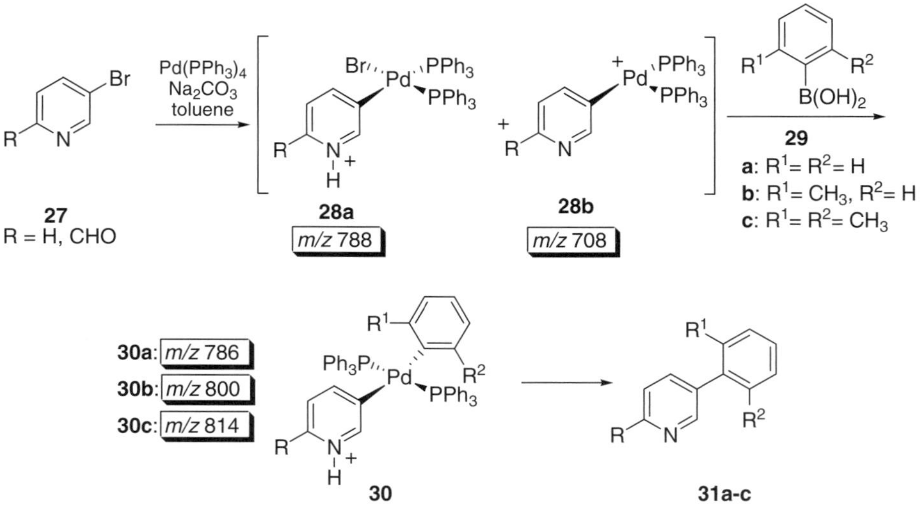

Scheme 5.12 Pd-catalyzed reaction (0.27 µmol L^{-1}) of pyridyl halides **27** (8.9 µmol L^{-1}) with phenylboronic acids **29a–c** (5.48 µmol L^{-1}) after dilution with MeOH for analysis.

studied by ESI-MS. In the catalytic cycle proposed (Scheme 5.13), the first step, the formation of Pd-allyl complexes **A** and **B**, is fast, while the second step, nucleophilic addition to the allylic system to give products **33a** and **33b**, is slower. The cationic intermediates **A** and **B** were observed in the ESI-MS spectra. The ratio **A** : **B** was attributed to the catalyst's ability to discriminate between two enantiomers with two different alkyl groups at the *para* position of the aryl group (Ar = 4-methylphenyl in **32a** and 4-ethylphenyl in **32b**). Five different catalyst precursors were tested with different stereoselectivities in homogeneous solution containing pseudoracemate **32a,b** and the anion of diethyl ethyl malonate as nucleophiles. All ten Pd-allyl intermediates were observed in the ESI-MS screening after two minutes of reaction time. A selectivity order was proposed from the ratio of ion intensities based on the ligand as **37** > **36** > **35** ≅ **34** ≅ **38**, the complex with ligand **37** being the most selective catalyst. The group proposed that the technique may be used with an unlimited number of catalysts in the screening simultaneously, as long as the signals do not overlap.

Roglans and coworkers [57] also studied the substitution of allyl acetate by acetylacetone in presence and absence of Pd catalyst by ESI-MS.

5.4.6
Stille Reaction

For a Stille reaction, on-line ESI(+)-MS(/MS) monitoring have allowed Santos and coworkers [58] to intercept and characterize (a) the actual catalytically active species Pd^0(PPh$_3$)$_2$ (Figure 5.10), (b) the oxidative addition product **41a**, and (c) the transmetalation intermediate **43a** and two products of this process **44a** and **45a**

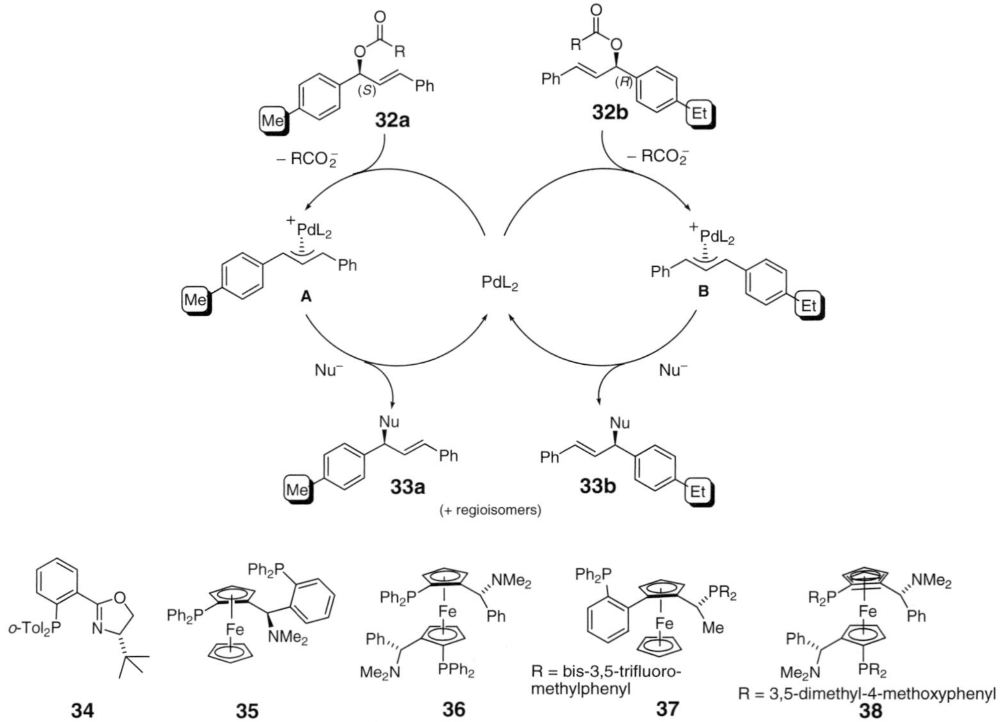

Scheme 5.13 Simultaneous screening of a mixture of five Pd catalysts (0.1 mmol L^{-1}) performed by Pfaltz. Compounds **32a,b** were used as pseudoracemates in solution as it was inferred by the authors that methyl and ethyl groups afforded no great differences in the reactivity of allyl groups toward Pd species, supposing almost the same reactivity behavior for both.

(Figure 5.11). Gas-phase reductive elimination (for **45b**$^+$) has also been observed. Therefore, for the first time, most (if not all) major intermediates of a Stille reaction have been intercepted, isolated, and characterized. Using ESI(−)-MS, the counteranion I$^-$ was the single species detected. Such straightforward experiments further illustrate the applicability of direct infusion ESI-MS(/MS) in revealing, elucidating, and helping to consolidate mechanisms of organic reactions as depicted in Scheme 5.14.

5.4.7
Alkynilation of Tellurides Mediated by Pd(II)

Under palladium dichloride catalysis, vinylic tellurides couple efficiently with alkynes with retention of the double-bond geometry. Looking for experimental support to validate the catalytic cycle proposed for this reaction, Raminelli and coworkers [59] decided to investigate the coupling reaction of vinylic tellurides with

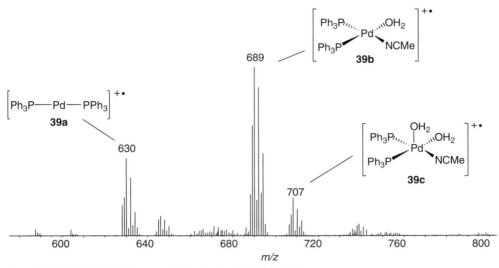

Figure 5.10 ESI(+)-MS of an acetonitrile solution of Pd(OAc)$_2$/PPh$_3$.

alkynes promoted by PdCl$_2$ by means of mass spectrometry techniques. ESI was applied to 'fish' Pd- and Te-containing cationic intermediates involved in the reaction described in Scheme 5.15 directly from the reaction medium to the gas phase for ESI-MS and ESI-MS/MS analysis. Reaction of telluride **47** with alkyne **48** was performed according to Scheme 5.15 to give **49**. The reaction mixture was electrosprayed via the ESI source operated in the positive ion mode. The reaction mixture was stirred for

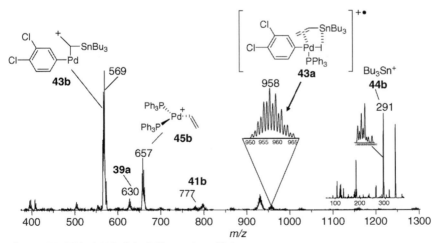

Figure 5.11 ESI(+)-MS of the Stille reaction of 3,4-dichloroiodobenzene (**40a**) and vinyltributyltin (**42**) in acetonitrile mediated by Pd(PPh$_3$)$_4$.

160 | *5 On-line Monitoring Reactions by Electrospray Ionization Mass Spectrometry*

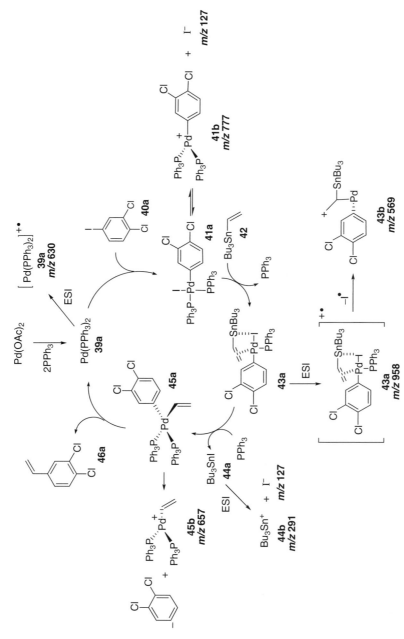

Scheme 5.14 Expanded mechanism by ESI(+)-MS of the Stille reaction of 3,4-dichloroiodobenzene (**40a**) and vinyltributyltin (**42**) in acetonitrile mediated by Pd(OAc)$_2$.

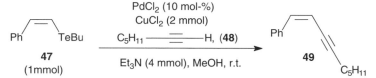

Scheme 5.15 Coupling of vinylic tellurides with alkynes catalyzed by PdCl$_2$ and CuCl$_2$.

1 h, and gave a number of ions that were attributed to the species shown in Scheme 5.16.

The most relevant data for the validation of the proposed catalytic cycle was the detection of three Te- and Pd-containing cationic complexes; **50** of m/z 717, **51** of m/z 499, and **52** of m/z 627. Ions **50** and **51** were suggested to be formed by solution ionization of the neutral species LnPdCl$_2$: that is, LnPdCl$_2$ → LnPdCl$^+$ + Cl$^-$. Cations analogous to **50** and **51** have been suggested [22, 55, 60] using NMR and cyclic voltammetry in reactions where ArPd$^+$ complexes were postulated. Herein, by

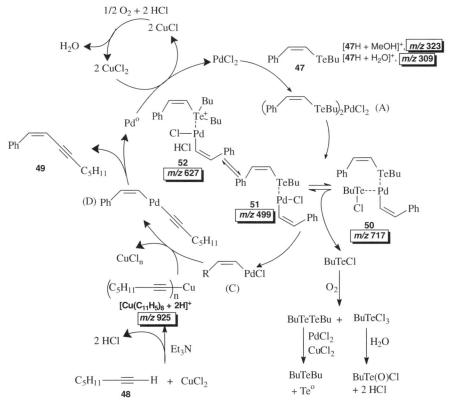

Scheme 5.16 Expanded mechanism based on ESI(+)-MS tandem MS/MS experiments for the reaction of **47** with alkyne **48** to give **49** using PdCl$_2$ and CuCl$_2$ in the absence of an inert atmosphere.

ESI fishing of **50** and **51** in their cationic forms, a solution equilibrium between these two species was suggested to exist in the palladium insertion process. It was also suggested that BuTeCl acted as a ligand that could stabilize **50**, which may be formed by coordination of two styryl butyl tellurides to $PdCl_2$ followed by transmetalation. Since the relative intensity of the cationic **50** and **51** remained nearly constant during up to 36 h of reaction, as shown by continuous ESI-MS monitoring, the ligand exchange equilibrium between **50** and **51** is likely dynamic. The mechanism in which the cationic **52** is formed was not so straightforwardly rationalized, but a proposed route to **52** involves **51** in a ligand exchange process, namely exchange of **47** by $[47 + Bu]^+$ and further association with neutral HCl present in the reaction medium. The species **50** and **51** are Pd-styryl complexes that can undergo β-hydrogen elimination to yield PhC≡CH. However, this acetylene was not observed as a by-product in the coupling reaction. It is known that cations such as Li^+, Ag^+, and Tl^+ can inhibit such β-hydrogen eliminations [61]. It is also known that cation association (**51** → **52**) enhances the solubility of metal complexes [62]. Organocopper cluster intermediates were also identified by ESI-MS and characterized as alkynylcopper species by characteristic Cu isotopic patterns and ESI-MS/MS structural analysis. An expanded catalytic cycle for the coupling of vinylic tellurides with alkynes catalyzed by palladium dichloride was proposed as depicted in Scheme 5.16 [63].

5.4.8
Lewis Acid-Catalyzed Additions

Ricci and coworkers [64] studied oxazoline moiety fused with a cyclopenta[β]thiophene as ligands on the copper-catalyzed enantioselective addition of Et_2Zn to chalcone. The structure of the active Cu species was determined by ESI-MS. Evans and coworkers [65] studied C_2-symmetric copper(II) complexes as chiral Lewis acids. The catalyst-substrate species were probed using electrospray ionization mass spectrometry. Comelles and coworkers studied Cu(II)-catalyzed Michael additions of β-dicarbonyl compounds to 2-butenone in neutral media [66]. ESI-MS studies suggested that copper enolates of the α-dicarbonyl formed *in situ* are the active nucleophilic species. Schwarz and coworkers investigated by ESI-MS iron enolates formed in solutions of iron(III) salts and β-ketoesters [67]. Studying the mechanism of palladium complex-catalyzed enantioselective Mannich-type reactions, Fujii and coworkers characterized a novel binuclear palladium enolate complex as intermediate by ESI-MS [68].

5.4.9
C—H Activation and Hydrogenations

Chen and Gerdes [69] performed a combination of ESI-MS and kinetic studies in the Pt^{II}-catalyzed H/D exchange (Scheme 5.17) and Ir-based hydrogenations [70]. Noyori used ESI(+)-MS to intercept the active species of the asymmetric hydrogenation reaction of acetophenone with chiral Ru-complex to give (R)-phenylethanol in 82% *ee* [71], Scheme 5.18.

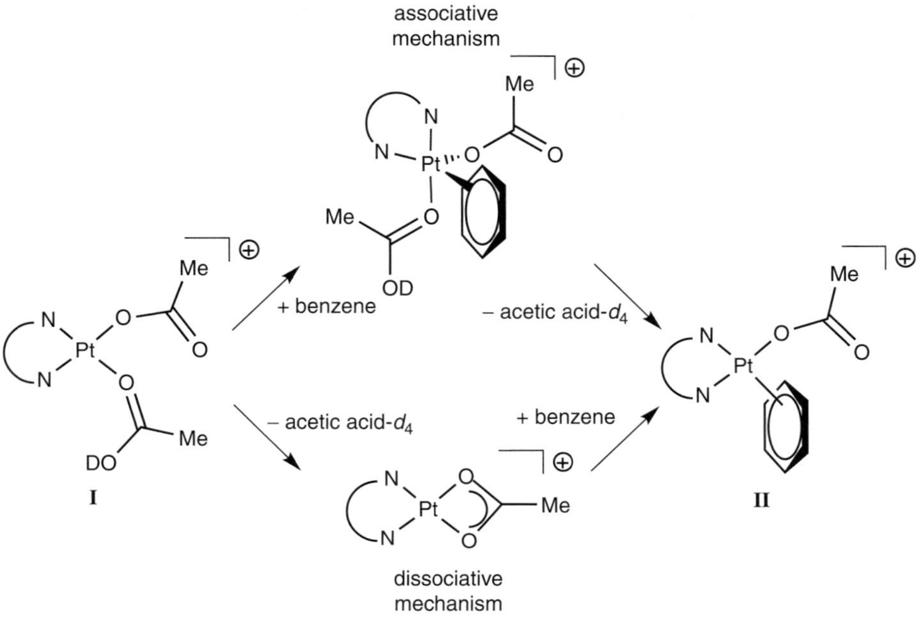

Scheme 5.17 Cationic Pt(II) carboxylato complexes in catalytic arene C—H activation.

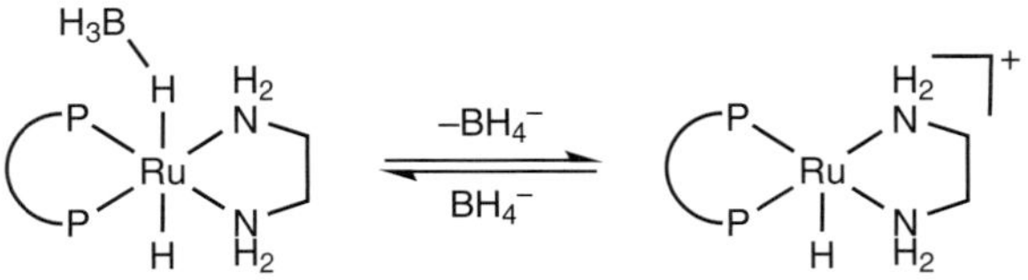

Scheme 5.18 Active Noyori species for asymmetric hydrogenation.

5.4.10
Oxidation Reactions

The (Schiff base)vanadium(V) complex **53** with tridentate imine auxiliaries acts as a catalyst for the oxidation of Br$^-$ with *tert*-butyl hydroperoxide (TBHP) in nonaqueous solvents. The principal objective of this study performed by Hartung and coworkers [72] was associated with the search for an adequate combination of bromide source and primary oxidant that would reduce the inherent propensity of (Schiff base) vanadium(V) complexes in the presence of TBHP to convert alkenols (**54**) directly into hydroxy-functionalized tetrahydrofurans (**56**) [73]. Reactivity-selectivity studies on the vanadium(V)-catalyzed oxidation of bromide were investigated through ESI (−)-MS^n in order to interpret crucial steps in the process such as peroxide activation and the generation of an active brominating reagent. The reactivity of *tert*-butylperoxy complex **55** was explored by ESI-MS analysis of solutions that were obtained by mixing (Schiff base)vanadium(V) complex **53** in CH_2Cl_2/CH_3CN with TBHP and py·HBr, and an intense ion of m/z 237 was observed in the negative ion mode and assigned as Br_3^- ion. Further ions detected in this experiment originated from $[VOL_8(OH) - H]^-$ (m/z 338), $[VOL_8(Br)_2]^-$ (m/z 480), and $[V_2O_3(L_8)_2Br]^-$ (m/z 739). The positive ion mode ESI spectrum of this sample pointed to the formation of the following cations: $[Py_2H_2Br]^+$ of m/z 239, $[VOL_8py]^+$ of m/z 401, $[V_2O_3(L_8)_2H]^+$ of m/z 661, and $[V_2O_3(L_8)_2pyH]^+$ of m/z 740. It should, however, be noted that these cations were also detected from a solution of **53** and pyHBr in CH_2Cl_2/CH_3CN without TBHP. The authors proposed a mechanism of the new method for bromocyclization of alkenols proceeding in three decisive steps as depicted in Scheme 5.19. In the first step, TBHP binds to a vanadium(V)-based catalyst to furnish the corresponding (*tert*-butylperoxy)(Schiff base)vanadium(V) complex (e.g., **55**). The activated peroxide is considered to be the oxidant for the conversion of Br^- into Br_2, which then serves as a brominating reagent for the subsequent bromocyclization of alkenols in a third vanadium(V)-independent step.

Smith and coworkers [74] studied the system 1,4,7-trimethyl-1,4,7-triazacyclononane (TMTACN), $MnSO_4$, and H_2O_2 in basic aqueous acetonitrile, and reported it was an effective system for the epoxidation of cinnamic acid. They used ESI(+)-MS to detect the manganese complexes, ligand and ligand oxidation products, and ESI(−)-MS to detect also the substrates and their oxidation products.

Itoh and coworkers [75] used ESI to identify a diiron complex of binaphthol-containing chiral ligand in the catalytic oxidation of alkane with *m*-CPBA to the corresponding alcohols. O'Hair *et al.* [76] reported an example that involves the oxidation of methanol to formaldehyde in the gas phase using binuclear molybdenum oxides, $[Mo_2O_6(OH)]^-$ and $[Mo_2O_5(OH)]^-$, as catalysts and nitromethane as the oxidant.

Aqueous ozonation of the 22 most common amino acids and some small peptides was studied through electrospray mass (ESI-MS) and tandem mass spectrometry by Eberlin and coworkers [77]. After 5 min of ozonation, only histidine, methionine, tryptophane, and tyrosine formed oxidation products clearly detectable by ESI-MS (Scheme 5.20).

5.4.11
Epoxidation

One of the most elegant methods for the selective formation of C—O bonds is the catalytic Jacobsen–Katsuki epoxidation, the enantioselective synthesis of optically active epoxides by oxygen-transfer reactions with chiral, nonracemic manganese oxo salen complexes. These complexes have been suggested as the catalytically active species in epoxidations catalyzed by metal-salen and porphyrin complexes [78]. One of these complexes was for the first time isolated and characterized by Feichtinger and Plattner through ESI-MS studies [79].

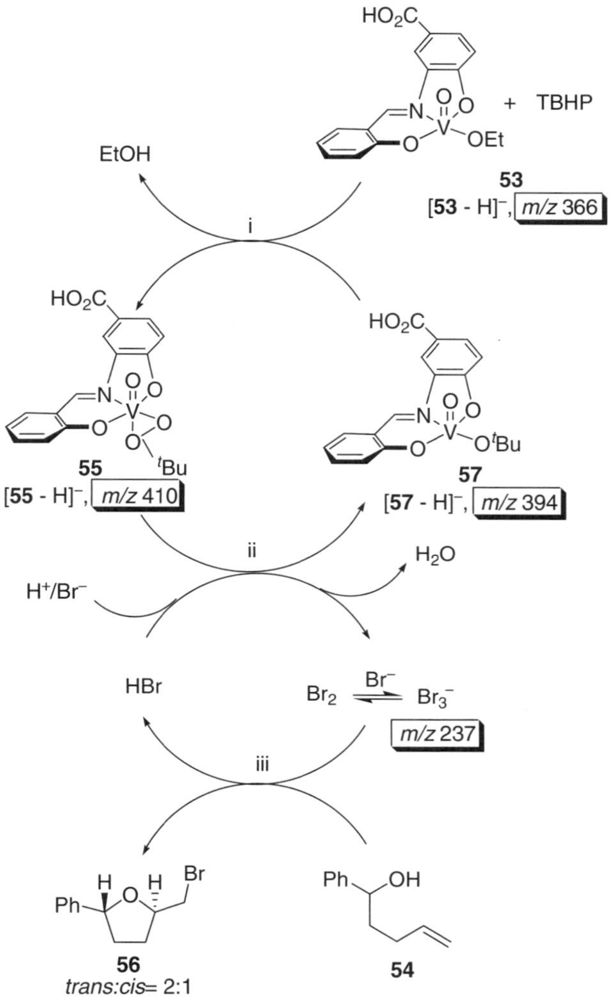

Scheme 5.19 Schematic presentation of reaction cycles leading to the formation of 2-(bromomethyl)-5-phenyltetrahydrofuran (**56**) from TBHP, pyHBr, 1-phenyl-4-penten-1-ol (**54**), via vanadium(V) compounds **55** and **57**. The solutions were diluted with CH_3CN before analysis to a final concentration of $1.0\,mmol\,L^{-1}$.

Adam and coworkers have characterized [Mn^{IV}(salen)] complexes by ESI-MS/MS [80]. The group proposed that cis and trans epoxide formation followed separate pathways, and the possibility that the reaction mixture had multiple, rapidly equilibrating oxidizing species, each of which preferentially followed one of the reaction pathways, could not be excluded. Herein, they suggested a bifurcation step in the catalytic cycle to account for the dependence of the diastereoselectivities on the oxygen source as depicted in Scheme 5.21 [80b].

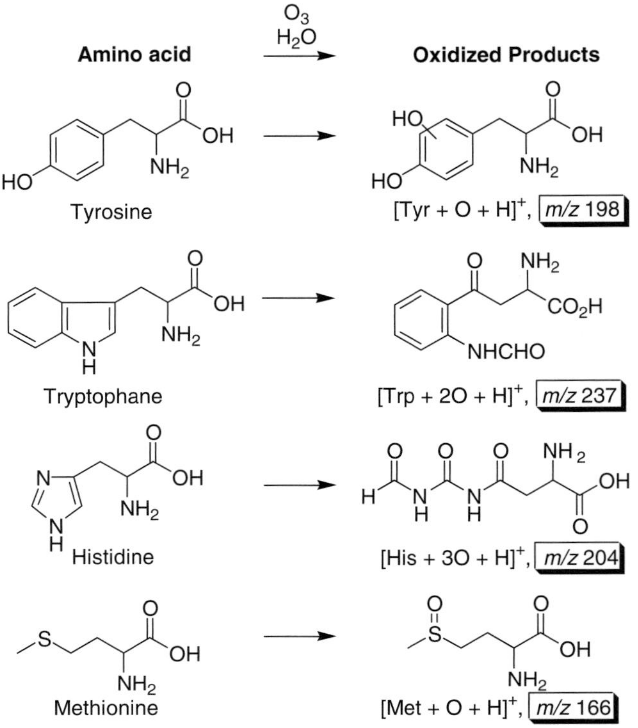

Scheme 5.20 Ozone oxidation products of amino acids screened by ESI(+)-MS.

5.4.12
The Baylis-Hillman Reaction

In a representative example of on-line monitoring the Baylis-Hillman reaction was studied by ESI-MS in both the positive and negative ion mode. The reactions were

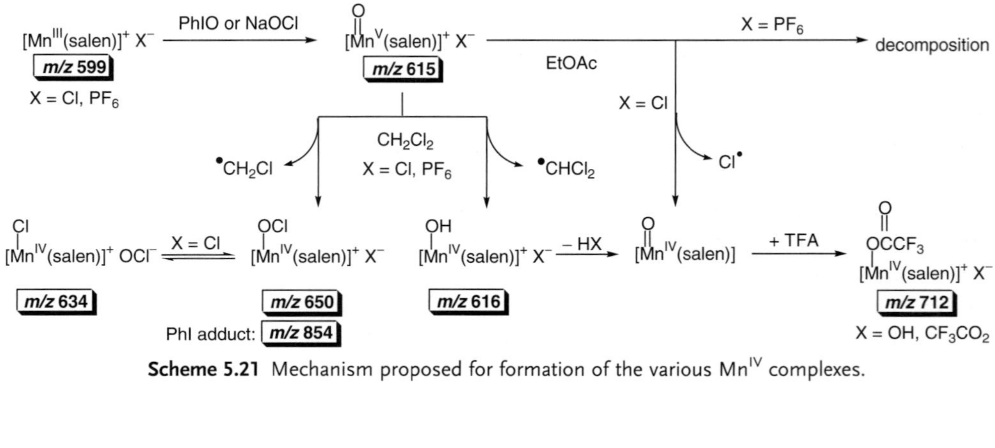

Scheme 5.21 Mechanism proposed for formation of the various Mn^{IV} complexes.

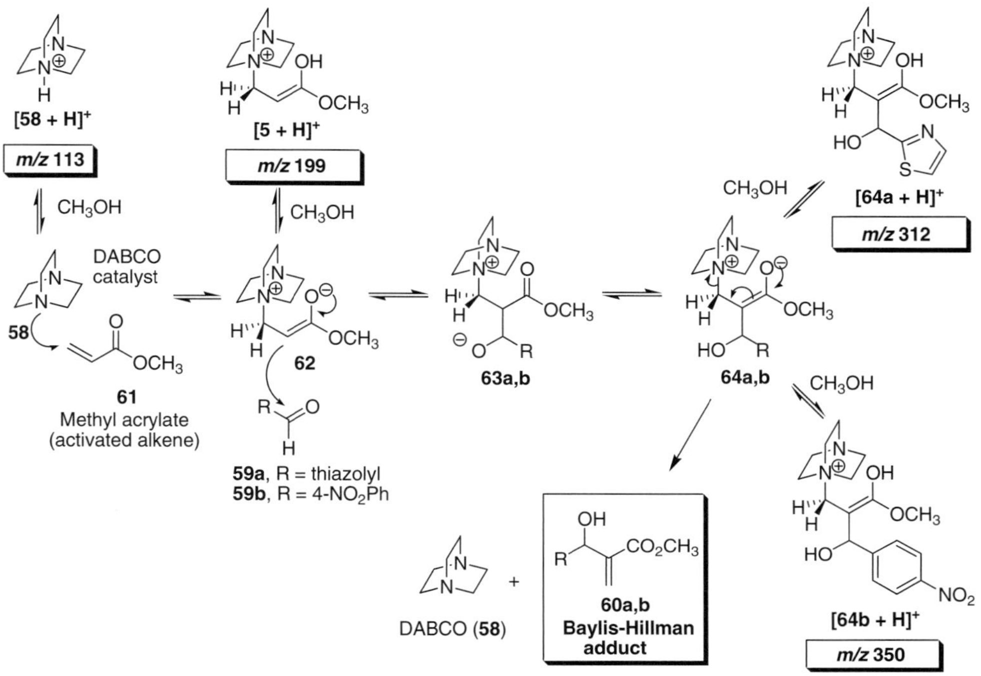

Scheme 5.22 Mechanism of Baylis–Hillman reaction of methyl acrylate and aldehydes catalyzed by DABCO. The protonated species expected to be intercepted and structurally characterized by ESI(+)-MS/MS, with their respective m/z ratios.

performed in a sealed syringe as the reaction medium directly coupled to the ion source [81]. The proposed intermediates for the catalytic cycle of the reaction (**58–64**, Scheme 5.22) were successfully intercepted and structurally characterized for the first time using ESI-MS and MS/MS. Strong evidence was collected corroborating the currently accepted mechanism [82].

5.4.13
The Baylis-Hillman Reaction Co-Catalyzed by Ionic Liquids

In a similar manner, on-line monitoring of Baylis-Hillman reactions co-catalyzed by ionic liquids was applied to gently fish from solution to the gas phase supramolecular species responsible for the co-catalytic role of ionic liquids in the reaction (Scheme 5.23) [83]. Several supramolecular species formed by coordination of reagents and products were trapped, identified, and characterized via MS analysis and MS/MS dissociations. Via competitive experiments, it was also reported that the efficiency order of different co-catalysts was $BMI.CF_3CO_2 > BMI.BF_4 > BMI.PF_6$, which was the opposite to that observed by Afonso [84] in the liquid phase. Based on the interception of these unprecedented supramolecular species, it was proposed

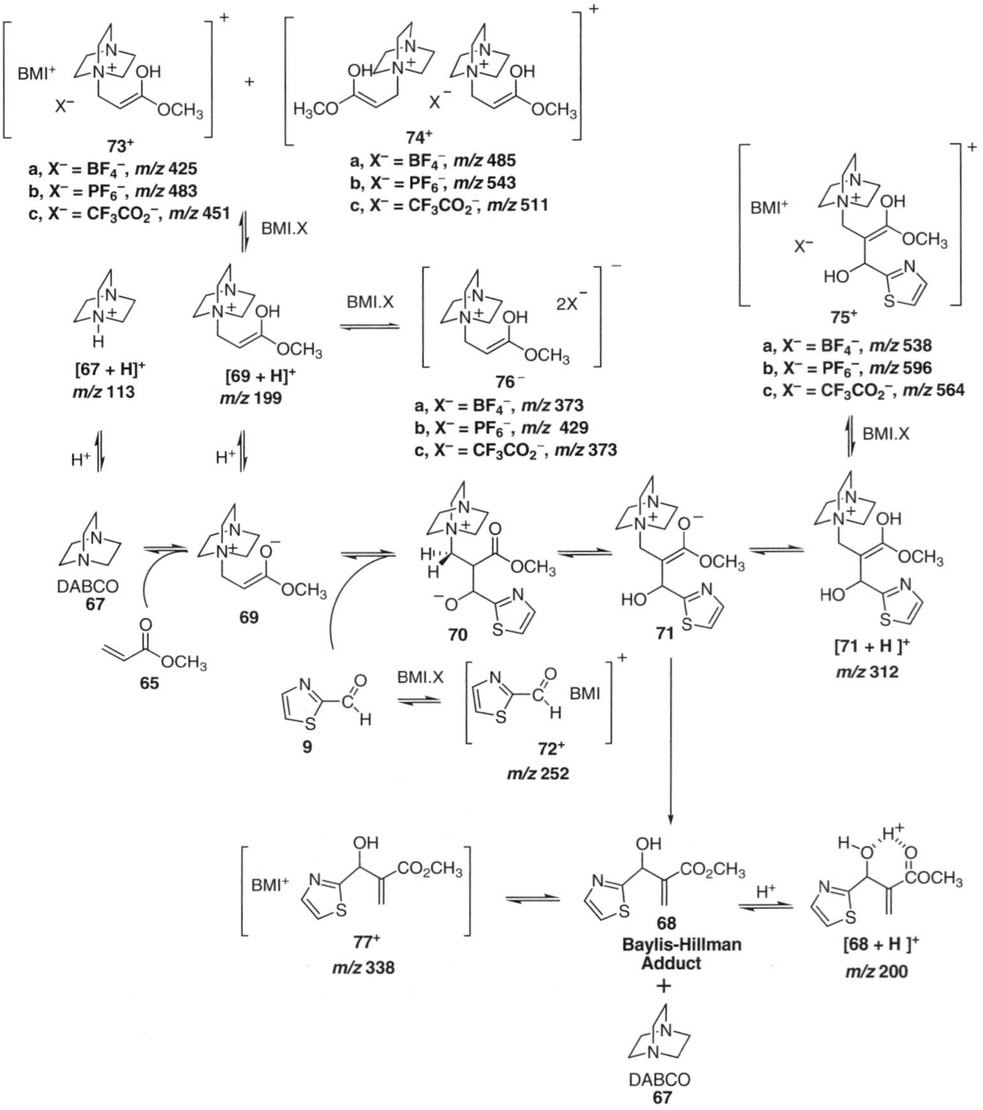

Scheme 5.23 Baylis–Hillman reaction of methyl acrylate (**65**) and 2-thiazolecarboxaldehyde (**66**) co-catalyzed by both DABCO (**67**) and an ionic liquid BMI.X (X = BF_4^-, PF_6^-, $CF_3CO_2^-$).

that 1,3-dialkylimidazolium ionic liquids function as efficient co-catalysts for the reaction by: (i) activating the aldehyde toward nucleophilic attack via BMI coordination (species 72^+) and (ii) by stabilizing the zwitterionic species that act as the main BH intermediates through supramolecular coordination (species 73^+, 74^+, 75^+, and 77^+).

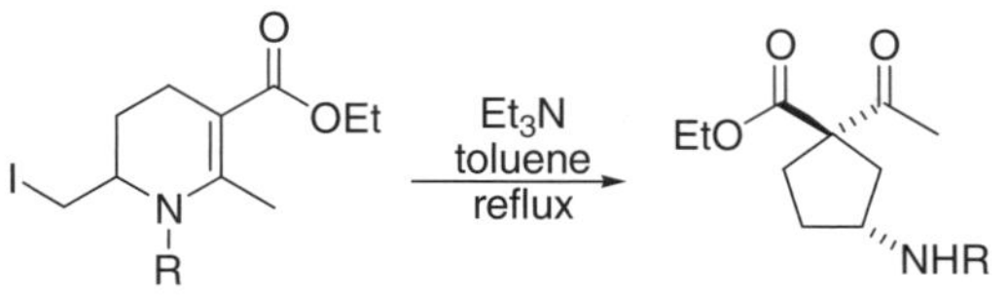

79a (R = Ph, 72 h)
79b (R = Bn, 24 h)
79c (R = p-MeOPh)

80a (73%)
80b (85%)
80c (77%)

Scheme 5.24 Ring contraction reaction.

5.4.14
Ring Contraction Reaction

Looking for experimental support to validate the mechanism of ring contraction for the reaction depicted in Scheme 5.24, on-line monitoring of the reaction by ESI-MS and ESI-MS/MS using the syringe as the reaction medium was performed [85]. A bicyclic iminium ion intermediate 81^+ was proposed to participate in the reaction (Scheme 5.25), but both reactant **79** and product **80** were neutral molecules. These neutral species were, however, expected to be in equilibrium in solutions of protic solvents such as methanol with their protonated forms. Therefore, ESI could transfer both reactants and products to the gas phase as $[M + H]^+$ species for MS analysis. Even an unfavored equilibrium was helpful considering the exceptionally high sensitivity of the ESI-MS technique. First, the ESI-MS monitoring of the reaction of iodo-β-enamino-esters **79a–c** with Et_3N as the base was performed. Iodo-β-enamino-esters **79** (1.0 equiv) and Et_3N (1.0 equiv) were mixed in 1:1 toluene/methanol (2 mL) at 25 °C, and the reaction was monitored by ESI-MS. Major ions were clearly detected by ESI-MS, corresponding to **79a**, the protonated reactant $[79a + H]^+$ of m/z 386, intermediate $81a^+$ of m/z 258, and the protonated product $[80a + H]^+$ of m/z 276. Likewise for **79b** and **79c**, the protonated reactants $[79b-c + H]^+$, the intermediates $81b-c^+$, and the final protonated products $[80b-c + H]^+$ were clearly detected. The detected cations, as shown by continuous ESI-MS monitoring, were the same from 1 to 60 min of reaction. Scheme 5.25 summarizes a general reaction pathway showing neutral and protonated reactants and protonated products, as well as the key bicyclic iminium ion intermediates $81a-c^+$ (with their respective m/z ratios) that have been intercepted and structurally characterized by ESI(−)-MS(/MS).

5.4.15
Nucleophilic Substitution Reactions – The Meisenheimer Complex

Another interesting approach using ESI(−)-MS was used to probe the mechanism of nucleophilic substitution reactions of methylated hydroxylamines, hydrazine and hydrogen peroxide with bis(2,4-dinitrophenyl)phosphate (BDNPP) [86]. ESI-MS screening of the reaction was performed to fish ionic intermediates and products directly from solution into the gas phase, and the key intermediates were fully

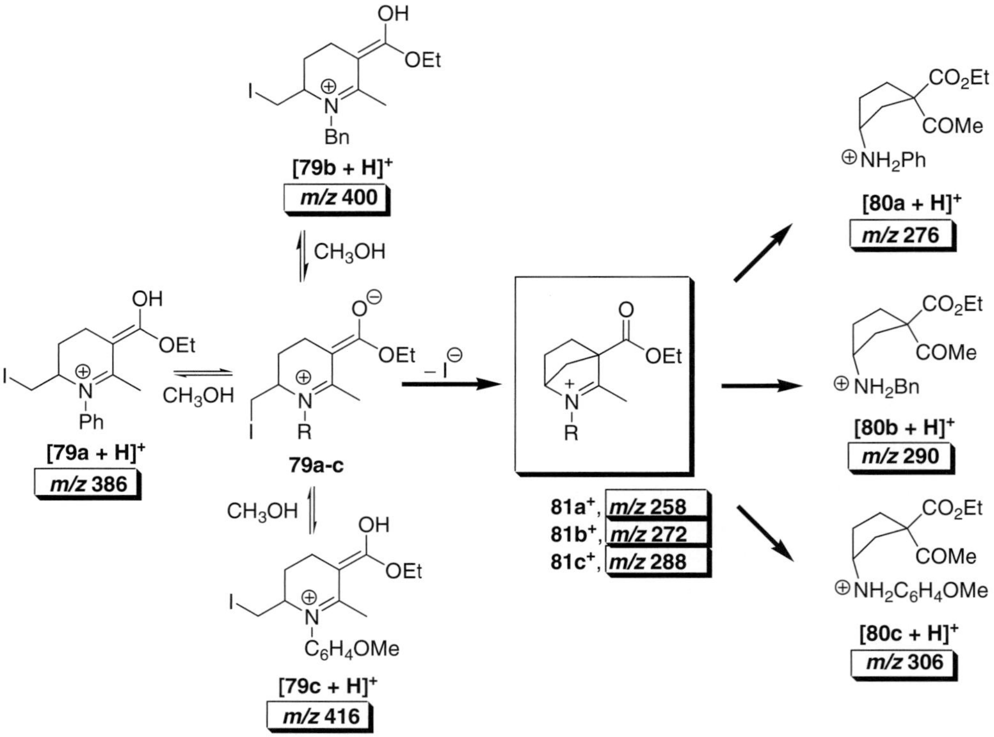

Scheme 5.25 Mechanism probed by ESI-MS for ring contraction reaction of iodo-β-enamino-esters. The screening was performed using **79a–c** (1 mmol), Et$_3$N (1 mmol) in toluene/methanol (1 : 1, 20 mL) at a flow rate of 0.01 mL min^{-1}.

characterized by ESI-MS/MS experiments. Based on the ESI(−)-MS(/MS) results, NMR and kinetic data, and the results on the reaction of NH$_2$OH with BDNPP, it was concluded that there was both aromatic substitution, generating DNPP, **84**, and **85**, and initial phosphorylation of the OH group of NHMeOH by BDNPP (Scheme 5.26, Figure 5.12). The latter generates dinitrophenyl oxide (DNP), forming intermediate **83**, which breaks down slowly by two distinct pathways: (a) aromatic nucleophilic substitution described above, giving **85** and **87**, or (b) spontaneous rearrangement whereby the terminal NHMe group attacks the dinitrophenyl moiety to form a transient cyclic Meisenheimer complex (**88**), as for reaction with NH$_2$OH. This complex **88** rapidly ring opens, giving **85** as a long-lived product, as through a mechanism proposed in Scheme 5.5. Ionic intermediates **83** and **86** (Schemes 5.26 and 5.27) ore the ESI-MS/MS of the ions of m/z 292 changed drastically with time. Samples taken after 10 min of reaction in solution showed that these ions dissociated in the mass spectrometer by CID mainly into four fragment ions of m/z 263, 183, 97, and 79. The fragment ions of m/z 263 and 183 were formed in the gas phase from **83** (Scheme 5.27a), whereas dissociation to ions of m/z 97 and 79 was evidence for the presence of **86** (Scheme 5.27b). Therefore, both **83** and **86** were present in the

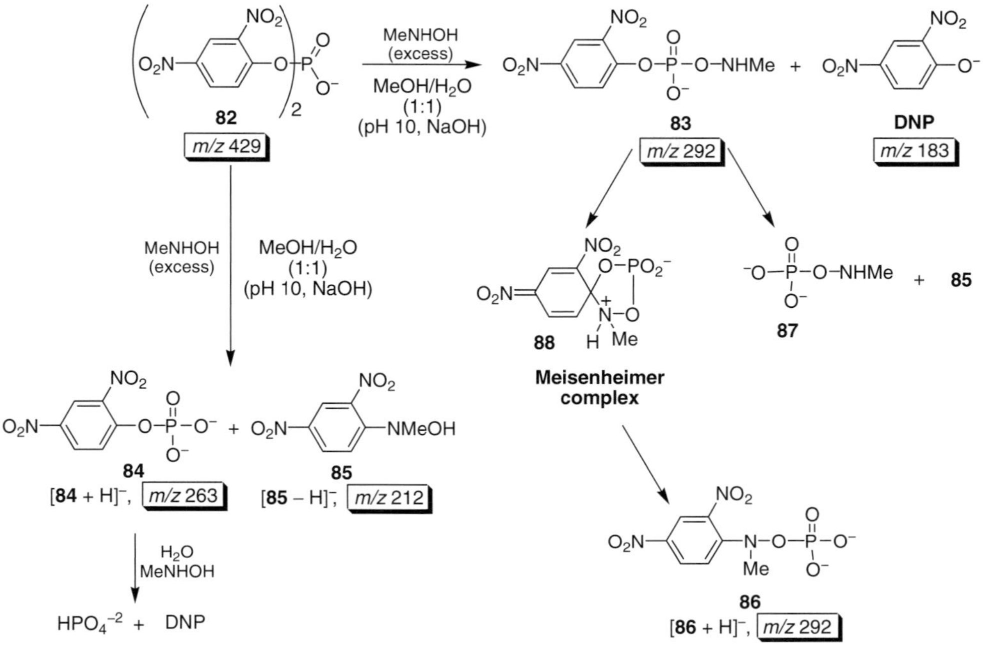

Scheme 5.26 Mechanism for the reaction of MeNHOH and **82** probed by ESI(−)-MS experiments, which corroborated kinetics studies.

reaction mixture at an early stage of the reaction. After 30 or 60 min of reaction, however, the ion of m/z 292 generated only the two fragment ions of m/z 97 (**89**) and 79. This temporal change in product ion distribution from the CID of ion of m/z 292 confirmed that, in a later stage of reaction, intermediate **86**, rather than **83**, became a dominant species through rearrangement from **83** in solution.

5.4.16
Oxidative Degradation of Caffeine

Caffeine is quickly and completely degraded under the oxidative conditions of the UV/H_2O_2, TiO_2/UV, and Fenton systems to CO_2, NH_3, and NH_2Me. Augusti and coworkers [87] reported in an interesting article the mechanism of this degradation (Scheme 5.26). The continuous on-line and real-time monitoring, by electrospray ionization mass spectrometric (ESI-MS) and tandem mass spectrometric experiments (ESI-MS/MS) as well as high-accuracy MS measurements, show that caffeine is first oxidized to N-dimethylparabanic acid, likely via initial OH insertion into the C4—C8 caffeine double bond. A second degradation intermediate, di(N-hydroxymethyl)parabanic acid, has been identified by ESI-MS and characterized by ESI-MS/MS and high-accuracy mass measurements (with errors around of 5–7 ppm). This polar and likely relatively unstable compound, which is not detected by other techniques, is likely formed via further oxidation of N-dimethylparabanic acid at

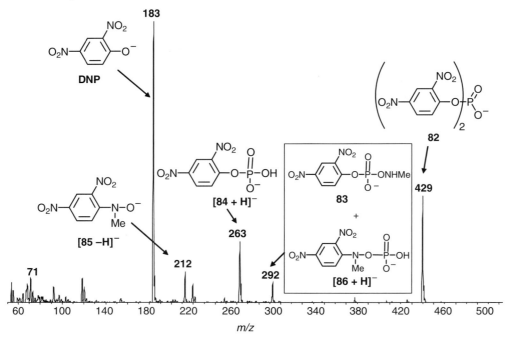

Figure 5.12 ESI-MS spectrum of the reaction mixture of 0.01 mol L^{-1} BDNPP with 0.1 mol L^{-1} NHMeOH, in aqueous methanol (50% v/v) at pH 10.

both of its *N*-methyl groups and constitutes an unprecedented intermediate in the degradation of caffeine.

After UV irradiation for 90 and 150 min, ESI was able to gently and efficiently fish directly from the solution to the gas phase for MS analysis two additional and

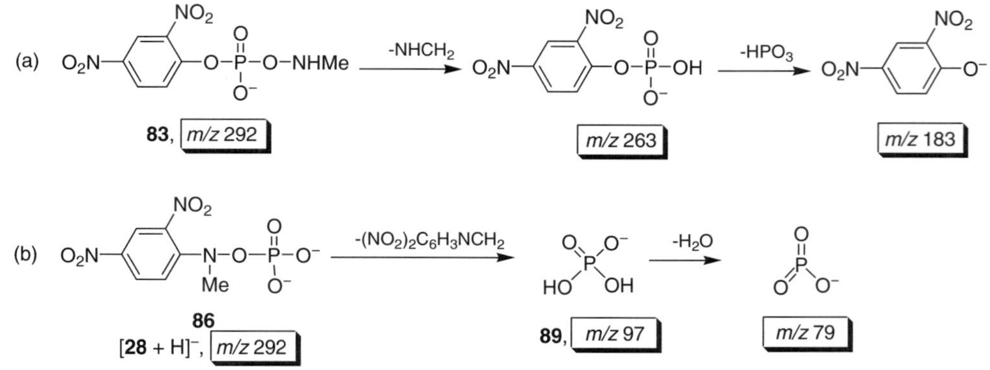

Scheme 5.27 Differentiation of the isomers **83** (a) and **86** (b) by fragmentation pathways (MS/MS experiments). The isomer **83** (m/z 292) was proposed to achieve **89** (m/z 292) through a Meisenheimer complex intermediate.

increasingly abundant ions of m/z 143 (dimethylparabanic acid, **91**) and m/z 175 (dihydroxymethylparabanic acid, **92**). The conversion of **90** to **91** and then to **92** (an unprecedented intermediate in the oxidation of caffeine) is further supported by the ESI tandem mass spectra of their protonated molecules, which display a series of structurally diagnostic fragment ions. The authors suggested that the degradation of organic compounds by the UV/TiO$_2$, H$_2$O$_2$/UV, and Fenton systems goes through the mechanism for the conversion of caffeine (**90**) to dimethylparabanic acid (**91**) involving initially the (fast) attack of the hydroxyl radicals to the C4=C8 double bond of caffeine (Scheme 5.28). After successive hydroxylations and oxidations, **91** and **92** are formed, whereas **92** is slowly mineralized to CO$_2$, NH$_3$, and NH$_2$Me.

5.4.17
Mimicking Atmospheric Oxidation of Isoprene

Photo-oxidation of isoprene in the atmosphere was thought to yield only lighter and more volatile products such as formaldehyde, methacrolein, and methyl vinyl ketone. Evidence for the formation of hygroscopic polar products such as **94** that can give rise to aerosols by gas-to-particle formation processes opened up a new and wide-range panorama for the isoprene role in the atmosphere. Under simulated atmospheric conditions, photo-oxidation of isoprene by ozone was shown to be relatively slow and to occur mainly via reaction with OH radicals. Furthermore, when the OH radical-initiated photo-oxidation of isoprene was performed in the absence of NO$_x$, 1,2-diol

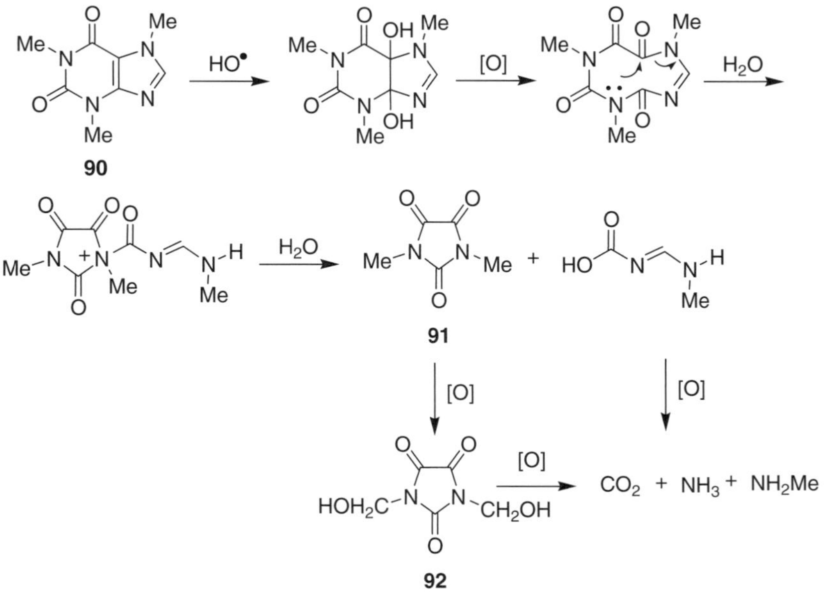

Scheme 5.28 Mechanism of degradation of caffeine by ESI-MS on-line monitoring through TiO$_2$/UV, H$_2$O$_2$/UV or Fenton systems.

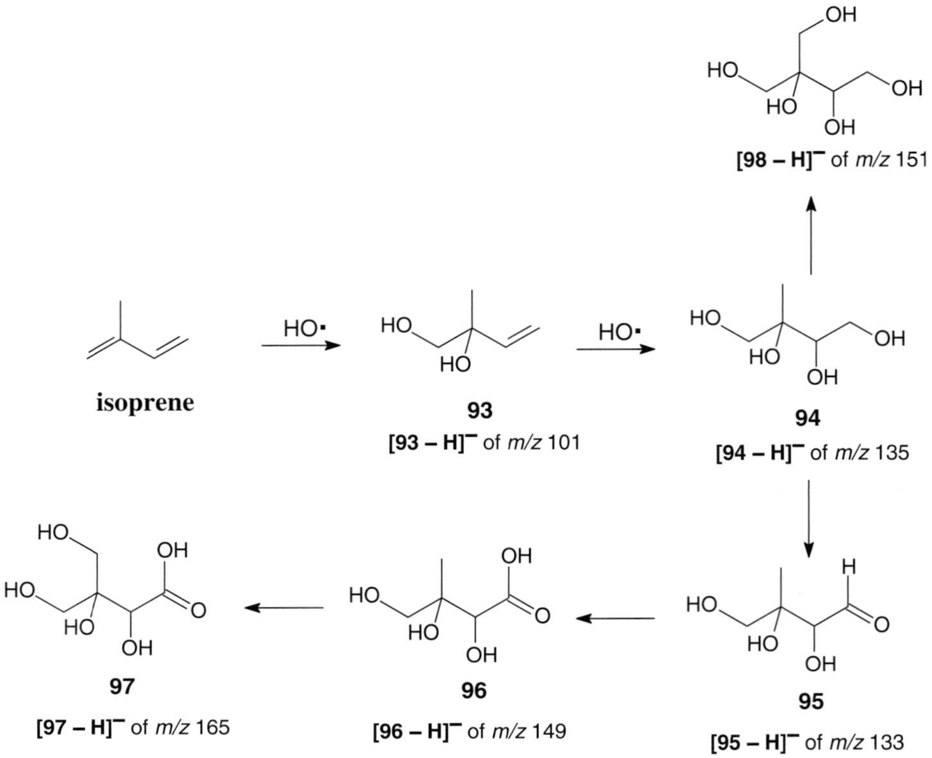

Scheme 5.29 Proposed route for the formation of the oxygenated products **93–98** in the AIBN-initiated reaction of isoprene with hydroxyl radicals from hydrogen peroxide.

derivatives (**93**) were formed (Scheme 5.29). Aiming at supporting the 'C$_5$ isoprene hypothesis' [88] for the formation of the secondary organic aerosol (SOA) marker compounds, the diastereoisomeric 2-methyltetrols **94**, Santos and coworkers [89] tried to mimic in solution the OH radical-mediated oxidation of isoprene and to monitor the process on-line and in real-time using ESI-MS(/MS) operating in the negative ion mode. In this article, the OH radical-mediated oxidation of isoprene by H$_2$O$_2$ in solution initiated either photochemically or by AIBN seems to mimic adequately the atmospheric oxidation of isoprene, as revealed by ESI(−)-MS(/MS) monitoring. The *in situ* fishing, detection and structural characterization of **94** as its deprotonated molecule supports the 'C$_5$ isoprene hypothesis' formulated by Claeys for the formation of the 2-methyltetrols **94** [88]. These diasteromeric polyols have been recently identified in atmospheric aerosols of the Amazon forest, and their highly hygroscopic nature likely enhances the capability of aerosols to act as cloud condensation nuclei. Other polyoxygenated compounds assigned as **93** (the diol precursors of **94**), **95–98**, have also been intercepted and structurally characterized. Products **95–98** are suggested to be formed by further oxidation via **93** and **94**, and they may therefore also be considered as plausible SOA components formed in normal or more extreme OH radical-mediated photo-oxidation of biogenic isoprene (Scheme 5.27). ESI(−)-MS(/MS) monitoring showed also promise for mimicking, and hence for studying, the OH radical-mediated oxidation of first-generation gas phase oxidation products of isoprene and α-pinene, such as, for example, methacrolein and pinic acid.

5.4.18
α-Methylenation of Ketoesters

Extensive studies have been devoted to optimizing the synthesis of α-methylene carbonyl compounds, as they are not only useful as synthetic intermediates but may also mimic biologically active natural products compounds and be used as potential antitumor drugs. Eberlin and coworkers reported a convenient one-pot method to prepare α-methylene ketoesters (Scheme 5.28) via direct α-methylenation of ketoesters as well a mechanistic study of this interesting reaction by electrospray mass spectrometric (ESI-MS) experiments [90]. They investigated via ESI(+)-MS monitoring the reaction of **103** with p-formaldehyde catalyzed by morpholine in acetic acid: acetonitrile solution (Scheme 5.28). As the key players of the catalytic cycle could be either ionic or neutral species, they hoped to intercept the protonated molecules of the neutral species from the acetic acid solutions, as achieved before [81]. Few examples of disfavored proton-transfer equilibrium are described, but it could be useful in intercepting neutral species because of the high sensitivity of ESI-MS in transferring ionic intermediates from solution to the gas phase.

The ESI-MS collected for the reaction was quite clean and mechanistically enlightening (Figure 5.13). As soon as the reaction mixture was formed and electrosprayed, ESI-MS detected the reactants, namely protonated morpholine **99** of m/z 88 and protonated ethyl benzoylacetate **103** of m/z 193, and additionally two key cationic intermediates: **101** of m/z 118 and the iminium ion **100** of m/z 100 (Scheme 5.29). Previously, only kinetic evidence for the intermediacy of iminium ions such as **100** in Mannich reactions was available. Another key piece of information was provided by continuous monitoring; after approximately 12 min of reaction, the ESI-MS changed considerably, detecting another key intermediate: the aldol **102** in its protonated form of m/z 292 (Figure 5.13b). The abundance of the starting reagents [**103** + H]$^+$ of m/z 193 and **99** (m/z 88) as well as intermediates **100** (m/z 100) and **101** (m/z 118) had decreased accordingly. The on-line screening depicted in Figure 5.13 exemplifies for [**103** + H]$^+$ of m/z 193 and [**102** + vH]$^+$ of m/z 292 that α-methylenations can therefore be continuously monitored by ESI-MS by the consumption of the starting carbonyl compound and the formation of the final product via its immediate aldol precursor. Scheme 5.30 summarizes the catalytic cycle proposed for the direct Mannich-type α-methylenation of ketoester **103** in the presence of **99**, which forms the α-methyleneketoester **104**, based on previous mechanistic interpretations but showing now the three cationic intermediates **100**–**102** intercepted by ESI-MS and structurally characterized by ESI-MS/MS.

Metzger recently showed a similar approach for L-proline catalyzed reactions (Scheme 5.31) [91].

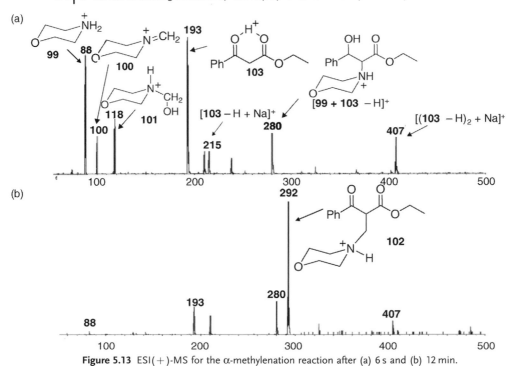

Figure 5.13 ESI(+)-MS for the α-methylenation reaction after (a) 6 s and (b) 12 min.

5.4.19
Transient Intermediates of Petasis and Tebbe Reagent

After confirming that methylenetitanocene (**106**) does in fact participate in Petasis reactions (Scheme 5.32) [92], Eberlin, Santos and coworkers decided to extend the methodology to other related reactions for which there is no conclusive mechanistic data. This was the case for Tebbe olefination reactions [93]. In APCI, neutral molecules are transferred from the drying spray droplets to the gas phase, where the molecules are ionized at atmospheric pressure via either electron abstraction, protonation or deprotonation (or a combination of these processes) due to ion/molecule reactions initiated by corona discharge. Therefore, the set of ions detected by APCI-MS is expected to closely reflect the solution composition of neutral species, and the gentle evaporation process is also likely to preserve the metal coordination spheres of organometallic species. Reagents, intermediates, and products are likely therefore to be transferred intact from solution to the gas phase, and then ionized gently and mass analyzed during APCI-MS of a reaction solution. APCI-MS in the positive ion mode has been used to efficiently protonate the postulated reaction intermediates **107** and **108**. The hypothesis that allylic titanocene complexes are also generated from non-allylic starting materials (vinyl ethers derivatives) in a related Tebbe-like reaction was also investigated. Key titanacycle intermediates **107** and **108** support both the classical mechanism of Ti-mediated Tebbe olefination reactions as

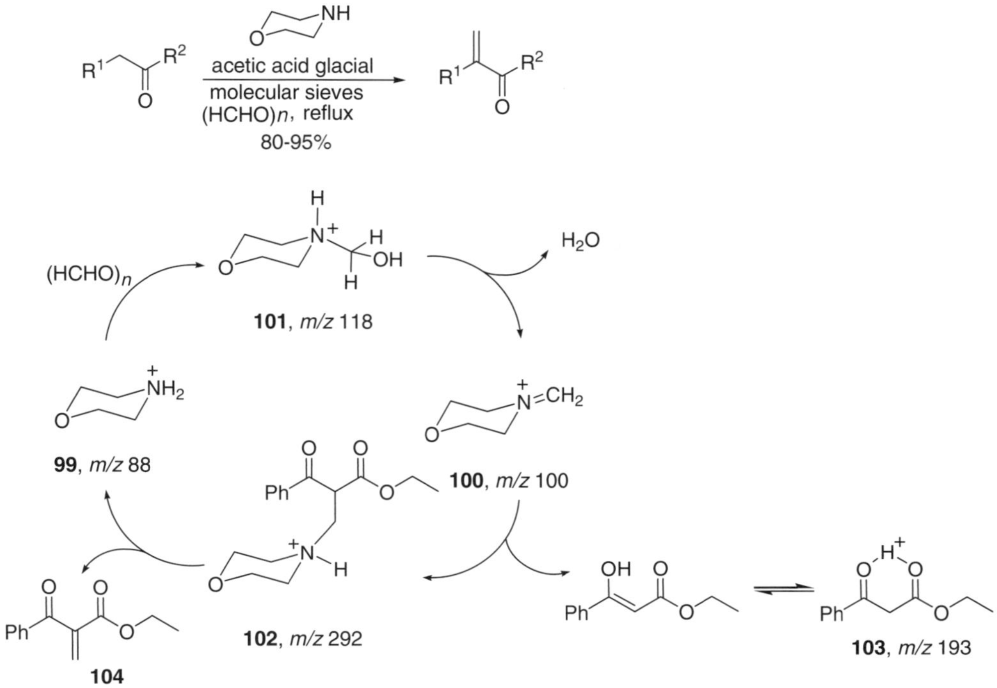

Scheme 5.30 Direct Mannich-type α-methylenation of carbonyl compounds.

postulated by Tebbe and the mechanism for the Tebbe-like [2 + 2] reaction postulated by Hanzawa [94]. As postulated by Tebbe, **105** generates **106**, which reacts further with the ketone to yield the oxatitanacyclic intermediates **107a–c**. To test the feasibility of this reaction toward ketones, the group applied APCI-MS to the screening of Tebbe olefination reactions and unequivocally detected [**107** + H]$^+$ as a characteristic cluster of isotopologous ions, the most abundant (for ^{48}Ti) being those of m/z

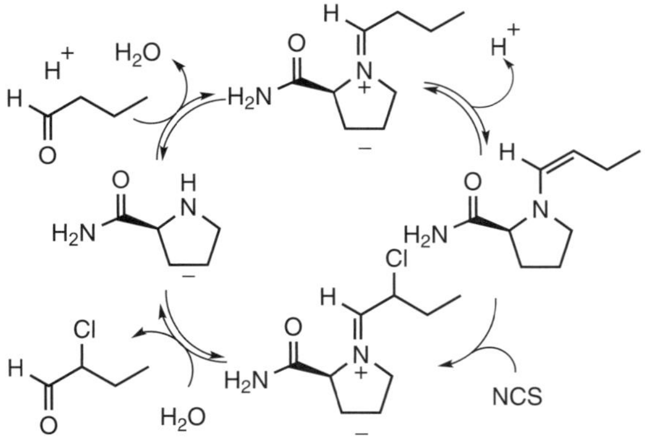

Scheme 5.31 Mechanism of the L-prolinamide-catalyzed α-chlorination of butanal with NCS.

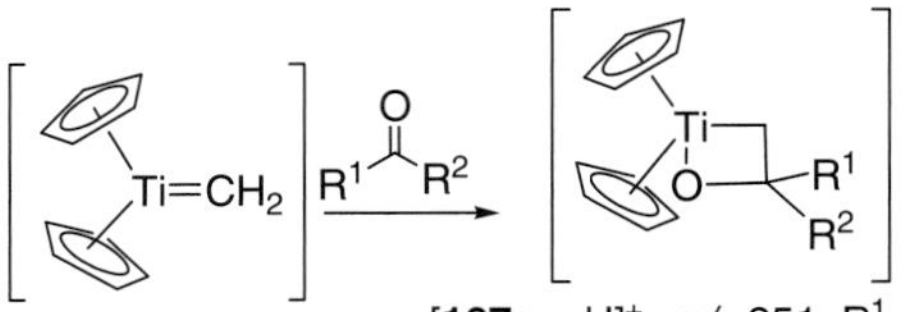

[**106** + H]$^+$, *m/z* 193 [**107a** + H]$^+$, *m/z* 251, $R^1 = R^2$ = Me
[**107b** + H]$^+$, *m/z* 265, R^1 = Me, R^2 = Et
[**107c** + H]$^+$, *m/z* 281, R^1 = Me, R^2 = OEt

Scheme 5.32 Transient intermediates of Petasis and Tebbe reagent reactions intercepted by APCI-MS.

251 for [**107a** + H]$^+$, *m/z* 265 for [**107b** + H]$^+$, and *m/z* 281 for [**107c** + H]$^+$. They then investigated the Tebbe-like [2 + 2] reaction of **105** with vinyl ethers known to afford allyl titanocenes **109** (Scheme 5.33). Using different vinyl ethers, the expected Ti-intermediates [**108a–c** + H]$^+$ were unambiguously detected by characteristic clusters of Ti-containing ions. This study further illustrates the broad potential of API mass spectrometry to study mechanisms of organometallic reactions via the fishing and structural characterization of their key intermediates, a vast but still little explored field.

5.4.20
On-Line Screening of the Ziegler–Natta Polymerization Reaction

Significant remarkable mechanistic investigation has been performed on the Ziegler–Natta polymerization reaction. Several mechanistic and kinetic studies have led to the generally accepted reaction mechanism of Ziegler–Natta polymerization, which is depicted in Scheme 5.34. It is seen that treatment of a toluene solution of Cp_2ZrCl_2 (**110**) with methylaluminoxane (MAO) leads to a rapid initial ligand exchange reaction that first generates the monomethyl complex Cp_2ZrCH_3Cl

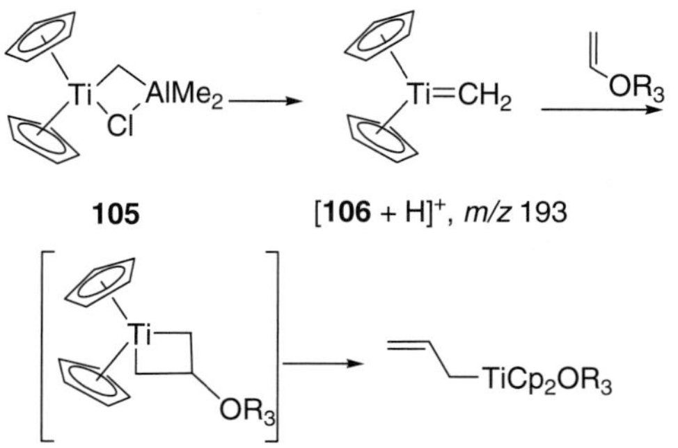

[**108a** + H]$^+$, *m/z* 265 [**109a** + H]$^+$, *m/z* 265, R^1 = Et
[**108b** + H]$^+$, *m/z* 279 [**109b** + H]$^+$, *m/z* 279, R^1 = Pr
[**108c** + H]$^+$, *m/z* 293 [**109c** + H]$^+$, *m/z* 293, R^1 = *t*Bu

Scheme 5.33 Transient intermediates from the reaction of Tebbe reagent and vinyl ethers intercepted by APCI-MS.

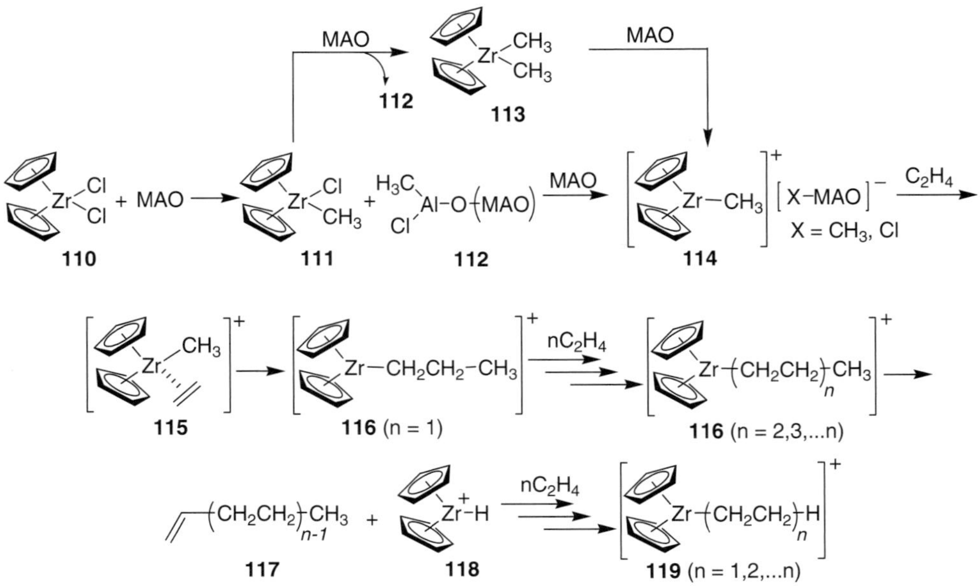

Scheme 5.34 Proposed mechanism of Ziegler–Natta polymerization of C_2H_4 using the homogenous catalyst Cp_2ZrCl_2/MAO.

(111) [95]. An excess of MAO leads to Cp_2ZrMe_2 (113) [95a]. Abstraction of chloride from 111 or methyl from 113 by MAO gives the catalytically active ion-paired species $[Cp_2ZrCH_3]^+$ (114) with the counter-ion $[X-Al(Me)O-]_n^-$ (X = Cl, Me) [95b] based on solid-state XPS [96a] and ^{13}C NMR [96b] studies as well as ^{91}Zr and ^{13}C NMR investigations of $Cp_2Zr(CH_3)_2$/MAO solutions [97]. The cation 114 in the presence of ethene gives, via π-complex 115, the insertion product 116 (n = 1) as the first intermediate of the polymerization process, which is followed by step-by-step insertion of ethane, achieving the cationic alkyl zirconocenes 116 (n = 2, 3 ... n). β-Elimination gives the uneven chain polymer 117 containing a terminal C=C double bond and cationic zirconocene hydride 118, which is able to start polymerization to give, via zirconocene cation 119, the even chain polymer [98]. However, there is some experimental evidence suggesting that the general classification of zirconocene/MAO catalyst systems as single-site catalysts may be an oversimplification [99].

Santos and Metzger [100] demonstrated that cation $Cp_2ZrCH_3^+$ (114), a species of great relevance to the polymer field, was easily detected in the reaction solution of Cp_2ZrCl_2 and MAO by ESI-MS. It was characterized by MS/MS, and the catalytic activity was directly demonstrated by ion/molecule reaction of 114 and ethene in the gas phase. Furthermore, for the first time, using a microreactor on-line coupled to the ESI source (Figure 5.5) they were able to intercept the intermediate alkyl zirconium cations 116 of the growing polymer chain of the homogeneous Ziegler–Natta polymerization of ethene directly from the solution (Figure 5.14), characterize them

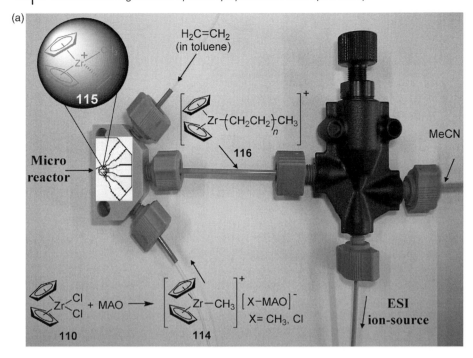

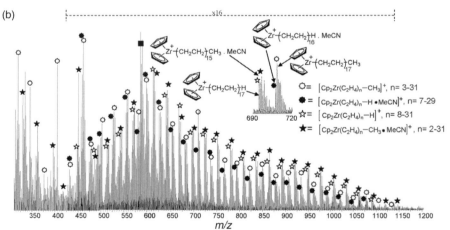

Figure 5.14 (a) Microreactor coupled on-line to ESI source. In the first micromixer (left), toluene solutions of the preformed catalyst [Cp$_2$ZrCl$_2$]/MAO (1 : 1.2 equiv) and of C$_2$H$_4$ were mixed continuously to initiate the polymerization. The reaction occurred in the capillary transferring the reacting solution to the second micromixer (right), where it is quenched by MeCN and from which it is fed directly to the ESI source. (b) Ziegler–Natta polymerization of ethene: Positive-mode ESI mass spectrum of the MeCN-quenched reaction solution of [Cp$_2$ZrCl$_2$]/MAO (1 : 1.2 equiv) and C$_2$H$_4$ (see Figure a). The spectrum shows the odd- and even-numbered-chain {Cp$_2$Zr}–alkyl species, as well as the MeCN adducts. The overall reaction time was approximately 1.7 s.

mass spectrometrically, and prove directly their catalytic activity by gas-phase reactions with ethene.

5.4.21
On-Line Screening of the Brookhart Polymerization Reaction

Employing the same device as that described above, Santos and Metzger reported the study of the Brookhart polymerization of ethane and of 1-butene with the homogeneous catalyst diimine Pd(Me)Cl (Scheme 5.35), with MAO or silver trifluoromethanesulfonate (AgOTf) as co-catalysts, using a microreactor coupled directly to the ESI source of a quadrupole time-of-flight (Q-TOF) mass spectrometer, focusing on the direct detection and mass spectrometric characterization of the transient cationic and catalytically active species involved, and on the direct demonstration of their catalytic activity [101]. The experiment reported here (Figure 5.15 and 5.16) has the advantages of high sensitivity (microgram quantities), very short assay times, direct analysis of all intermediates in the reaction mixture (active species that are being formed and consumed with time, and no quenching), new methodology to intercept highly sensitive intermediates with short lifetimes, and the proving of further activity of several species trapped from solution through ion/molecule reactions with neutrals in the collision cell. This report is another example of the successful application of API in revealing, elucidating, and helping to consolidate previously proposed reaction mechanisms.

5.4.22
TeCl$_4$ Addition to Propargyl Alcohols

TeCl$_4$ is found to display reactivity toward alkynes similar to that of p-methoxyphenyltellurium trichloride, whereas it reacts with aromatic and 3-hydroxyalkynes by different mechanisms, as shown by characteristic stereochemistries of the products. The complete anti stereospecificity of the additions of TeCl$_4$ to all propargyl alcohols studied is consistent with a cyclic chelated telluronium ion intermediate (122) in this reaction. Using ESI-MS and ESI-MS/MS, Santos and coworkers [102] have been able to intercept and characterize the active electrophile TeCl$_3^+$ in THF solution of TeCl$_4$, as well as its THF complex and several TeCl$_x$(OH)$_y^+$ derivatives (Figure 5.17). For the first time also, on-line ESI-MS(/MS) monitoring permitted key Te(IV) cationic intermediates of the electrophilic addition of TeCl$_4$ to alkynes to be captured from the solution and to be gently and directly transferred to the gas phase for mass measurement, determination of isotopic patterns, and structural investigation via collision-induced dissociation (Scheme 5.36). Two of the reaction products reported herein (the cyclic telluranes) have recently been found to act as cysteine protease inhibitors [103]; hence the mechanistic aspects of such reactions revealed in this study may assist the design of new members of this class of potential anti-metastasis agents. To date, despite their chemical and biological interest, only a few tellurium compounds have been investigated by mass spectrometry [104].

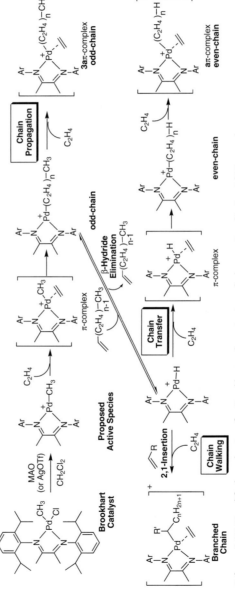

Scheme 5.35 Simplified mechanism for Brookhart polymerization of ethene using diimine-Pd(II) complex.

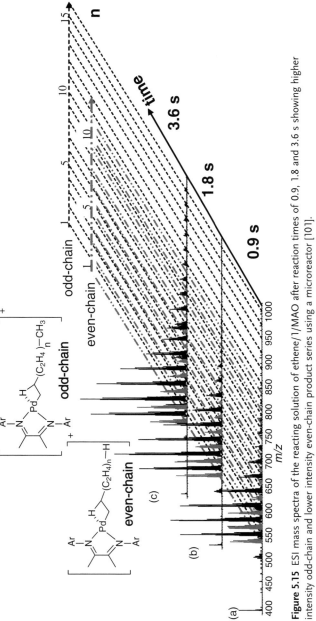

Figure 5.15 ESI mass spectra of the reacting solution of ethene/1/MAO after reaction times of 0.9, 1.8 and 3.6 s showing higher intensity odd-chain and lower intensity even-chain product series using a microreactor [101].

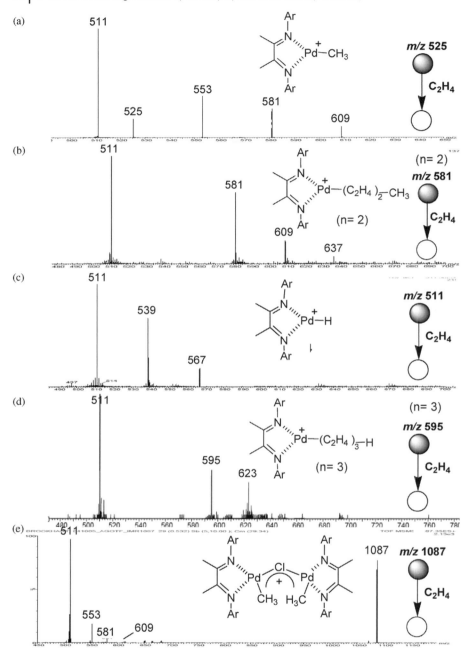

Figure 5.16 Proof of catalytic activity in the gas phase of various previously proposed intermediates. Product ion mass spectra for reactions of ethene with several precursor ions [101].

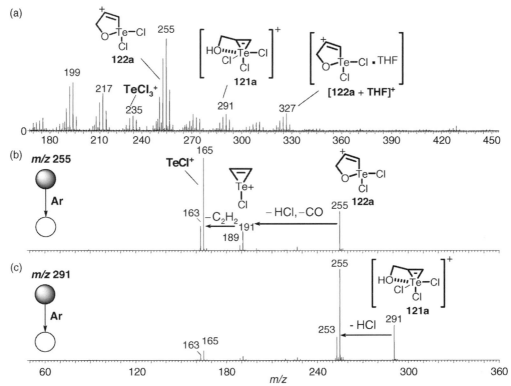

Figure 5.17 (a) ESI(+)-MS for on-line screening of a solution of **120a**/TeCl$_4$ (1.0 : 0.9 equiv) in THF. ESI-MS/MS for CID of the intermediate species (b) **122a** of m/z 255 and (c) **121a** of m/z 291 intercepted from the reaction solution [102].

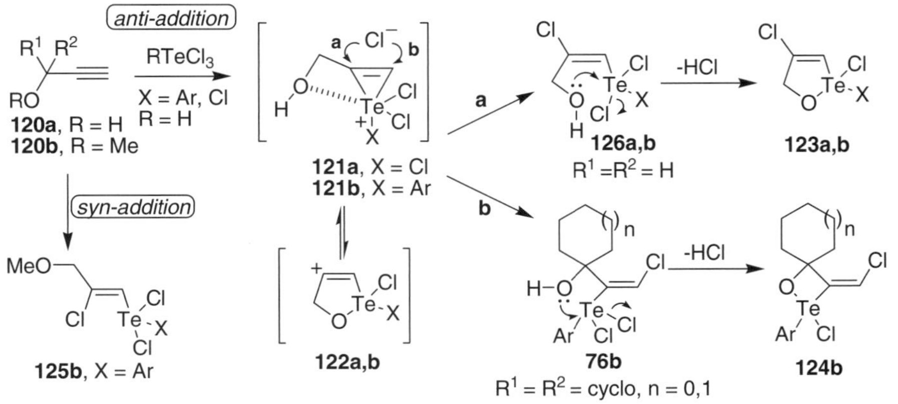

Scheme 5.36 Propose mechanism of TeCl$_4$ addition to propargyl alcohols.

5.4.23
Mechanism of Tröger's Base Formation

In general, it is now accepted that the mechanism for Tröger's base formation involves a series of *in situ* Friedel–Crafts reactions of **130**, but the detailed course of the reaction has not yet been fully established [105]. In trying to elucidate major mechanistic aspects, different methylene sources as well as different anilines have been used to form Tröger's bases by Abella and coworkers [106]. In this study, direct infusion ESI-MS/MS was used to monitor Tröger's base formation from different anilines (*p*-toluidine and 4-aminoveratrol), using both formaldehyde and urotropine (**128**) as the methylene source. To gain further evidence for the methylene transfer step, gas-phase ion/molecule reactions of protonated urotropine with two volatile amines were also performed. A hybrid linear ion-trap equipment was used in which N_2 gas was replaced with reactive neutrals by needle valve adaptors that allowed the introduction of reactive gases into the collision cell. Interestingly, urotropine of m/z 141 was indeed found to react with ethylamine and aniline to form directly the respective iminium ions of m/z 56 and 106 (Figure 5.18). Figure 5.18a shows that the imine $C_2H_5-NH=CH_2^+$ (as for **130**, Scheme 5.36) of m/z 56 is formed in high yield

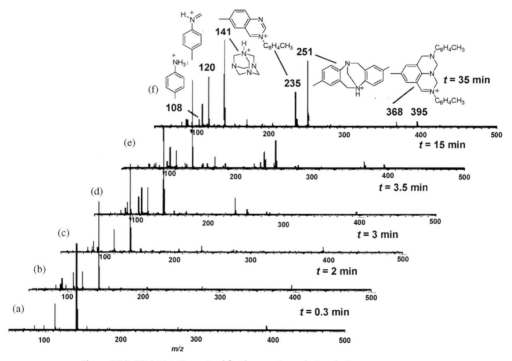

Figure 5.18 ESI(+)-MS acquired for the reaction solution during the on-line monitoring of the reaction of *p*-toluidine and urotropine in net TFA after (a) 20 s; (b) 2 min; (c) 3 min; (d) 3.5 min; (e) 15 min; and (f) 35 min of reaction.

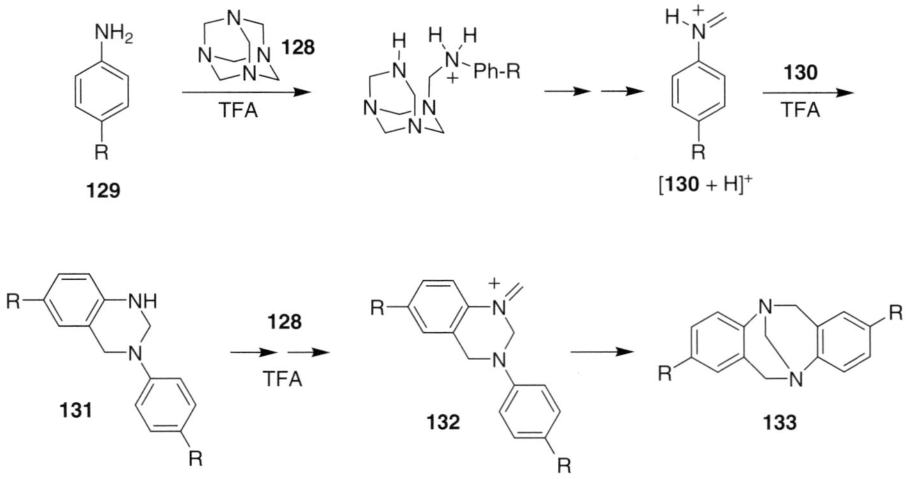

Scheme 5.37 Mechanism for the formation of Tröger's base as probed by ESI(+)-MS(/MS) and ion/molecule reactions.

in the ion/molecule reactions. Herein, the adduct of *m/z* 186, the first transient intermediate leading to methylene transfer, is obtained from nucleophilic addition of the amine to protonated urotropine. Furthermore, in reactions with aniline (Figure 5.16b), the corresponding imine of *m/z* 106 (as for **130**, Scheme 5.37), that is, Ph−NH=CH$_2^+$, is formed as an abundant ion, and the transient adduct of *m/z* 234 is detected as a low-abundance ion. On the basis of the mechanistic data collected from on-line ESI-MS(/MS) monitoring (Figures 5.18 and 5.19) and the gas-phase ion/molecule reactions, an experimentally probed mechanism for the formation of Tröger's bases using either formaldehyde or, more particularly, urotropine as the source of methylene was presented (Scheme 5.37). This mechanism proposes the participation of all of the intermediates intercepted by ESI-MS and properly characterized by ESI-MS/MS.

5.5
Conclusion

Over the past few years, there has been a rapid increase in the application of on-line API-MS monitoring techniques to the study of organic reaction mechanisms, mainly to intercept for the first time reactive intermediates from these reactions. The ability to isolate ions direct from crude reaction mixtures has a variety of outstanding features and advantages that allow new applications of transient species in mechanistic chemistry. Undoubtedly, API-MS is now an important tool for future studies of labile and sensitive intermediaries in solution, with no need for prior purification or isolation for further characterization of active species, intermediates of reactions, and products due to on-line purifications. Such new MS techniques are also

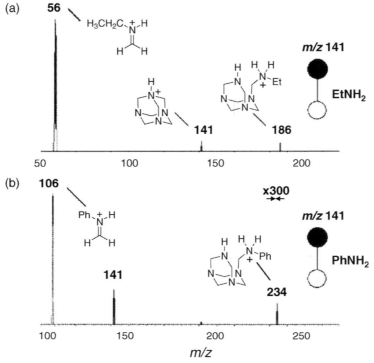

Figure 5.19 ESI-MS/MS for gas phase reactions of protonated urotropine of m/z 141 with (a) ethylamine and (b) aniline. The insert ×300 in (b) shows that the intensity of the ion of m/z 234 was increased by a factor of 300.

beginning to be used to transfer the reaction intermediates directly from solution to the gas phase to probe mechanisms of reactions of fundamental importance and practical application. Furthermore, using API-MS many intermediates were intercepted, isolated, detected, and then structurally characterized.

References

1 (a) Whitehouse, C.M., Dreyer, R.N., Yamashita, M., and Fenn, J.B. (1985) Electrospray interface for liquid chromatographs and mass spectrometers. *Anal. Chem.*, **57**, 675–679; (b) Fenn, J.B., Mann, M., Meng, C.K., Wong, S.F., and Whitehouse, C.M. (1989) Electrospray ionization for mass-spectrometry of large biomolecules. *Science*, **246**, 64–71; (c) Cole, R.B. (ed.) (1997) *Electrospray Ionization Mass Spectrometry – Fundamentals, Instrumentation, and Applications*, John Wiley & Sons, Inc., New York; (c) Fenn, J.B. (1993) Ion formation from charged droplets - roles of geometry, energy, and time. *J. Am. Soc. Mass Spectrom.*, **4**, 524–535.

2. Adlhart, C. and Chen, P. (2000) Fishing for catalysts: mechanism-based probes for active species in solution. *Helv. Chim. Acta*, **83**, 2192–2196.

3. (a) Plattner, D.A. (2001) Electrospray mass spectrometry beyond analytical chemistry: studies of organometallic catalysis in the gas phase. *Int. J. Mass Spectrom.*, **207**, 125–144; (b) Plattner, D.A. (2003) Metalorganic chemistry in the gas phase: insight into catalysis. *Top. Curr. Chem.*, **225**, 153–203.

4. (a) Santos, L.S., Knaack, L., and Metzger, J.O. (2005) Investigation of chemical reactions in solution using API-MS. *Int. J. Mass Spectrom.*, **246**, 84–104; (b) Chen, P. (2003) Electrospray ionization tandem mass spectrometry in high-throughput screening of homogeneous catalysts. *Angew. Chem. Int. Ed.*, **42**, 2832–2847; (c) Eberlin, M.N. (2007) Electrospray ionization mass spectrometry: a major tool to investigate reaction mechanisms in both solution and the gas phase. *Eur. J. Mass Spectrom.*, **13**, 19–28.

5. (a) Fenn, J.B., Mann, M., Meng, C.K., Wong, S.F., and Whitehouse, C.M. (1990) Electrospray ionization-principles and practice. *Mass Spectrom. Rev.*, **9**, 37–70; (b) Fenn, J.B. (2003) Electrospray wings for molecular elephants (Nobel lecture). *Angew. Chem. Int. Ed.*, **42**, 3871–3894.

6. Kebarle, P. and Tang, L. (1993) From ions in solution to ions in the gas-phase – the mechanism of electrospray mass-spectrometry. *Anal. Chem.*, **65**, 972–986.

7. Gaskell, S.J. (1997) Electrospray: principles and practice. *J. Mass Spectrom.*, **32**, 677–688.

8. Jones, J.L., Dongre, A.R., Somogyi, A., and Wysocki, V.H. (1994) Sequence dependence of peptide fragmentation efficiency curves determined by electrospray-ionization surface-induced dissociation mass-spectrometry. *J. Am. Chem. Soc.*, **116**, 8368–8369.

9. Kebarle, P. and Ho, Y. (1997) On the mechanism of electrospray ionization, in *Electrospray Ionization Mass Spectrometry* (ed. R.B. Cole), John Wiley & Sons, Inc., New York, pp. 3–63.

10. de la Mora, J.F., Van Berckel, G.J., Enke, C.G., Cole, R.B., Martinez-Sanchez, M., and Fenn, J.B. (2000) Electrochemical processes in electrospray ionization mass spectrometry – discussion. *J. Mass Spectrom.*, **35**, 939–952.

11. Van Berkel, G.J. (1997) The electrolytic nature of electrospray, in *Electrospray Ionization Mass Spectrometry* (ed. R.B. Cole) John Wiley & Sons, Inc., New York, pp. 65–105.

12. KaneMaguire, L.A.P., Kanitz, R., and Sheil, M.M. (1996) Electrospray mass spectrometry of neutral pi-hydrocarbon organometallic complexes. *Inorg. Chimica Acta*, **245**, 209–214.

13. Vandell, V.E. and Limbach, P.A. (1998) Electrospray ionization mass spectrometry of metalloporphyrins. *J. Mass Spectrom.*, **33**, 212–220.

14. Lavanant, H., Hecquet, E., and Hoppilliard, Y. (1999) Complexes of L-histidine with Fe^{2+}, Co^{2+}, Ni^{2+}, Cu^{2+}, Zn^{2+} studied by electrospray ionization mass spectrometry. *Int. J. Mass Spectrom.*, **187**, 11–23.

15. (a) Van Berkel, G.J., Asano, K.G., and Kertesz, V. (2002) Enhanced study and control of analyte oxidation in electrospray using a thin-channel, planar electrode emitter. *Anal. Chem.*, **74**, 5047–5056; (b) Van Berkel, G.J., Asano, K.G., and Gragner, M.C. (2004) Controlling analyte electrochemistry in an electrospray ion source with a three-electrode emitter cell. *Anal. Chem.*, **76**, 1493–1499; (c) Van Berkel, G.J., Asano, K.G., and Schnier, P.D. (2001) Electrochemical processes in a wire-in-a-capillary bulk-loaded, nano-electrospray emitter. *J. Am. Soc. Mass Spectrom.*, **12**, 853–862; (d) Van Berkel, G.J., Giles, G.E., Bullock, J.S., and Gray, L.J. (1999) Computational simulation of redox reactions within a metal electrospray emitter. *Anal. Chem.*, **71**, 5288–5296.

16 Zhang, K., Zimmerman, D.M., Chung-Phillips, A., and Cassady, C.J. (1993) Experimental and ab initio studies of the gas-phase basicities of polyglycines. *J. Am. Chem. Soc.*, **115**, 10812–10822.

17 Blades, A.T., Jayaweera, P., Ikonomou, M.G., and Kebarle, P. (1990) Studies of alkaline-earth and transition-metal M^{++} gas-phase ion chemistry. *J. Chem. Phys.*, **92**, 5900–5906.

18 Gianelli, L., Amendola, V., Fabbrizzi, L., Pallavicini, P., and Mellerio, G.G. (2001) Investigation of reduction of Cu(II) complexes in positive-ion mode electrospray mass spectrometry. *Rapid Commun. Mass Spectrom.*, **15**, 2347–2353.

19 (a) Vachet, R.W., Hartman, J.A.R., and Callahan, J.H. (1998) Ion-molecule reactions in a quadrupole ion trap as a probe of the gas-phase structure of metal complexes. *J. Mass Spectrom.*, **33**, 1209–1225; (b) Vachet, R.W., Hartman, J.R., Gertner, J.W., and Callahan, J.H. (2001) Investigation of metal complex coordination structure using collision-induced dissociation and ion-molecule reactions in a quadrupole ion trap mass spectrometer. *Int. J. Mass Spectrom.*, **204**, 101–112; (c) Hartman, J.R., Vachet, R.W., and Callahan, J.H. (2000) Gas, solution, and solid state coordination environments for the nickel(II) complexes of a series of aminopyridine ligands of varying coordination number. *Inorg. Chim. Acta*, **297**, 79–87; (d) Hartman, J.R., Vachet, R.W., Pearson, W., Wheat, R.J., and Callahan, J.H. (2003) A comparison of the gas, solution, and solid state coordination environments for the copper(II) complexes of a series of aminopyridine ligands of varying coordination number. *Inorg. Chim. Acta*, **343**, 119–132.

20 (a) Schäfer, A., Fischer, B., Paul, H., Bosshard, R., Hesse, M., and Viscontini, M. (1992) Pterin chemistry 94. electrospray-ionization mass-spectrometry - detection of a radical cation present in solution - new results on the chemistry of (tetrahydropteridinone)-metal complexes. *Helv. Chim. Acta*, **75**, 1955–1964; (b) Schäfer, A., Paul, H., Fischer, B., Hesse, M., and Viscontini, M. (1995) Reaction of 5,6,7,8-tetrahydropterin with iron(III) acetylacetonate - detection of radical cations by electrospray-ionization mass-spectrometry. *Helv. Chim. Acta*, **78**, 1763–1776.

21 Wilson, S.R., Perez, J., and Pasternak, A. (1993) ESI-MS detection of ionic intermediates in phosphine-mediated reactions. *J. Am. Chem. Soc.*, **115**, 1994–1997.

22 Aliprantis, A.O. and Canary, J.W. (1994) Observation of catalytic intermediates in the Suzuki reaction by electrospray mass-spectrometry. *J. Am. Chem. Soc.*, **116**, 6985–6986.

23 Moreno-Manas, M., Perez, M., and Pleixats, R. (1996) Palladium-catalyzed Suzuki-type self-coupling of arylboronic acids. A mechanistic study. *J. Org. Chem.*, **61**, 2346–2351.

24 Hinderling, C. and Chen, P. (1999) Rapid screening of olefin polymerization catalyst libraries by electrospray ionization tandem mass spectrometry. *Angew. Chem. Int. Ed.*, **38**, 2253–2556.

25 Hinderling, C. and Chen, P. (2000) Mass spectrometric assay of polymerization catalysts for combinational screening. *Int. J. Mass Spectrom.*, **195**, 377–383.

26 Hambitzer, G. and Heitbaum, J. (1986) Electrochemical thermospray mass spectrometry. *Anal. Chem.*, **58**, 1067–1070.

27 Sam, J.W., Tang, X.J., Magliozzo, R.S., and Peisach, J. (1995) Electrospray mass-spectrometry of iron bleomycin-II - investigation of the reaction of Fe(III) bleomycin with iodosylbenzene. *J. Am. Chem. Soc.*, **117**, 1012–1218.

28 Griep-Raming, J., Meyer, S., Bruhn, T., and Metzger, J.O. (2002) Investigation of reactive intermediates of chemical reactions in solution by electrospray ionization mass spectrometry: radical

chain reactions. *Angew. Chem. Int. Ed.*, **41**, 2738–2742.
29 Meyer, S. and Metzger, J.O. (2003) Use of electrospray ionization mass spectrometry for the investigation of radical cation chain reactions in solution: detection of transient radical cations. *Anal. Bioanal. Chem.*, **377**, 1108–1114.
30 Wilson, D.J. and Konermann, L. (2003) A capillary mixer with adjustable reaction chamber volume for millisecond time-resolved studies by electrospray mass spectrometry. *Anal. Chem.*, **75**, 6408–6414.
31 (a) Arakawa, R., Jian, L., Yoshimura, A., Nozaki, K., Ohno, T., Doe, H., and Matsuo, T. (1995) Online mass analysis of reaction-products by electrospray-ionization - photosubstitution of ruthenium(II) diimine complexes. *Inorg. Chem.*, **34**, 3874–3878; (b) Arakawa, R., Mimura, S., Matsubayashi, G., and Matsuo, T. (1996) Photolysis of (diamine) bis(2,2′-bipyridine)ruthenium(II) complexes using on-line electrospray mass spectrometry. *Inorg. Chem.*, **35**, 5725–5729.
32 Brum, J. and Dell'Orco, P. (1998) On-line mass spectrometry: real-time monitoring and kinetics analysis for the photolysis of idoxifene. *Rapid Commun. Mass Spectrom.*, **12**, 741–745.
33 Dell'Orco, P., Brum, J., Matsuoka, R., Badlani, M., and Muske, K. (1999) Monitoring process-scale reactions using API mass spectrometry. *Anal Chem.*, **71**, 5165–5170.
34 (a) Ding, W., Johnson, K.A., Amster, I.J., and Kutal, C. (2001) Identification of photogenerated intermediates by electrospray ionization mass spectrometry. *Inorg. Chem.*, **40**, 6865–6866; (b) Ding, W., Johnson, K.A., Kutal, C., and Amster, I.J. (2003) Mechanistic studies of photochemical reactions with millisecond time resolution by electrospray ionization mass spectrometry. *Anal. Chem.*, **75**, 4624–4630.
35 Modestov, A.D., Gun, J., Savotine, I., and Lev, O. (2004) On-line electrochemical–mass spectrometry study of the mechanism of oxidation of N,N-dimethyl-p-phenylenediamine in aqueous electrolytes. *J. Electroanal. Chem.*, **565**, 7–19.
36 Diehl, G. and Karst, U. (2002) On-line electrochemistry - MS and related techniques. *Anal. Bioanal. Chem.*, **373**, 390–398.
37 (a) Van Berkel, G.J. and Zhou, F. (1995) Characterization of an electrospray ion source as a controlled-current electrolytic cell. *Anal. Chem.*, **67**, 2916–2923; (b) Vanberkel, G.J., McLuckey, S.A., and Glish, G.L. (1992) Electrochemical origin of radical cations observed in electrospray ionization mass-spectra. *Anal. Chem.*, **64**, 1586–1593; (c) Van Berkel, G.J. and Zhou, F. (1995) Electrospray as a controlled-current electrolytic cell: electrochemical ionization of neutral analytes for detection by electrospray mass spectrometry. *Anal. Chem.*, **67**, 3958–3964; (d) Van Berkel, G.J. (2000) Insights into analyte electrolysis in an electrospray emitter from chronopotentiometry experiments and mass transport calculations. *J. Am. Soc. Mass Spectrom.*, **11**, 951–960.
38 (a) Stirk, K.M., Kiminkinen, M.L.K., and Kenttamaa, H.I. (1992) Ion molecule reactions of distonic radical cations. *Chem. Rev.*, **92**, 1649–1665; (b) Brodbelt, J.S. (1997) Analytical applications of ion-molecule reactions. *Mass Spectrom. Rev.*, **16**, 91–110; (c) Eberlin, M.N. (1997) Triple-stage pentaquadrupole (QqQqQ) mass spectrometry and ion/molecule reactions. *Mass Spectrom. Rev.*, **16**, 113–144; (d) Gronert, S. (2001) Mass spectrometric studies of organic ion/molecule reactions. *Chem. Rev.*, **101**, 329–360; (e) Speranza, M. (2004) Enantioselectivity in gas-phase ion-molecule reactions. *Int. J. Mass Spectrom.*, **232**, 277–317; (f) Cooks, R.G., Eberlin, M.N., Zheng, X., Chen, H., and Tao, W.A. (2006) Polar acetalization and

transacetalization in the gas phase: the eberlin reaction. *Chem. Rev.*, **106**, 188–211; (g) Green, M.K. and Lebrilla, C.B. (1997) Ion-molecule reactions as probes of gas-phase structures of peptides and proteins. *Mass Spectrom. Rev.*, **16**, 53–71.

39 Hess, T.F., Renn, T.S., Watts, R.J., and Paszczynski, A.J. (2003) Studies on nitroaromatic compound degradation in modified Fenton reactions by electrospray ionization tandem mass spectrometry (ESI-MS-MS). *Analyst*, **128**, 156–160.

40 (a) Meyer, S., Koch, R., and Metzger, J.O. (2003) Investigation of reactive intermediates of chemical reactions in solution by electrospray ionization mass spectrometry: radical cation chain reactions. *Angew. Chem. Int. Ed.*, **42**, 4700–4703; (b) Furmeier, S. and Metzger, J.O. (2004) Detection of transient radical cations in electron transfer-initiated Diels-Alder reactions by electrospray ionization mass spectrometry. *J. Am. Chem. Soc.*, **126**, 14485–14992; (c) Zhang, X., Wang, H.Y., and Guo, Y.L. (2006) Interception of the radicals produced in electrophilic fluorination with radical traps (TEMPO, DMPO) studied by electrospray ionization mass spectrometry. *Rapid. Commun. Mass Spectrom.*, **20**, 1877–1882; (d) Zhang, X., Liao, Y.X., Qian, R., Wang, H.Y., and Guo, Y.L. (2005) Investigation of radical cation in electrophilic fluorination by ESI-MS. *Org. Lett.*, **7**, 3877–3880.

41 Moura, F.C.C., Araujo, M.H., Dalmazio, I., Alves, T.M.A., Santos, L.S., Eberlin, M.N., Augusti, R., and Lago, R.M. (2006) Investigation of reaction mechanisms by electrospray ionization mass spectrometry: characterization of intermediates in the degradation of phenol by a novel iron/magnetite/hydrogen peroxide heterogeneous oxidation system. *Rapid Commun. Mass Spectrom.*, **20**, 1859–1863.

42 Griep-Raming, J. and Metzger, J.O. (2000) An electrospray ionization source for the investigation of thermally initiated reactions. *Anal. Chem.*, **72**, 5665–5668.

43 Fürmeier, S., Griep-Raming, J., Hayen, A., and Metzger, J.O. (2005) Chelation-controlled radical chain reactions studied by electrospray ionization mass spectrometry. *Chem. Eur. J.*, **11**, 5545–5554.

44 (a) Hayen, A., Koch, R., and Metzger, J.O. (2000) 1,3-stereoinduction in radical reactions. *Angew. Chem. Int. Ed.*, **39**, 2758–2761; (b) Hayen, A., Koch, R., Saak, W., Haase, D., and Metzger, J.O. (2000) 1,3-Stereoinduction in radical reactions: Radical additions to dialkyl 2-alkyl-4-methyleneglutarates. *J. Am. Chem. Soc.*, **112**, 12458–12468.

45 Arakawa, R., Tachiyashiki, S., and Matsuo, T. (1995) Detection of reaction intermediates - photosubstitution of (polypyridine)ruthenium(II) complexes using online electrospray mass-spectrometry. *Anal. Chem.*, **67**, 4133–4138.

46 Arakawa, R., Matsuda, F., Matsubayashi, G.-E., and Matsuo, T. (1997) Structural analysis of photo-oxidized (ethylenediamine)bis(2,2'-bipyridine) ruthenium(II) complexes by using on-line electrospray mass spectrometry of labeled compounds. *J. Am. Soc. Mass Spectrom.*, **8**, 713–717.

47 Kimura, K., Mizutani, R., Yokoyama, M., Arakawa, R., Matsubayashi, G.-E., Okamoto, M., and Doe, H. (1997) All-or-none type photochemical switching of cation binding with malachite green carrying a bis(monoazacrown ether) moiety. *J. Am. Chem. Soc.*, **119**, 2062–2063.

48 Arakawa, R., Lu, J., Mizuno, K., Inoue, H., Doe, H., and Matsuo, T. (1997) On-line electrospray mass analysis of photoallylation reactions of dicyanobenzenes by allylic silanes via photoinduced electron transfer. *Int. J. Mass Spectrom. Ion Processes*, **160**, 371–376.

49 Schuster, D.I., Cao, J.R., Kaprinidis, N., Wu, Y.H., Jensen, A.W., Lu, Q.Y., Wang,

H., and Wilson, S.R. (1996) [2 + 2] photocycloaddition of cyclic enones to C-60. *J. Am. Chem. Soc.*, **118**, 5639–5647.

50 (a) Gill, T.P. and Mann, K.R. (1983) Photochemistry of [(eta. -C₅H₅)Fe(. eta. -p-xyl)]PF₆ in acetonitrile solution. Characterization and reactivity of [(eta. -C₅H₅)Fe(MeCN)₃]⁺. *Inorg. Chem.*, **22**, 1986–1991; (b) McNair, A.M., Schrenk, J.L., and Mann, K.R. (1984) Effect of arene substituents and temperature on the arene replacement reactions of [(eta.5-C₅H₅)Fe(. eta. 6-arene)]⁺ and [(eta. 5-C₅H₅)Ru(. eta. 6-arene)]⁺. *Inorg. Chem.*, **23**, 2633–2640; (c) Chrisope, D.R., Park, K.M., and Schuster, G.B. (1989) Photodissociation of cyclopentadienyliron(II) arene cations - detection and characterization of reactive intermediates by means of time-resolved laser spectroscopy. *J. Am. Chem. Soc.*, **111**, 6195–6201; (d) Jakubek, V. and Lees, A.J. (2000) Quantitative wavelength-dependent photochemistry of the [CpFe (eta(6)-ipb)]PF₆(ipb = isopropylbenzene) photoinitiator. *Inorg. Chem.*, **39**, 5779–5786.

51 Keizer, S.P., Han, W.J., and Stillman, M.J. (2002) Photochemically Induced Radical Reactions of Zinc Phthalocyanine. *Inorg. Chem.*, **41**, 353–358.

52 Sabino, A.A., Machado, A.H.L., Correia, C.R.D., and Eberlin, M.N. (2004) Probing the mechanism of the Heck reaction with arene diazonium salts by electrospray mass and tandem mass spectrometry. *Angew. Chem. Int. Ed.*, **43**, 2514–2518.

53 Masllorens, J., Moreno-Manãs, M., Pla-Quintana, A., and Roglans, A. (2003) First Heck reaction with arenediazonium cations with recovery of Pd-triolefinic macrocyclic catalyst. *Org. Lett.*, **5**, 1559–1561.

54 Moreno-Manãs, M., Pleixats, R., Sebastian, R.M., Vallribera, A., and Roglans, A. (2004) Organometallic chemistry of 15-membered tri-olefinic macrocycles: catalysis by palladium(0) complexes in carbon-carbon bond-forming reactions. *J. Organomet. Chem.*, **689**, 3669–3684.

55 Armendía, M.A., Lafont, F., Moreno-Manãs, M., Pleixats, R., and Roglans, A. (1999) Electrospray ionization mass spectrometry detection of intermediates in the palladium-catalyzed oxidative self-coupling of areneboronic acids. *J. Org. Chem.*, **64**, 3592–3594.

56 Markert, C. and Pfaltz, A. (2004) Screening of chiral catalysts and catalyst mixtures by mass spectrometric monitoring of catalytic intermediates. *Angew. Chem. Int. Ed.*, **43**, 2498–2500.

57 Chevrin, C., Le Bras, J., Hénin, F., Muzart, J., Pla-Quintana, A., Roglans, A., and Pleixats, R. (2004) Allylic substitution mediated by water and palladium: Unusual role of a palladium(II) catalyst and ESI-MS analysis. *Organometallics*, **23**, 4796–4799.

58 Santos, L.S., Rosso, G.B., Pilli, R.A., and Eberlin, M.N. (2007) The mechanism of the Stille reaction investigated by electrospray ionization mass spectrometry. *J. Org. Chem.*, **72**, 5809–5812.

59 Raminelli, C., Prechtl, M.H.G., Santos, L.S., Eberlin, M.N., and Comasseto, J.V. (2004) Coupling of vinylic tellurides with alkynes catalyzed by palladium dichloride: evaluation of synthetic and mechanistic details. *Organometallics*, **23**, 3990–3996.

60 (a) Ripa, L. and Hallberg, A. (1996) Controlled double-bond migration in palladium-catalyzed intramolecular arylation of enamidines. *J. Org. Chem.*, **61**, 7147–7155; (b) Brown, J.M. and Hii, K.K. (1996) Characterization of reactive intermediates in palladium-catalyzed arylation of methyl acrylate (Heck reaction). *Angew. Chem. Int. Ed.*, **35**, 657–659; (c) Hii, K.K., Claridge, T.D.W., and Brown, J.M. (1997) Intermediates in the intermolecular, asymmetric Heck arylation of dihydrofurans. *Angew. Chem. Int. Ed.*, **36**, 984–987.

61 (a) Wang, Z., Zhang, Z.G., and Lu, X.Y. (2000) Effect of halide ligands on the reactivity of carbon-palladium bonds: implications for designing catalytic reactions. *Organometallics*, **19**, 775–780; (b) Zhang, Z.G., Lu, X.Y., Zang, Q.H., and Han, X.L. (2001) Role of halide ions in divalent palladium-mediated reactions: competition between beta-heteroatom elimination and beta-hydride elimination of a carbon-palladium bond. *Organometallics*, **20**, 3724–3728; (c) Liu, G.S. and Lu, X.Y. (2002) Palladium(II)-catalyzed coupling reactions of alkynes and allylic compounds initiated by intramolecular carbopalladation of alkynes. *Tetrahedron Lett.*, **43**, 6791–6794; (d) de Meijere, A. and Meyer, F.E. (1995) Fine feathers make fine birds - the Heck reaction in modern garb. *Angew. Chem. Int. Ed.*, **33**, 2379–2411, and references therein.

62 Alexakis, A., Berlan, J., and Besace, Y. (1986) Organocopper conjugate addition reaction in the presence of trimethylchlorosilane. *Tetrahedron Lett.*, **27**, 1047–1050.

63 Other examples for Pd studies, see: (a) Guo, H., Qian, R., Liao, Y.X., Ma, S.M., and Guo, Y.L., (2005) ESI-MS studies on the mechanism of Pd(0)-catalyzed three-component tandem double addition-cyclization reaction. *J. Am. Chem. Soc.*, **127**, 13060–13064; (b) Qian, R., Guo, H., Liao, Y.X., Guo, Y.L., and Ma, S.M. (2005) Probing the mechanism of the palladium-catalyzed addition of organoboronic acids to allenes in the presence of AcOH by ESI-FTMS. *Angew. Chem. Int. Ed.*, **44**, 4771–4774.

64 Bonini, B.F., Capito, E., Comes-Franchini, M., Ricci, A., Bottoni, A., Bernardi, F., Miscione, G.P., Giordano, L., and Cowley, A.R. (2004) Diastereoselective synthesis of thieno [3′,2′: 4, 5]cyclopenta[1,2-d][1,3]-oxazolines - New ligands for the copper-catalyzed asymmetric conjugate addition of diethylzinc to enones. *Eur. J. Org. Chem.*, 4442–2451.

65 Evans, D.A., Kozlowski, M.C., Murry, J.A., Burgey, C.S., Campos, K.R., Connell, B.T., and Staples, R.J. (1999) C-2-symmetric copper(II) complexes as chiral Lewis acids. Scope and mechanism of catalytic enantioselective aldol additions of enolsilanes to (benzyloxy)acetaldehyde. *J. Am. Chem. Soc.*, **121**, 669–685.

66 Comelles, J., Moreno- Manãs, M., Perez, E., Roglans, A., Sebastian, R.M., and Vallribera, A. (2004) Ionic and covalent copper(II)-based catalysts for Michael additions. The mechanism. *J. Org. Chem.*, **69**, 6834–6842.

67 Trage, C., Schröder, D., and Schwarz, H. (2005) Coordination of iron(III) cations to beta-keto esters as studied by electrospray mass spectrometry: implications for iron-catalyzed Michael addition reactions. *Chem. Eur. J.*, **11**, 619–627.

68 Fujii, A., Hagiwara, E., and Sodeoka, M. (1999) Mechanism of palladium complex-catalyzed enantioselective Mannich-type reaction: characterization of a novel binuclear palladium enolate complex. *J. Am. Chem. Soc.*, **121**, 5450–5458.

69 (a) Gerdes, G. and Chen, P. (2003) Comparative gas-phase and sollution-phase investigations of the mechanism of C—H activation by $[(\text{N-N})\text{Pt}(CH_3)(L)]^+$. *Organometallics*, **22**, 2217–2225; (b) Gerdes, G. and Chen, P. (2004) Cationic platinum(II) carboxylato complexes are competent in catalytic arene C—H activation under mild conditions. *Organometallics*, **23**, 3031–3036.

70 (a) Hinderling, C., Plattner, D.A., and Chen, P. (1997) Direct observation of a dissociative mechanism for C—H activation by a cationic iridium(III) complex. *Angew. Chem. Int. Ed.*, **36**, 243–244; (b) Hinderling, C., Feichtinger, D., Plattner, D.A., and Chen, P. (1997) A combined gas-phase, solution-phase, and computational study of C—H activation by cationic iridium(III) complexes. *J. Am. Chem. Soc.*, **119**, 10793–10804; (c) Dietiker, R. and Chen, P. (2004)

Gas-phase reactions of the [(PHOX)IrL$_2$]$^+$ ion olefin-hydrogenation catalyst support an Ir-I/Ir-III. *Angew. Chem. Int. Ed.*, **43**, 5513–5516.

71 Sandoval, C.A., Ohkuma, T., Muniz, K., and Noyori, R. (2003) Mechanism of asymmetric hydrogenation of ketones catalyzed by BINAP/1,2-diamine-ruthenium(II) complexes. *J. Am. Chem. Soc.*, **125**, 13490–13503.

72 Greb, M., Hartung, J., Köhler, F., Spehar, K., Kluge, R., and Csuk, R. (2004) The (Schiff base)vanadium(V) complex catalyzed oxidation of bromide - A new method for the *in situ* generation of bromine and its application in the synthesis of functionalized cyclic ethers. *Eur. J. Org. Chem.*, 3799–3812.

73 Hartung, J. and Greb, M. (2002) Transition metal-catalyzed oxidations of bishomoallylic alcohols. *J. Organomet. Chem.*, **661**, 67–84; (b) Hartung, J., Drees, S., Greb, M., Schmidt, P., Svoboda, I., Fuess, H., Murso, A., and Stalke, D. (2003) (Schiff-base)vanadium(V) complex-catalyzed oxidations of substituted bis(homoallylic) alcohols - Stereoselective synthesis of functionalized tetrahydrofurans. *Eur. J. Org. Chem.*, 2388–2408.

74 Gilbert, B.C., Smith, J.R.L., Payeras, A.M.I., Oakes, J., and Prats, R.P.I. (2004) A mechanistic study of the epoxidation of cinnamic acid by hydrogen peroxide catalysed by manganese 1,4,7-trimethyl-1,4,7-triazacyclononane complexes. *J. Mol. Catal. A: Chem.*, **219**, 265–272.

75 Nagataki, T., Tachi, Y., and Itoh, S. (2005) Synthesis, characterization, and catalytic oxygenation activity of dinuclear iron(III) complex supported by binaphthol-containing chiral ligand. *J. Molec. Cat. A: Chem.*, **225**, 103–109.

76 Waters, T., O'Hair, R.A.J., and Wedd, A.G. (2000) Probing the catalytic oxidation of alcohols via an anionic dimolybdate centre using multistage mass spectrometry. *Chem. Commun.*, 225–226.

77 Kotiaho, T., Eberlin, M.N., Vainiotalo, P., and Kostiainen, R. (2000) Electrospray mass and tandem mass spectrometry identification of ozone oxidation products of amino acids and small peptides. *J. Am. Soc. Mass Spectrom.*, **11**, 526–535.

78 (a) Jacobsen, E.N. (1993) Asymmetric Catalytic Epoxidation of Unfunctionalized Olefins. *Catalytic Asymmetric Synthesis* (ed. I. Ojima) VCH Publishers, New York, pp. 159–202; (b) Katsuki, T. (1996) Mn-salen catalyst, competitor of enzymes, for asymmetric epoxidation. *J. Mol. Catal. A: Chem.*, **113**, 87–107; (c) Srinivasan, K., Michaud, P., and Kochi, J.K. (1986) Epoxidation of olefins with cationic (salen) manganese(III) complexes. The modulation of catalytic activity by substituents. *J. Am. Chem. Soc.*, **108**, 2309–2320; (d) Siddall, T.L., Miyaura, N., Huffman, J.C., and Kochi, J.K. (1983) Isolation and molecular structure of unusual oxochromium(V) cations for the catalytic epoxidation of alkenes. *J. Chem. Soc. Chem. Commun.*, 1185–1186; (e) Samsel, E.G., Srinivasan, K., and Kochi, J.K. (1985) Mechanism of the chromium-catalyzed epoxidation of olefins. Role of oxochromium(V) cations. *J. Am. Chem. Soc.*, **107**, 7606–7617; (f) Srinivasan, K. and Kochi, J.K. (1985) Synthesis and molecular structure of oxochromium(V) cations. Coordination with donor ligands. *Inorg. Chem.*, **24**, 4671–4679.

79 Feichtinger, D. and Plattner, D.A. (2001) Probing the reactivity of oxomanganese-salen complexes: an electrospray tandem mass spectrometric study of highly reactive intermediates. *Chem. Eur. J.*, **7**, 591–599.

80 (a) Adam, W., Mock-Knoblauch, C., Saha-Moller, C.R., and Herderich, M. (2000) Are Mn-IV species involved in Mn(salen)-catalyzed Jacobsen-Katsuki epoxidations? A mechanistic elucidation of their formation and reaction modes by EPR spectroscopy, mass-spectral analysis, and product studies: chlorination versus

oxygen transfer. *J. Am. Chem. Soc.*, **122**, 9685–9691; (b) Adam, W., Roschmann, K.J., Saha-Moller, C.R., and Seebach, D. (2002) cis-stilbene and (1 alpha,2 beta,3 alpha)-(2-ethenyl-3-methoxycyclopropyl) benzene as mechanistic probes in the Mn-III(salen)-catalyzed epoxidation: influence of the oxygen source and the counterion on the diastereoselectivity of the competitive concerted and radical-type oxygen transfer. *J. Am. Chem. Soc.*, **124**, 5068–5073.

81 Santos, L.S., Pavam, C.H., Almeida, W.P., Coelho, F., and Eberlin, M.N. (2004) Probing the mechanism of the Baylis-Hillman reaction by electrospray ionization mass and tandem mass spectrometry. *Angew. Chem. Int. Ed.*, **43**, 4330–4333.

82 (a) Hill, J.S. and Isaacs, N.S. (1990) Mechanism of alpha-substitution reactions of acrylic derivatives. *J. Phys. Org. Chem.*, **3**, 285–288; (b) Hoffmann, H.M.R. and Rabe, J. (1983) A new, efficient and stereocontrolled synthesis of trisubstituted alkenes via functionalized acrylic esters. *Angew. Chem. Int. Ed.*, **22**, 796–797; (c) Bode, M.L. and Kaye, P.T. (1991) A Kinetic and mechanistic study of the Baylis-Hillman reaction. *Tetrahedron Lett.*, **32**, 5611–5614; (d) Fort, Y., Berthe, M.C., and Caubere, P. (1992) The Baylis-Hillman reaction-mechanism and applications revisited. *Tetrahedron*, **48**, 6371–6384.

83 Santos, L.S., DaSilveira Neto, B.A., Consorti, C.S., Pavam, C.H., Almeida, W.P., Coelho, F., Dupont, J., and Eberlin, M.N. (2006) The role of ionic liquids in co-catalysis of Baylis-Hillman reaction: interception of supramolecular species via electrospray ionization mass spectrometry. *J. Phys. Org. Chem.*, **19**, 731–736.

84 Rosa, J.N., Afonso, C.A.M., and Santos, A.G. (2001) Ionic liquids as a recyclable reaction medium for the Baylis-Hillman reaction. *Tetrahedron*, **57**, 4189–4193.

85 Ferraz, H.M.C., Pereira, F.L.C., Goncalo, E.R.S., Santos, L.S., and Eberlin, M.N. (2005) Unexpected synthesis of conformationally restricted analogues of gamma-amino butyric acid (GABA): mechanism elucidation by electrospray ionization mass spectrometry. *J. Org. Chem.*, **70**, 110–114.

86 (a) Domingos, J.B., Longhinotti, E., Brandao, T.A.S., Bunton, C.A., Santos, L.S., Eberlin, M.N., and Nome, F. (2004) Mechanisms of nucleophilic substitution reactions of methylated hydroxylamines with bis(2,4-dinitrophenyl)phosphate. Mass spectrometric identification of key intermediates. *J. Org. Chem.*, **69**, 6024–6033; (b) Domingos, J.B., Longhinotti, E., Brandao, T.A.S., Santos, L.S., Eberlin, M.N., Bunton, C.A., and Nome, F. (2004) Reaction of bis(2, 4-dinitrophenyl) phosphate with hydrazine and hydrogen peroxide. Comparison of O- and N-phosphorylation. *J. Org. Chem.*, **69**, 7898–7905.

87 Dalmazio, I., Santos, L.S., Lopes, R.P., Eberlin, M.N., and Augusti, R. (2005) Advanced oxidation of caffeine in water: on-line and real-time monitoring by electrospray ionization mass spectrometry. *Env. Sci. Technol.*, **39**, 5982–5988.

88 Claeys, M., Graham, B., Vas, G., Wang, W., Vermeylen, R., Pashynska, V., Cafmeyer, J., Guyon, P., Andreae, M.O., Artaxo, P., and Maenhaut, W. (2004) Formation of secondary organic aerosols through photooxidation of isoprene. *Science*, **303**, 1173–1176.

89 Santos, L.S., Dalmazio, I., Eberlin, M.N., Claeys, M., and Augusti, R. (2006) Mimicking the atmospheric OH-radical-mediated photooxidation of isoprene: formation of cloud-condensation nuclei polyols monitored by electrospray ionization mass spectrometry. *Rapid Commun. Mass Spectrom.*, **20**, 2104–2108.

90 Milagre, C.D.F., Milagre, H.M.S., Santos, L.S., Lopes, M.L.A., Rodrigues, J.A.R., Moran, P.J.S., and Eberlin, M.N. (2007) Probing the mechanism of direct

Mannich-type alpha-methylenation of ketoesters via electrospray ionization mass spectrometry. *J. Mass Spectrom.*, **42**, 1287–1293.

91 (a) Marquez, C.A., Fabbretti, F., and Metzger, J.O. (2007) Electrospray ionization mass spectrometric study on the direct organocatalytic alpha-halogenation of aldehydes. *Angew. Chem. Int. Ed.*, **46**, 6915–6917; (b) Marquez, C. and Metzger, J.O. (2006) ESI-MS study on the aldol reaction catalyzed by L-proline. *Chem. Commun.*, 1539–1541. For similar studies, see: (c) Wang, H.Y., Zhang, X., Guo, Y.L., and Lu, L. (2005) Mass spectrometric studies of the gas phase retro-Michael type fragmentation reactions of 2-hydroxybenzyl-N-pyrimidinylamine derivatives. *J. Am. Soc. Mass Spectrom.*, **16**, 1561–1573; (d) Zhang, X. and Guo, Y.L. (2006) Electrospray ionization mass spectrometric study on the alpha-fluorination of aldehydes. *Rapid Commun. Mass Spectrom.*, **20**, 3477–3480.

92 Meurer, E.C., Santos, L.S., Pilli, R.A., and Eberlin, M.N. (2003) Probing the mechanism of the petasis olefination reaction by atmospheric pressure chemical ionization mass and tandem mass spectrometry. *Org. Lett.*, **5**, 1391–1394.

93 Meurer, E.C., Rocha, L.L., Pilli, R.A., Eberlin, M.N., and Santos, L.S. (2006) Transient intermediates of the Tebbe reagent intercepted and characterized by atmospheric pressure chemical ionization mass spectrometry. *Rapid Commun. Mass Spectrom.*, **20**, 2626–2629.

94 Hanzawa, Y., Kowase, N., Momose, S., and Taguchi, T. (1998) A Cp_2TiCl_2-Me_3Al (1:4) reagent system: an efficient reagent for generation of allylic titanocene derivatives from vinyl halides, vinyl ethers and carboxylic esters. *Tetrahedron*, **54**, 11387–11398.

95 (a) Kaminsky, W. and Steiger, R. (1988) Polymerization of olefins with homogeneous zirconocene alumoxane catalysts. *Polyhedron*, **7**, 2375–2381; (b) Cam, D. and Giannini, U. (1992) Concerning the reaction of zirconocene dichloride and methylalumoxane - homogeneous Ziegler-Natta catalytic-system for olefin polymerization. *Makromol. Chem.*, **193**, 1049–1055.

96 XPS: (a) Gassman, P.G. and Callstrom, M.R. (1987) Isolation, and partial characterization by XPS, of two distinct catalysts in the Ziegler-Natta polymerization of ethylene. *J. Am. Chem. Soc.*, **109**, 7875–7876; ^{13}C NMR: (b) Sishta, C., Hathorn, R.M., and Marks, T.J., (1992) Group-4 metallocene alumoxane olefin polymerization catalysts - CPMAS NMR spectroscopic observation of cation-like zirconocene alkyls. *J. Am. Chem. Soc.*, **114**, 1112–1114.

97 (a) Siedle, A.R., Lamanna, W.M., Newmark, R.A., and Schroepfer, J.N. (1998) Mechanism of olefin polymerization by a soluble zirconium catalyst. *J. Mol. Catal. A*, **128**, 257–271; (b) Tritto, I., Li, S.X., Sacchi, M.C., Locatelli, P., and Zannoni, G. (1995) *Macromolecules*, **28**, 5358; (c) Tritto, I., Li, S., Sacchi, M.C., and Zannoni, G. (1993) H-1 and C-13 NMR spectroscopic study of titanium metallocene-aluminoxane catalysts for olefin polymerizations. *Macromolecules*, **26**, 7111–7115.

98 (a) Kaminsky, W. (1996) New polymers by metallocene catalysis. *Macromol. Chem. Phys.*, **197**, 3907–3945; (b) Kaminsky, W. and Strubel, C. (1998) Hydrogen transfer reactions of supported metallocene catalysts. *J. Mol. Catal. A*, **128**, 191–200.

99 (a) Coevoet, D., Cramail, H., and Deffieux, A. (1999) Activation of iPr(CpFluo)ZrCl2 by methylaluminoxane, 3 Kinetic investigation of the syndiospecific hex-1-ene polymerization in hydrocarbon and chlorinated media. *Macromol. Chem. Phys.*, **200**, 1208–1214; (b) Pedeutour, J.-N., Coevoet, D., Cramail, H., and Deffieux, A. (1999) Activation of iPr (CpFluo)ZrCl$_2$ by methylaluminoxane, 4

UV visible spectroscopic study in hydrocarbon and chlorinated media. *Macromol. Chem. Phys.*, **200**, 1215–1221; (c) Babushkin, D.E., Semikolenova, N.V., Zakharov, V.A., and Talsi, E.P. (2000) Mechanism of dimethylzirconocene activation with methylaluminoxane: NMR monitoring of intermediates at high Al/Zr ratios. *Macromol. Chem. Phys.*, **201**, 558–567.

100 Santos, L.S. and Metzger, J.O. (2006) Study of homogeneously catalyzed Ziegler-Natta polymerization of ethene by ESI-MS. *Angew. Chem. Int. Ed.*, **45**, 977–981.

101 Santos, L.S. and Metzger, J.O. (2008) On-line monitoring of Brookhart polymerization by electrospray ionization mass spectrometry. *Rapid Commun. Mass Spectrom.*, **22**, 898–904.

102 Santos, L.S., Cunha, R.L.O.R., Comasseto, J.V., and Eberlin, M.N. (2007) Electrospray ionization mass spectrometric characterization of key Te(IV) cationic intermediates for the addition of $TeCl_4$ to alkynes. *Rapid Commun. Mass Spectrom.*, **21**, 1479–1484.

103 Cunha, R.L.O.R., Urano, M.E., Chagas, J.R., Almeida, P.C., Bincoletto, C., Tersariol, I.L.S., and Comasseto, J.V. (2005) Tellurium-based cysteine protease inhibitors: evaluation of novel organotellurium(IV) compounds as inhibitors of human cathepsin B. *Bioorg. Med. Chem. Lett.*, **15**, 755–760.

104 (a) Cojocaru, M., Elyashiv, I., and Albeck, M. (1997) Mass spectrometric study of some organotellurium(IV) compounds. *J. Mass Spectrom.*, **31**, 705–713; (b) Williams, F.D. and Dunbar, F.X. (1968) Preparation of Tritelluroformaldehyde. *Chem. Commun.*, 459; (c) Duffield, A.M., Budzikiewicz, H., and Djerassi, C. (1965) Mass spectrometry in structural and stereochemical problems .71. A study of influence of different heteroatoms on mass spectrometric fragmentation of 5-membered heterocycles. *J. Am. Chem. Soc.*, **87**, 2920; (d) Wieber, M. and Kaunzinger, E. (1977) Organotellurium (IV) compounds - esters of diorganotellurium hydroxide. *J. Organomet. Chem.*, **129**, 339–346.

105 (a) Wagner, E.C. (1935) Condensations of aromatic amines with formaldehyde in media containing acid III the formation of Troeger's base. *J. Am. Chem. Soc.*, **57**, 1296–1298; (b) Miller, T.R. and Wagner, E.C. (1941) Some analogs of Troeger's base and related compounds. *J. Am. Chem. Soc.*, **63**, 832–836.

106 Abella, C.A.M., Benassi, M., Santos, L.S., Eberlin, M.N., and Coelho, F. (2007) The mechanism of Troeger's base formation probed by electrospray ionization mass spectrometry. *J. Org. Chem.*, **72**, 4048–4054.

6
Gas Phase Ligand Fragmentation to Unmask Reactive Metallic Species
Richard A. J. O'Hair

6.1
Introduction and Scope of the Review

The newer ionization methods of electrospray ionization, cold-spray ionization [1], atmospheric pressure chemical ionization [2], and matrix-assisted laser desorption [3] are making possible the mass spectrometry-based analysis of a diverse range of metal-based species [4]. With the very recent coupling of ESI to a glovebox [5], even challenging organometallic and inorganic species should be examinable via MS. Electrospray ionization has been particularly useful in the direct interception of reactive intermediates in solution. Pioneering work in the area includes that of Wilson and Canary in the early 1990s, who studied the intermediates formed in a number of textbook organic reactions including the Wittig, Mitsunobu, and Staudinger reactions [6] and C—C bond-coupling reactions [7]. Since this topic has been recently reviewed [8, 9] and also forms the basis of chapters 3, 4, 5 and 7 of this book, it is not discussed further here.

The focus of this chapter is on the combined used of ESI and collision-induced dissociation (CID) as a means of 'synthesizing' reactive intermediates so that their fundamental gas-phase reactivity may be examined via, for example, the use of ion-molecule reactions (IMR) [10][1]. The motivation for such studies is to gain fundamental insights into metal-containing species that may have relevance to catalysis and metal-mediated reactions [13–18]. This field builds upon a wealth of previous metal ion chemistry studies in which other ionization techniques were utilized [14, 19–28]. Indeed, the potential for using CID to generate important (and sometimes novel) organic, organoelement and transition metal reactive intermediates was recognized some time ago. For example, Squires used decarboxylation of carboxylates and/or loss of aldehydes from alkoxide anions to study the gas-phase chemistry of naked alkyl anions [29–32]. Thus, the prototypical methyl anion,

---

1) It is worth noting that photolysis [11] and thermolysis of metal complexes have a rich history and have been used as a route to reactive intermediates and new materials [12].

Reactive Intermediates: MS Investigations in Solution. Edited by Leonardo S. Santos.
Copyright © 2010 WILEY-VCH Verlag GmbH & Co. KGaA, Weinheim
ISBN: 978-3-527-32351-7

formed via Eqs. (6.1) and (6.2), readily reacts with a range of neutral reagents, undergoing electron transfer with O_2 (Eq. (6.3)), proton transfer with weak acids such as ammonia (Eq. (6.4)), and an addition/elimination reaction with propionaldehyde (Eq. (6.5)) [30].

$$CH_3CO_2^- \rightarrow CH_3^- + CO_2 \qquad (6.1)$$

$$CH_3CH_2O^- \rightarrow CH_3^- + CH_2O \qquad (6.2)$$

$$CH_3^- + O_2 \rightarrow O_2^{-\bullet} + CH_3\bullet \qquad (6.3)$$

$$CH_3^- + NH_3 \rightarrow NH_2^- + CH_4 \qquad (6.4)$$

$$CH_3^- + CH_3CH_2CHO \rightarrow CH_3CH=C(CH_3)O^- + H_2 \qquad (6.5)$$

CID on organosilicon ions [33] has been used to generate and study the ion-molecule reactions of a range of unusual species such as the silaacetone enolate anion [34], the silaformamide anion [35] and $HSiX^-$ (where X=O [36], S and NH [37]). Finally, Freiser pioneered the use of CID to generate a range of novel transition metal ions, including metal cluster dimers [38] and metal hydride cations [39]. For example, by taking advantage of the multistage mass spectrometry capabilities of a Fourier-transform mass spectrometer, Carlin et al. 'synthesized' Group VIII metal deuteride ions ($[FeD]^+$, $[CoD]^+$ and $[NiD]^+$) in the gas phase via the combined sequence of laser-desorption of monoatomic metal cations (Eq. (6.6)), ion-molecule reactions of these cations with CD_3NO_2 or CD_3ONO (e.g., Eq. (6.7)), and CID of the resultant metal alkoxide cations (Eq. (6.8)) [39]. The resultant metal deuteride cations were allowed to react with simple hydrocarbons. The major reaction pathway for $[MetalD]^+$ with alkanes was dehydrogenation, yielding either metal-alkyl or metal-allyl product ions. In contrast to the bare metal ions, these reactions occur via initial oxidative insertion into C—H bonds. $[NiD]^+$ and $[CoD]^+$ were rendered more reactive by the hydride ligand, both reacting with ethane and $[NiD]^+$ reacting with methane (Eqs. (6.9) and (6.10)).

$$Metal(rod) + h\nu \rightarrow Metal^+ \qquad (6.6)$$

$$Metal^+ + CD_3ONO \rightarrow [MetalOCD_3]^+ + NO \qquad (6.7)$$

$$[MetalOCD_3]^+ \rightarrow [MetalD]^+ + CD_2O \qquad (6.8)$$

$$[NiD]^+ + CH_4 \rightarrow [NiH]^+ + CH_3D \qquad (6.9)$$

$$\rightarrow [NiCH_3]^+ + HD \qquad (6.10)$$

The aim of this book chapter is neither to review all past work on ESI/MS of metallic species [40–45] nor the CID of metal containing species as a means of forming reactive intermediates. Rather the focus is on the fairly recent combination of ESI and CID as a means of generating reactive metallic species [10].[2)] A number of these studies have also benefited from the use of theoretical calculations [46], and these will be discussed where appropriate. The ESI process will not be reviewed here, nor will the instrumental aspects of the experiments (for reviews see: Refs. [47–51]). Finally, the engineering of ligands to carry a charge 'handle' and thus make otherwise neutral complexes 'electrospray and MS active' has recently been reviewed [52].

6.2
Unmasking Reactive Metallic Intermediates via Collision-Induced Dissociation

Ligands coordinated to metal centers can undergo a number of reactions under CID conditions. These can include loss of the intact ligand as a neutral species, as a protonated species, or as a radical ion. The latter reaction has been exploited as a means of generating a diverse range of radical ions of biomolecules [53, 54], including radical cations and anions of peptides [55, 56] and radical cations of nucleobases [57]. Since these reactions essentially only unmask a vacant coordination site at the metal center, they are not reviewed here [58]. Instead, the focus will be on reactions in which the ligand itself undergoes fragmentation. Once again, there are a number of ligand fragmentation reactions that may occur. For ligands with a connectivity of X-Y-Z these can include loss of YZ (Eq. (6.11)) or a rearrangement with concomitant loss of XY (Eq. (6.12)). Each of these reactions can give rise to a range of intermediates and can occur via heterolytic or homolytic processes. In the next sections, the focus is on the formation and reactions of specific classes of intermediates.

$$[(L)_n \text{MetalXYZ}]^{+/-} \rightarrow (L)_n \text{MetalX}]^{+/-} + YZ \qquad (6.11)$$

$$\rightarrow [(L)_n \text{MetalZ}]^{+/-} + XY \qquad (6.12)$$

6.2.1
Formation and Reactivity of Organometallics (Eq. (6.12), Z = C)

Decarboxylation of metal carboxylate ions appears to be a general way of synthesising organometallic ions in the gas phase (for related reactions in solution see: [59, 60], some of which are related to textbook organometallic species used in organic synthesis (e.g., Grignard [61] and Gilman [62] reagents). In fact, CID of the parent metal acetate ions can proceed via a number of different pathways including decarboxylation (Eq. (6.13)), acetate anion loss (Eq. (6.14)), oxidation of the acetate

---

[2)] CID of ESI-generated ions can be performed on mass-selected ions, or it can occur in the interface region directly after the electrospray source (also known as 'in-source' CID).

anion with concomitant reduction of the metal (Eq. (6.15)), loss of ketene (Eq. (6.16)), loss of water (Eq. (6.17)), and loss of the auxiliary ligand, L^- (Eq. (6.18)).

$$[CH_3CO_2Metal(L)_n]^- \rightarrow [CH_3Metal(L)_n]^- + CO_2 \quad (6.13)$$

$$\rightarrow CH_3CO_2^- + [Metal(L)_n] \quad (6.14)$$

$$\rightarrow [Metal(L)_n]^- + CH_3CO_2 \quad (6.15)$$

$$\rightarrow [HOMetal(L)_n]^- + CH_2CO \quad (6.16)$$

$$\rightarrow [HCCOMetal\ L)_n]^- + H_2O \quad (6.17)$$

$$\rightarrow L^- + [CH_3CO_2Metal(L)_{n-1}] \quad (6.18)$$

Which of these reactions is favored will depend on the type of metal, the auxiliary ligand, L^-, and the oxidation state of the metal. While no comprehensive survey of main group and transition metal acetates with a range of auxiliary ligands has been carried out, some trends have emerged from studies on alkali and alkaline earth metals [63, 64], group VI oxometalate anions [65], and copper and silver species [66–68]:

- Alkali metal acetate ions $[Metal(O_2CCH_3)_2]^-$ (where Metal = lithium, sodium, potassium, rubidium, and cesium, $L = CH_3CO_2$, and $n = 1$) all fragment via loss of the acetate anion (Eq. (6.14)) [63] with virtually no formation of the organometalate. In contrast, the alkaline earth acetate ions (Metal = magnesium, calcium, strontium, and barium $L = CH_3CO_2$, and $n = 2$) not only fragment via loss of the acetate anion (Eq. (6.14)), but also all fragment to form the organometalates (Eq. (6.13)) [63].
- Oxometalate anions favor the ketene loss channel (Eq. (6.16)), with the selectivity being best for Mo and W [65].
- Decarboxylation (Eq. (6.13)) readily proceeds for the Cu(I) ion $[(CH_3CO_2)_2Cu]^-$, but the redox reaction (Eq. (6.15)) occurs for the Cu(II) ion $[(CH_3CO_2)_3Cu]^-$ [66].

The formation and reactivity of specific organometallic ions are discussed in the following sections.

6.2.1.1 Formation and Reactions of Organolithium Ions

As noted above, $[Li(O_2CCH_3)_2]^-$ mainly fragments via loss of the acetate anion (Eq. (6.14)) [63]. One way of overcoming this problem is to prepare precursor complexes which possess only one carboxylate ligand bound to Li. Since simple carboxylates such as the acetate anion posses a charge of -1, this means that the complex $Li(O_2CCH_3)$ would have a net charge of 0 and thus not be observable by mass spectrometry. This issue can be solved by engineering [52] the carboxylate ligand to either be a dianion, as shown by Hare et al. [69] (Scheme 6.1) or a zwitterion (Scheme 6.2) [70].

The product of the decarboxylation of betaine is formally a lithiated ylide (**1**, Scheme 6.2). This ylide, $[(CH_3)_3NCH_2Li]^+$, was shown to exhibit different

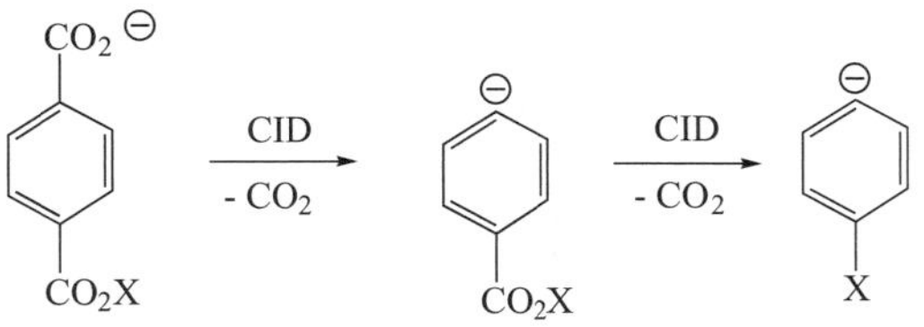

Scheme 6.1 Formation of organoalkalis (X = H, Na, K) by engineering the carboxylate ligand to be a dianion (see Ref. [69]).

reactivity from that of its amine isomer, $[(CH_3CH_2)(CH_3)_2NLi]^+$. Thus the ylide reacts with water via an acid-base reaction (Eq. (6.19)), while the amine reacts via adduct formation (Eq. (6.20)). Collisional activation of $[(CH_3CH_2)(CH_3)_2NLi]^+$ resulted in only loss of ion signal, while the ylide $[(CH_3)_3NCH_2Li]^+$ fragmented via loss of a methyl radical, LiMe, and ethane [70].

$$[(CH_3)_3NCH_2Li]^+ + H_2O \rightarrow (CH_3)_4N^+ + LiOH \tag{6.19}$$

$$[(CH_3CH_2)(CH_3)_2NLi]^+ + H_2O \rightarrow [(CH_3CH_2)(CH_3)_2NLi(OH_2)]^+ \tag{6.20}$$

6.2.1.2 Formation and Reactions of Alkaline Earth Organometalates

As noted above, alkaline earth acetate ions all fragment via decarboxylation (Eq. (6.13)), albeit to various extents. $[CH_3CO_2MgL_2]^-$ ions readily fragment under collisional activation conditions, leading to exclusive decarboxylation (Eq. (6.13)) when L = Cl but compete with acetate ion formation (Eq. (6.14)) when L = CH_3CO_2 [64]. These experimental observations are consistent with DFT calculations, which predict that the reactions of Eqs.(6.13) and (6.14) have similar endothermicities for L = CH_3CO_2 (47.7 kcal mol^{-1} and 51.7 kcal mol^{-1} respectively), but that acetate loss is considerably more endothermic for L = Cl (78.8 kcal mol^{-1} for reaction 13 versus 48.9 kcal mol^{-1} for reaction 14). DFT calculations also provide insights into the transition states for decarboxylation. In both cases these are four centered transition states in which the CH_3 group migrates to the Mg atom. All $[CH_3MetalL_2]^-$ (where Metal = Mg, Ca, Sr and Ba; L = CH_3CO_2) were successfully synthesized, but due to the low mass cut-off of the ion trap, it is not possible to comment on the relative yields of the acetate loss channel (Eq. (6.14)) [63]. DFT calculations, however, suggest that the reactions of Eqs.(6.13) and (6.14) have similar endothermicities for all these alkaline earth acetates.

The ability to synthesize organoalkaline earths $[CH_3MetalL_2]^-$ and study their ion-molecule reactions provides a unique opportunity to establish how reactivity is

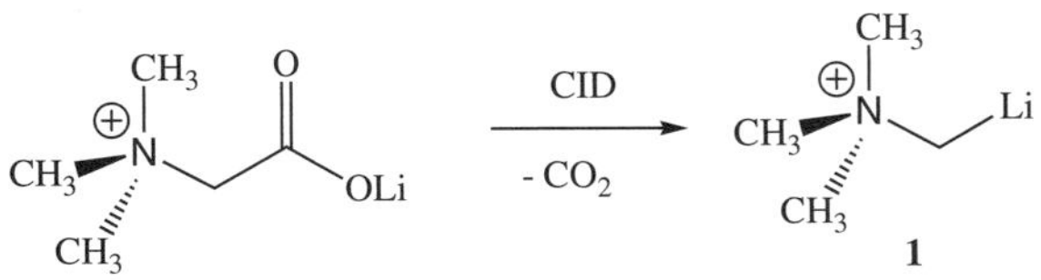

Scheme 6.2 Formation of organolithium by engineering the carboxylate ligand to be a zwitterion (see Ref. [70]).

controlled by the auxiliary ligand, the nature of the metal, and the substrate [63, 64]. $[CH_3MgL_2]^-$ (L = Cl and = O_2CCH_3) exhibits some of the reactivity of Grignard reagents, reacting with neutral species containing an acidic proton (AH) via addition with concomitant elimination of methane to form $[AMgL_2]^-$ ions (Eq. (6.21)). Kinetic measurements, combined with DFT calculations, provided clear evidence of an influence of the auxiliary ligand on reactivity of the organomagnesates $[CH_3MgL_2]^-$. Thus $[CH_3Mg(O_2CCH_3)_2]^-$ exhibited reduced reactivity toward water. The DFT calculations suggest that this may arise from the bidentate binding mode of acetate, which induces overcrowding of the Mg coordination sphere.

$$[CH_3MetalL_2]^- + AH \rightarrow [AMetalL_2]^- + CH_4 \quad (6.21)$$

The substrate also plays a key role in the reactivity of the $[CH_3MgL_2]^-$ ions. This is dramatically illustrated for the reaction of aldehydes containing enolizable protons, which reacted via enolization (Eq. (6.21)) rather than via the Grignard reaction (Eq. (6.22)). This is consistent with DFT calculations on the competition between enolization and the Grignard reaction for $[CH_3MgCl_2]^-$ ions reacting with acetaldehyde, which suggest that while the latter has a smaller barrier, it is entropically disfavored.

$$[CH_3MgL_2]^- + RCHO \rightarrow [RC(CH_3)OMgL_2]^- \quad (6.22)$$

Interestingly, this study provides an additional example of the role of the substrate in the reactions of $[CH_3MgL_2]^-$ (Eq. (6.21)). Thus, while acetaldehyde is much more acidic than water (by over 20 kcal mol^{-1}), it reacts much less readily with $[CH_3MgCl_2]^-$. This illustrates a dramatic impact of metal coordination on the effective acidity of the reagent, further confirmed by the DFT-calculated energetics, which suggest both a kinetic barrier and a thermodynamic effect (the overall enthalpy changes of the water and acetaldehyde reactions are essentially the same). Finally, when acetic acid is the substrate, the $[CH_3MgL_2]^-$ ions complete a two-step catalytic cycle for the decarboxylation of acetic acid (Eq. (6.23), Scheme 6.3) [64].

$$CH_3CO_2H \rightarrow CH_4 + CO_2 \quad (6.23)$$

Given that the acetate ligand appears to tame the reactivity of organomagnesates, in the next study this ligand was used to probe the reactivity of other organoalkaline

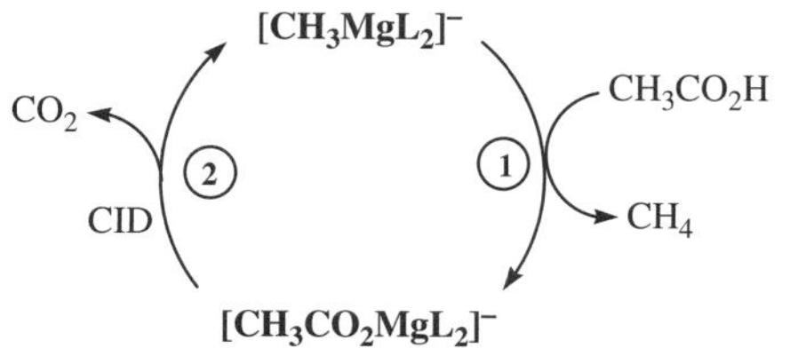

Scheme 6.3 Gas-phase catalytic cycles for the metal-mediated decarboxylation of acetic acid [64].

metalates [63]. Each of the organometalates [CH$_3$Metal(O$_2$CCH$_3$)$_2$]$^-$ react with water via addition with concomitant elimination of methane to form the metal hydroxide [HOMetal(O$_2$CCH$_3$)$_2$]$^-$ ions (Eq. (6.21)), with a relative reactivity order of: [CH$_3$Ba(O$_2$CCH$_3$)$_2$]$^-$ ≈ [CH$_3$Sr(O$_2$CCH$_3$)$_2$]$^-$ > [CH$_3$Ca(O$_2$CCH$_3$)$_2$]$^-$ > [CH$_3$Mg(O$_2$CCH$_3$)$_2$]$^-$. The DFT-predicted reaction exothermicities for these reactions generally supported the reaction trends observed experimentally, [CH$_3$Mg(O$_2$CCH$_3$)$_2$]$^-$ being the least reactive.

6.2.1.3 Formation and Reactions of Organocuprates and Organoargentates

Experiment and theory have been used to study the formation of organoargentates [67] and organocuprates [68] and to examine the reactivity of the dimethyl metalate ions (CH$_3$)$_2$Metal$^-$ (where Metal = Cu and Ag) with one of the simplest and most reactive alkyl halides, CH$_3$I [66]. These experiments require two steps of CID to promote double decarboxylation (Scheme 6.4).

In the first study, CID of [(CH$_3$CO$_2$)$_2$Ag]$^-$, 2, resulted in loss of CO$_2$ to yield [CH$_3$CO$_2$AgCH$_3$]$^-$, 3, as the only charged species (Eq. (6.13) where Metal = Ag, L = CH$_3$CO$_2$ and n = 1), with loss of acetate anion to yield neutral silver acetate, [AgO$_2$CCH$_3$] not being observed (Eq. (6.14) where Metal = Ag, L = CH$_3$CO$_2$ and n = 1). In contrast, CID of [CH$_3$CO$_2$AgCH$_3$]$^-$, 3, revealed the presence of two competing fragmentation pathways: loss of CO$_2$ to form [(CH$_3$)$_2$Ag]$^-$, 4, (Eq. (6.13) where Metal = Ag, L = CH$_3$ and n = 1), and the loss of acetate anion to yield neutral methylsilver [AgCH$_3$] (Eq. (6.14) where Metal = Ag, L = CH$_3$, and n = 1). These results are consistent with the DFT-predicted thermochemistry, which suggests that Eq. (6.13) is preferred over Eq. (6.14) for the first stage of CID on [(CH$_3$CO$_2$)$_2$Ag]$^-$ (+24.3 kcal mol^{-1} versus +51.4 kcal mol^{-1}), but that both equations are energetically similar for the second stage of CID on [CH$_3$CO$_2$AgCH$_3$]$^-$ (+36.5 kcal mol^{-1} for Eq. (6.13) versus +33.9 kcal mol^{-1} for Eq. (6.14).

The most recent study has used a combination of experiment and theory to examine the general applicability of the double decarboxylation strategy as a means of 'synthesizing' a wide array of organocuprates [68]. Thus, the homocuprates [(Me)$_2$Cu]$^-$ and [(Et)$_2$Cu]$^-$ were generated in the gas phase by double decarboxylation of the copper carboxylate centers [(MeCO$_2$)$_2$Cu]$^-$ and [(EtCO$_2$)$_2$Cu]$^-$, respectively. The same strategy was explored for generating the heterocuprates [MeCuR]$^-$ from [MeCO$_2$CuO$_2$CR]$^-$ (R = Et, Pr, iPr, tBu, allyl, benzyl, Ph). A number of side reactions were observed to be in competition with the second stage of decarboxylation, as discussed further below. Detailed DFT calculations were carried out on the potential energy surfaces for the first and second decarboxylation reactions of all homo- and heterocuprates, as well as possible competing reactions.

$$(CH_3CO_2)_2\text{Metal}^{\ominus} \xrightarrow[-CO_2]{\text{CID (MS}^2)} CH_3CO_2\text{MetalCH}_3^{\ominus} \xrightarrow[-CO_2]{\text{CID (MS}^3)} (CH_3)_2\text{Metal}^{\ominus}$$
$$\quad\quad\quad 2 \quad\quad\quad\quad\quad\quad\quad\quad\quad\quad 3 \quad\quad\quad\quad\quad\quad\quad\quad\quad\quad 4$$

Scheme 6.4 Double decarboxylation strategy to form dimethylmetalates via two stages of CID. Metal = Ag and Cu.

CID of $[(CH_3CO_2)_2Cu]^-$, **2**, resulted in loss of CO_2 to yield $[CH_3CO_2CuCH_3]^-$, **3**, as the only charged species (Eq. (6.13) where Metal = Cu, L = CH_3CO_2, and n = 1), with no observation of loss of acetate anion to yield neutral copper acetate $[CuO_2CCH_3]$ (Eq. (6.14) where Metal = Cu, L = CH_3CO_2 and n = 1). In contrast, CID of $[CH_3CO_2CuCH_3]^-$, **3**, revealed the presence of two competing fragmentation pathways: loss of CO_2 to form $[(CH_3)_2Cu]^-$, **4**, (Eq. (6.13) where Metal = Cu, L = CH_3 and n = 1), and loss of acetate anion to yield neutral methylcopper $[CuCH_3]$ (Eq. (6.14) where Metal = Cu, L = CH_3, and n = 1). These results are consistent with DFT calculations. Thus CID on $[(CH_3CO_2)_2Cu]^-$, **2**, is predicted to have a decarboxylation (Eq. (6.13)) transition state energy of +38.5 kcal mol^{-1} (with a final endothermicity of 24.2 kcal mol^{-1}), while loss of the acetate anion (Eq. (6.14)) is endothermic by 61.8 kcal mol^{-1}. The transition state energy for decarboxylation (Eq. (6.13)) and the reaction endothermicity for acetate anion loss (Eq. (6.14)) are closer for CID on $[CH_3CO_2CuCH_3]^-$, **3**, being 38.5 and 42.9 kcal mol^{-1} respectively.

The double decarboxylation strategy for the synthesis of higher homologs of $[(Me)_2Cu]^-$ suffers from additional fragmentation pathways in the second stage of CID, which involve the alkyl ligand. Thus CID of $[EtCO_2CuEt]^-$ proceeds via three competing channels: (i) loss of CO_2 to form $[(Et)_2Cu]^-$ (Eq. (6.13) where Metal = Cu, L = Et, and n = 1), loss of propionate anion to yield neutral ethylcopper [CuEt] (Eq. (6.14) where Metal = Cu, L = Et and n = 1), and loss of C_2H_4 to form $[EtCO_2CuH]^-$, which occurs via a β-hydride elimination reaction (Eq. (6.24)). Support for the β-hydride mechanism came from two different deuterium-labeling experiments: (i) CID of $[CH_3CD_2CO_2CuCD_2CH_3]^-$ yielded $[CH_3CD_2CO_2CuH]^-$, (Eq. (6.25)), and (ii) CID of $[CD_3CH_2CO_2CuCH_2CD_3]^-$ yielded $[CD_3CH_2CO_2CuD]^-$ (Eq. (6.25)). Interestingly, this β-hydride elimination reaction is directly related to the thermal decomposition reaction of $Bu_3PCuCH_2CD_2CH_2CH_3$ studied by Whitesides over 30 years ago (Eq. (6.27)) [71].

$$[EtCO_2CuEt]^- \rightarrow [EtCO_2CuH]^- + C_2H_4 \quad (6.24)$$

$$[CH_3CD_2CO_2CuCD_2CH_3]^- \rightarrow [CH_3CD_2CO_2CuH]^- + CD_2=CD_2 \quad (6.25)$$

$$[CD_3CH_2CO_2CuCH_2CD_3]^- \rightarrow [CD_3CH_2CO_2CuD]^- + CH_2=CD_2 \quad (6.26)$$

$$Bu_3PCuCH_2CD_2CH_2CH_3 \rightarrow Bu_3PCuD + CH_2=CDCH_2CH_3 \quad (6.27)$$

There are four possible competing fragmentation pathways for the heterocarboxylate systems $[MeCO_2CuO_2CR]^-$: decarboxylation from either of the two inequivalent carboxylate groups respectively (Eqs. (6.28) and (6.29)), or loss of either of the corresponding carboxylate anions $MeCO_2^-$ or RCO_2^- to yield neutral CuO_2CR or CuO_2CMe, respectively (Eqs. (6.30) and (6.31)). While loss of the carboxylate anions (Eqs. (6.30) and (6.31)) can be distinguished due to their different m/z values, the decarboxylation reactions produce isomeric species $[MeCuO_2CR]^-$ and/or $[MeCO_2CuR]^-$ of the same m/z value. In order to distinguish between these isomers, a reactive probe was developed. Thus, the product(s) of decarboxylation were allowed

Table 6.1 DFT-predicted transition state energies for competing decarboxylation (Eqs. (6.28) and (6.29)) reactions for $[MeCO_2CuO_2CR]^-$ versus experimental products as determined via ion-molecule reactions with allyl iodide (Eqs. (6.32) and (6.33)).

R =	TS energy for $[MeCuO_2CR]^-$ + CO_2 (Eq. (6.28))[a]	Is $[ICuO_2CR]^-$ formed (Eq. (6.32))?	TS energy for $[MeCO_2CuR]^-$ + CO_2 (Eq. (6.29))[a]	Is $[MeCO_2CuI]^-$ formed (Eq. (6.33))?
Et	38.5	Yes, major	40.4	Yes, minor
Pr	38.7	Yes, major	40.1	Yes, minor
iPr	38.7	Yes, major	42.9	Yes, minor
tBu	38.5	Yes, major	47.3	No
Allyl	38.7	No	37.8	Yes, major
Benzyl	38.7	Yes, minor	36.7	Yes, major
Ph	38.7	Yes, minor	35.3	Yes, major

[a] In kcal mol^{-1}.

to undergo C—C bond coupling reactions with allyliodide. Isomer $[MeCuO_2CR]^-$ reacts to yield $[ICuO_2CR]^-$ (Eq. (6.32)), while isomer $[MeCO_2CuR]^-$ produces $[MeCO_2CuI]^-$ instead (Eq. (6.33)). Since $[ICuO_2CR]^-$ and $[MeCO_2CuI]^-$ have different m/z values, they can be used to probe the populations of the different $[MeCuO_2CR]^-$ and $[MeCO_2CuR]^-$ isomers formed upon CID (Eqs. (6.28) and (6.29)). The experimental results were checked via the use of DFT calculations on the potential energy surfaces associated with Eqs. (6.28)–(6.31). Decarboxylation is always favored over loss of the carboxylate anions. Table 6.1 highlights the agreement between experiment and theory for the sites of decarboxylation of $[MeCO_2CuO_2CR]^-$. In all cases, the lowest transition state energy for decarboxylation corresponds to the formation of the major product. Furthermore, some interesting trends emerge. For example, when R = an aliphatic alkyl group, the transition state energy follows the steric demand of the alkyl group: Et < iPr < tBu. Decarboxylation at the RCO_2^- ligand of $[MeCO_2CuO_2CR]^-$ was preferred when R = Ph, benzyl and allyl, indicating that the site of decarboxylation was dependent on additional factors such as Cu—C bond strengths.

$$[MeCO_2CuO_2CR]^- \rightarrow [MeCuO_2CR]^- + CO_2 \quad (6.28)$$

$$\rightarrow [MeCO_2CuR]^- + CO_2 \quad (6.29)$$

$$\rightarrow MeCO_2^- + CuO_2CR \quad (6.30)$$

$$\rightarrow MeCO_2Cu + RCO_2^- \quad (6.31)$$

$$[MeCuO_2CR]^- + C_3H_5I \rightarrow [ICuO_2CR]^- + MeC_3H_5 \quad (6.32)$$

$$[MeCO_2CuR]^- + C_3H_5I \rightarrow [MeCO_2CuI]^- + RC_3H_5 \quad (6.33)$$

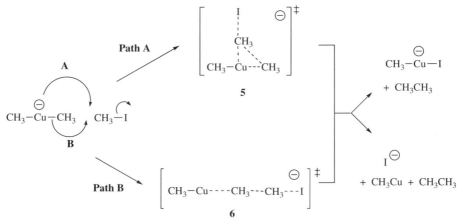

Scheme 6.5 Competing mechanisms for C–C bond coupling [66].

The dimethylcuprate ion (M = Cu) reacts with CH_3I via C–C bond cross coupling (Eqs. (6.34) and (6.35)) at a modest rate (corresponding to a reaction efficiency of 3 out of every 100 collisions) [66]. In contrast, its silver congener is unreactive. A comparison of the rate constants for the reaction of $(CH_3)_2Cu^-$ with CH_3I and CD_3I revealed that breaking the C–I bond involves an inverse isotope effect ($k_H/k_D = 0.88 \pm 0.40$). Comparison of the ion abundances of CH_3CuI^- and CD_3CuI^- in the reaction of $CD_3CuCH_3^-$ with CH_3I revealed an inverse β-deuterium isotope effect ($k_H/k_D = 0.82 \pm 0.05$) for the Cu–C bond. That two isotope effects are operating in the cross-coupling reaction was further confirmed by the combined isotope effect of $k_H/k_D = 0.66 \pm 0.05$ determined from a comparison of the ion abundances of CH_3CuI^- and CD_3CuI^- in the reactions of $CD_3CuCH_3^-$ with CD_3I.

$$[(CH_3)_2Cu]^- + CH_3I \rightarrow [CH_3CuI]^- + CH_3CH_3 \tag{6.34}$$

$$\rightarrow I^- + CH_3Cu + CH_3CH_3 \tag{6.35}$$

The experimental results are consistent with MP2/6–31++G** (with ECPs for the metal) *ab initio* calculations on two different mechanisms (Scheme 6.5) for cross coupling of $(CH_3)_2Metal^-$ (where Metal = Cu and Ag) with CH_3I. Path A involves the formation of a 'T-shaped' transition state, **5**, while Path B involves a 'side-on' S_N2 transition state, **6**. The *ab initio* calculations reveal that the former pathway has the lower barrier, with an energy below that of the separated reactants. In contrast, the transition state energies for both pathways for the silver congener are above the energy-separated reactants. This is consistent with the experimental observation that $(CH_3)_2Cu^-$ reacts with CH_3I but $(CH_3)_2Ag^-$ does not.

6.2.1.4 Formation of Metal Carbenes

Julian *et al.* have examined the gas-phase 'synthesis' of copper(I) and silver(I) Fischer carbenes via CID of metal complexes of diazomalonates, **7** (Scheme 6.6, where L = acetonitrile). The initial loss of N_2 generates a metastable Fischer carbene

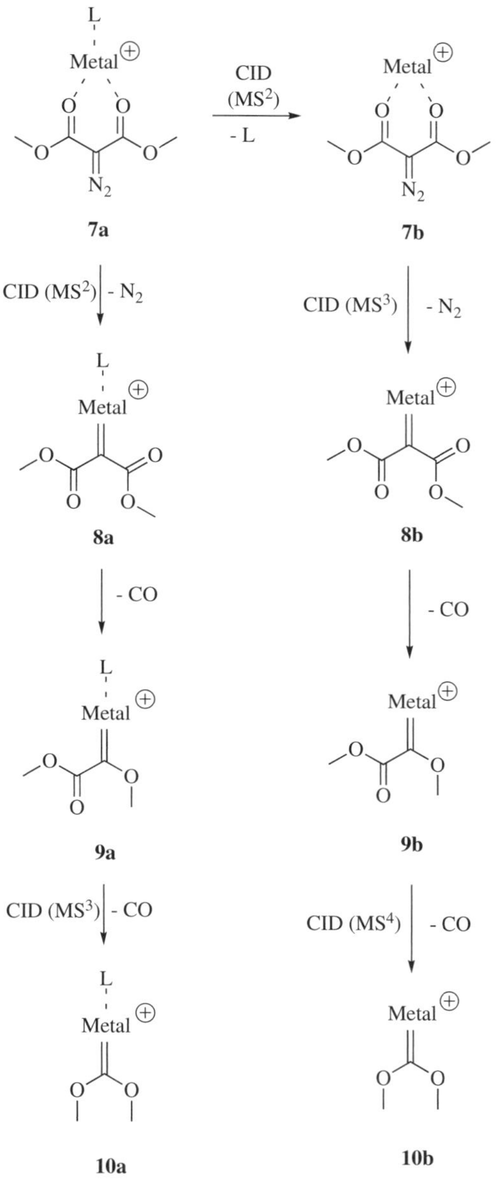

Scheme 6.6 Fischer carbene formation [72].

complex, **8**, which subsequently undergoes a Wolff rearrangement with loss of CO to give another Fischer carbene complex, **9**. Another stage of CID leads to the loss of a second CO molecule and the generation of a stable Fischer carbene, **10** [72]. There are subtle differences in the behavior of the silver and copper complexes. In the case of copper, the first stage of CID proceeds via loss of N_2 and CO to yield the overall sequence: **7a → 8a → 9a → 10a**. In contrast, for the silver complex, loss of the acetonitrile ligand precedes the losses of N_2 and CO, to yield the overall sequence: **7a → 7b → 8b → 9b → 10b**. Finally, the oxidation state of the metal appears to be crucial, with Cu(II) and Ni(II) complexes failing to undergo Fischer carbene formation.

6.2.2
Formation and Reactivity of Metal Hydrides (Eq. (6.12), Z = H)

6.2.2.1 Mononuclear Metal Hydrides

Early gas-phase studies of metal hydrides have been reviewed [73]. Metal hydrides can be formed via CID on a number of different precursor metal complexes via fragmentation of the following types of coordinated ligands: formate; alkoxide; and alkyl (recall Eq. (6.24) above). Magnesium hydride anions $HMgL_2^-$ (L = Cl and HCO_2) can be formed via CID on the formate precursors $HCO_2MgL_2^-$ (Eq. (6.36)) [74]. These undergo acid-base reactions with neutral acids (Eq. (6.37)). Note that when formic acid is used, the original formate, $HCO_2MgL_2^-$ is regenerated, thus closing a catalytic cycle for the decarboxylation of formic acid (Eq. (6.38)), which is related to that for the decarboxylation of acetic acid (Eq. (6.23), Scheme 6.3).

$$HCO_2MgL_2^- \to HMgL_2^- + CO_2 \qquad (6.36)$$

$$HMgL_2^- + AH \to AMgL_2^- + H_2 \quad (A=HO, CH_3O \text{ and } HCO_2) \qquad (6.37)$$

$$HCO_2H \to H_2 + CO_2 \qquad (6.38)$$

As noted in the introduction, metal hydrides can be formed via CID on metal alkoxides (Eq. (6.8)). Metal alkoxides generated via ESI also fragment to form metal hydrides as demonstrated by a number of studies [75, 76]. When there are other ligands present in the metal alkoxide, alternative pathways can also occur [77]. In fact, in some instances metal hydride formation may be undesirable. For example, in a study on the catalytic oxidation of methanol to formaldehyde (Eq. (6.39), Scheme 6.7) mediated by the molybdenum oxide anion, $[Mo_2O_6(OH)]^-$, the key CID step which liberates formaldehyde from the methoxide ligand produces two isomeric products in a ratio of 80% isomer **A** to 20% isomer **B** [77]. Isomer **A** is a reduced metal species with a hydroxide ligand, which can be reoxidised to the original catalyst (step 3 of Scheme 6.7), thereby closing the catalytic cycle. In contrast, formation of isomer **B**, which is the metal hydride, 'poisons' the catalyst since it is unreactive toward either methanol or the oxidant.

$$CH_3OH + CH_3NO_2 \to CH_2O + H_2O + CH_3NO \qquad (6.39)$$

Schwarz's group has employed in-source CID to generate a range of metal hydrides for reactivity studies [78–80]. For example, the influence of additional ligands (L = CO, H_2O, HO) on the reactivity of nickel hydride cluster cations, NiH^+ toward methane has been explored (Eq. (6.40)) [78]. All systems react with methane via a metathesis reaction (Eq. (6.40)), albeit at a slower rate than NiH^+. A related

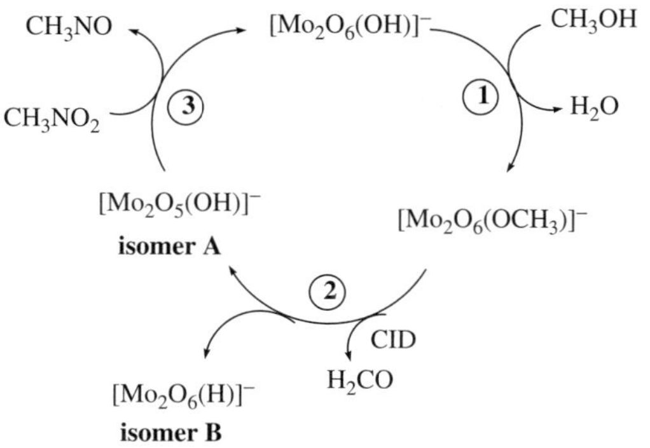

Scheme 6.7 Gas-phase catalytic cycles for the oxidation of methanol to formaldehyde. Reaction 2 is a key CID reaction of the methoxide ligand, which produces two isomeric products [77].

metathesis reaction has been observed for PdH^+ (Eq. (6.41)) [79].

$$[Ni(H)L]^+ + CH_4 \rightarrow [Ni(CH_3)L]^+ + H_2 \qquad (6.40)$$

$$PdH^+ + CH_4 \rightarrow PdCH_3^+ + H_2 \qquad (6.41)$$

In a recent study, the reactions of $[Metal(H)]^+$, $[Metal(H)(H_2O)]^+$, and $[Metal(H)(OH)]^+$ with molecular oxygen have been examined [80]. The bare hydrides (in fact deuterides were used to avoid issues of isobaric impurities) react sluggishly with molecular oxygen, with H transfer (Eq. (6.42)) competing with O–O bond activation (Eqs. (6.43) and (6.44)) for all three metals examined (Fe, Co and Ni). Solvation by a water molecule enhances the reactivity and also leads to products arising from H transfer (Eq. (6.45), only for Metal = Ni), formal ligand displacement (Eq. (6.46) for all three metals, Fe, Co and Ni), and O–O bond activation (Eqs. (6.47) and (6.48)) for all three metal examined (Fe, Co and Ni). Introduction of a hydroxide ligand changes the formal oxidation state of the metal center and also enhances the reactivity of the metal hydride, with H transfer becoming a minor channel for Ni only (Eq. (6.49)) and O–O bond activation (Eqs. (6.50)–(6.52)) dominating for all three metals examined (Fe, Co and Ni). Deuterium and ^{18}O labeling studies reveal that the mechanisms of these reactions are complex, but point to the key involvement of metal peroxides via insertion of O_2 into the Metal–H bond.

$$[Metal(D)]^+ + O_2 \rightarrow [Metal]^+ + DO_2 \qquad (6.42)$$

$$\rightarrow [Metal(O)]^+ + DO \qquad (6.43)$$

$$\rightarrow [Metal, D, O]^+ + O \qquad (6.44)$$

$$[Metal(H)(H_2O)]^+ + O_2 \rightarrow [Metal, H_2, O]^+ + HO_2 \qquad (6.45)$$

$$\rightarrow [Metal, H, O_2]^+ + H_2O \qquad (6.46)$$

$$\rightarrow [\text{Metal}, H_2, O_2]^+ + HO \qquad (6.47)$$

$$\rightarrow [\text{Metal}, H_3, O_2]^+ + O \qquad (6.48)$$

$$[\text{Metal}(H)(OH)]^+ + O_2 \rightarrow [\text{Metal}, H, O]^+ + HO_2 \qquad (6.49)$$

$$\rightarrow [\text{Metal}, O_2]^+ + H_2O \qquad (6.50)$$

$$\rightarrow [\text{Metal}, H, O_2]^+ + HO \qquad (6.51)$$

$$\rightarrow [\text{Metal}, H_2, O_2]^+ + O \qquad (6.52)$$

6.2.2.2 Multinuclear Metal Hydrides

The Ag_2H^+ and Ag_4H^+ cluster cations can be 'synthesized' via multiple stages of CID of silver-amino acid clusters [81, 82]. Since the mechanisms of these reactions are complex, only the pathways to the formation of these clusters are shown in Scheme 6.8.

The ion-molecule reactions of these silver hydrides have been examined with a number of neutral substrates and exhibit a number of bond activation/bond formation reactions of relevance to the chemistry of silver surfaces [83] and nanoparticles [84]. In the first study, the ion-molecule reaction of Ag_4H^+ was found to mediate C—C bond coupling with allyl bromide, CH_2=$CHCH_2Br$ [82], a reaction that mimics such processes on silver surfaces and nanoparticles (Eq. (6.53)). The gas-phase process is a multi-step reaction that proceeds via the following sequence: allyl bromide reacts with Ag_4H^+ via a metathesis reaction to yield Ag_4Br^+ (Eq. (6.54)), which subsequently reacts with a second molecule of allyl bromide to form the ion $Ag_4Br_2(C_3H_5)^+$ (Eq. (6.55)), which in turn reacts with a third molecule of allyl bromide to form the silver organometallic ion $Ag(C_3H_5)_2^+$ in combination with the

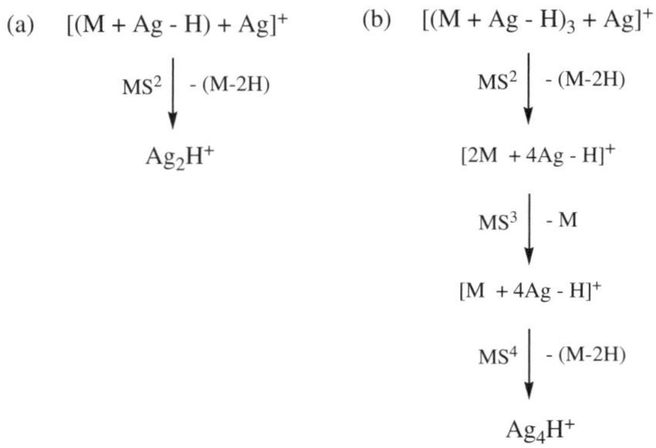

(a) $[(M + Ag - H) + Ag]^+$

$MS^2 \downarrow -(M-2H)$

Ag_2H^+

(b) $[(M + Ag - H)_3 + Ag]^+$

$MS^2 \downarrow -(M-2H)$

$[2M + 4Ag - H]^+$

$MS^3 \downarrow -M$

$[M + 4Ag - H]^+$

$MS^4 \downarrow -(M-2H)$

Ag_4H^+

Scheme 6.8 Fragmentation of amino acid clusters $[(M + Ag - H)_n + Ag]^+$ to form: (a) Ag_2H^+ (where M = glycine), and (b) Ag_4H^+ (where M = N,N-dimethylglycine).

neutral cluster Ag_3Br_3 (Eq. (6.56)). Overall, these equations combine to give the C−C bond-coupling reaction (Eq. (6.57)). CID of $Ag(C_3H_5)_2^+$ results in the sole formation of Ag^+ (Eq. (6.58)), providing supporting evidence that C−C bond coupling has fioccurred. Furthermore, the identity of the C−C bond-coupled C_6H_{10} product was shown to be 1,5-hexadiene via a comparison of the energy-resolved CID spectrum of the $Ag(C_6H_{10})^+$ ion with a range of ions of 'authentic' structures.

$$Ag_n + nCH_2=CHCH_2Br \rightarrow (AgBr)_2 + n/2(CH_2=CHCH_2)_2 \quad (6.53)$$

$$Ag_4H^+ + CH_2=CHCH_2Br \rightarrow Ag_4Br^+ + CH_2=CHCH_3 \quad (6.54)$$

$$Ag_4Br^+ + CH_2=CHCH_2Br \rightarrow Ag_4Br_2(C_3H_5)^+ \quad (6.55)$$

$$Ag_4Br_2(C_3H_5)^+ + CH_2=CHCH_2Br \rightarrow Ag(C_3H_5)_2^+ + Ag_3Br_3 \quad (6.56)$$

$$Ag_4H^+ + 3CH_2=CHCH_2Br \rightarrow Ag(C_3H_5)_2 + Ag_3Br_3 + CH_2=CHCH_3 \quad (6.57)$$

$$Ag(C_3H_5)_2^+ \rightarrow Ag^+ + (C_3H_5)_2 \quad (6.58)$$

In the second study, the reactions of the Ag_2H^+ and Ag_4H^+ clusters with 2-propanol and 2-butylamine were examined [81]. Both hydride clusters only undergo sequential ligation reactions with 2-propanol. In contrast, the clusters react with 2-butylamine via silver hydride displacement (Eq. (6.59)), sequential ligation (Eq. (6.60)), C−H, and C−N bond activation after sufficient ligands have been added to the cluster (Eq. (6.61)). For Ag_2H^+, two amine ligands are added before the third undergoes C−H and C−N bond activation (Eq. (6.61), n = 2, x = 2), whereas for Ag_4H^+ three amine ligands are added before the fourth undergoes C−H and C−N bond activation (n = 4, x = 3).

$$Ag_nH^+ + L \rightarrow Ag_{n-1}(L)^+ + AgH \quad (6.59)$$

$$Ag_nH^+ + xL \rightarrow Ag_nH(L)_x^+ \quad (6.60)$$

$$Ag_nH(L)_x^+ + CH_3CH_2CH(CH_3)NH_2 \rightarrow Ag_nH(C_4H_6)(L)_x^+ + NH_3 + H_2 \quad (6.61)$$

Two subsequent studies have examined the Ag_2H^+ and Ag_4H^+ mediated C−I bond activation reactions of 2-iodoethanol, ICH_2CH_2OH [85], and allyliodide, $CH_2=CHCH_2I$ [86]. Ag_2H^+ and Ag_4H^+ solely undergo ligand addition reactions with 2-iodoethanol (Eq. (6.60)), which is in stark contrast to the combined C−I and C−OH bond activation reaction observed for Ag_5^+ (Eq. (6.62)) [85]. Ag_2H^+ reacts with allyliodide via the metathesis reaction shown in Eq. (6.63) to yield Ag_2I^+ ion [86]. Interestingly, the related metathesis reaction is not observed for Ag_2H^+ reacting with ICH_2CH_2OH [85], highlighting the importance of the substrate structure. Ag_4H^+ yields a rich spectrum arising from multiple reactions of allyliodide

(Eqs. (6.64)–(6.68)). The primary product is Ag_4I^+, which arises from a metathesis reaction (Eq. (6.64)), which is related to reactions for the formation of Ag_4Br^+ (Eq. (6.54)) and Ag_2I^+ (Eq. (6.63)). Ag_4I^+ undergoes the secondary reactions shown in Eqs. (6.65) and (6.66). Each of these secondary products also reacts further with allyliodide to give the products shown in Eqs. (6.67) and (6.68). The following silver iodide clusters are formed in the sequence of reactions: (i) the stoichiometric silver iodide cluster $Ag_4I_3^+$ (Eqs. (6.64) → (6.65) → (6.67) and Eqs. (6.64) → (6.66) → (6.68)), and (ii) the non-stoichiometric silver iodide clusters Ag_4I^+ (Eq. (6.64)) and $Ag_4I_2^+$ (Eqs. (6.64) → (6.65)). At longer reaction time, $Ag_4I_3^+$ is further ligated, and some minor, smaller silver fragments are observed.

$$Ag_5^+ + ICH_2CH_2OH \rightarrow Ag_5HIO^+ + C_2H_4 \tag{6.62}$$

$$Ag_2H^+ + CH_2=CHCH_2I \rightarrow Ag_2I^+ + CH_2=CHCH_3 \tag{6.63}$$

$$Ag_4H^+ + CH_2=CHCH_2I \rightarrow Ag_4I^+ + CH_2=CHCH_3 \tag{6.64}$$

$$Ag_4I^+ + CH_2=CHCH_2I \rightarrow Ag_4I_2^+ + CH_2=CHCH_2^\bullet \tag{6.65}$$

$$Ag_4I^+ + CH_2=CHCH_2I \rightarrow Ag_4I_2(CH_2=CHCH_2)^+ \tag{6.66}$$

$$Ag_4I_2^+ + CH_2=CHCH_2I \rightarrow Ag_4I_3^+ + CH_2=CHCH_2^\bullet \tag{6.67}$$

$$Ag_4I_2(CH_2=CHCH_2)^+ + CH_2=CHCH_2I \rightarrow Ag_4I_3^+ + (CH_2=CHCH_2)_2 \tag{6.68}$$

Finally, the related gold hydride clusters $(R_3PAu)_2H^+$ (R = Me and Ph) have been formed by CID of an amino acid cluster (Eq. (6.69)) [87]. Unlike the 'bare' Ag_2H^+, $(R_3PAu)_2H^+$ is unreactive with a range of neutral molecules (H_2O, MeOH, 2-propanol, MeCN, pyridine, allyl iodide, O_2, N_2O, CH_3NO_2, DMSO). When subjected to CID, $(R_3PAu)_2H^+$ does not undergo either ligand loss (Eq. (6.70)) or cleavage of the cluster via loss of R_3PAuH (Eq. (6.71)), but rather undergoes cluster fragmentation with migration of a phosphine ligand to lose AuH (Eq. (6.72)). DFT calculations on a simple model system (R = H) reveal that the barrier to ligand migration and AuH loss (Eq. (6.72)) is below the endothermicities for the other competing reactions, consistent with the experimental results.

$$[(Me_2NCH_2CO_2) + 2AuPR_3]^+ \rightarrow (R_3PAu)_2H^+ + [H_7, C_4, N, O_2] \tag{6.69}$$

$$(R_3PAu)_2H^+ \rightarrow Au_2H(PR_3)^+ + PR_3 \tag{6.70}$$

$$\rightarrow Au(PR_3)^+ + R_3PAuH \tag{6.71}$$

$$\rightarrow Au(PR_3)_2^+ + AuH \tag{6.72}$$

6.2.3
Formation and Reactivity of Metal Oxides (Eq. (6.11), X = O; Eq. (6.12), Z = O)

The gas phase-chemistry of metal oxide ions has been the subject of several studies, and this area has been reviewed [16, 17, 88]. The focus here is on the 'synthesis' and reactions of metal oxides via heterolysis or homolysis of coordinated ligands with X—O bonds.

6.2.3.1 Bond Heterolysis

One of the first studies to have combined the use of ESI and CID as a means of generating transition metal oxide coordination complexes involved the use of CO_2 loss from coordinated carbonato ligands. In these studies three different metal complexes with the auxiliary ligands **11**, **12** and **13** from Scheme 6.9 were

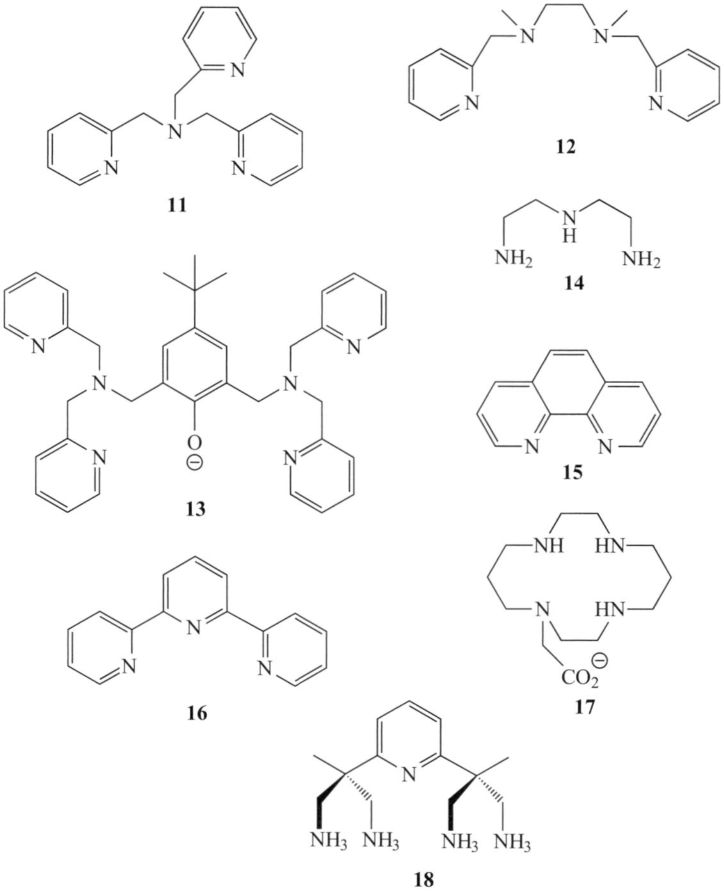

Scheme 6.9 Auxiliary ligands used in metal complexes used to generate metal oxides and nitrenes.

used. Both mononuclear and binuclear complexes underwent decarboxylation (Eqs. (6.73)–(6.75)).

$$[(L = 11)CoCO_3]^+ \rightarrow [(L = 11)CoCO_3]^+ + CO_2 \tag{6.73}$$

$$[((L = 12)Fe)_2(O)CO_3]^+ \rightarrow [((L+12)Fe)_2(O)_2]^+ + CO_2 \tag{6.74}$$

$$[((L = 13)Cu)_2(O)CO_3]^+ \rightarrow [((L = 13)Cu)_2(O)_2]^+ + CO_2 \tag{6.75}$$

CID of oxides coordinated to manganese(salen), [(salen)Mn(OX)]$^+$, provides a way of generating oxomanganese-salen complexes (Eq. (6.76)). This reaction is in competition with loss of the oxide, XO (Eq. (6.77)) [89–91]. The relative yields of these two fragmentation channels depend on the nature of the oxide, XO. Iodosobenzene mainly gives the oxomanganese-salen complex (Eq. (6.76)), pyridine-N-oxide mainly gives ligand loss (Eq. (6.77)), while pCN-C$_6$H$_4$NMe$_2$O gives both products. The oxomanganese-salen complexes undergo gas phase O atom transfer reactions with 2,3-dihydrofuran and sulfides (Eq. (6.78)) [92].

$$[(salen)Mn(OX)]^+ \rightarrow [(salen)Mn(O)]^+ + X \tag{6.76}$$

$$\rightarrow [(salen)Mn]^+ + XO \tag{6.77}$$

$$[(salen)Mn(O)]^+ + X \rightarrow [(salen)Mn]^+ + XO \tag{6.78}$$

The lithium oxide anion has been synthesized in the gas phase via two stages of CID on the lithium oxalate anion [93]. In the first stage, decarboxylation occurs to give the lithium salt of doubly deprotonated formic acid (Eq. (6.79)). In the second stage, decarbonylation occurs to give the lithium oxide anion (Eq. (6.80)). The lithium oxide anion is an extremely reactive species, which reacts with background water and oxygen via proton (Eq. (6.81)) and electron transfer (Eq. (6.82)) respectively. High level theoretical calculations suggest that LiO$^-$ is the strongest gas phase anionic base, stronger even than the bare methyl anion.

$$[LiO_2CCO_2]^- \rightarrow [LiO_2C]^- + CO_2 \tag{6.79}$$

$$[LiO_2C]^- \rightarrow LiO^- + CO \tag{6.80}$$

$$LiO^- + H_2O \rightarrow HO^- + LiOH \tag{6.81}$$

$$LiO^- + O_2 \rightarrow O_2^{-\bullet} + LiO^{\bullet} \tag{6.82}$$

6.2.3.2 Bond Homolysis of Metal Nitrites and Nitrates

One of the first studies to demonstrate bond homolysis involved an examination of the CID reactions of [Pt(L)NO$_2$]$^+$, where L = dien, 4 (Scheme 6.8) [94]. Ligand fragmentation via loss of NO (Eq. (6.83)) was observed to dominate over loss of the protonated ligand (Eq. (6.84)).

$$[\text{Pt(dien)NO}_2]^+ \rightarrow [\text{Pt(dien)O}]^+ + \text{NO} \tag{6.83}$$

$$\rightarrow [\text{Pt(dien)-H}]^+ + \text{HNO}_2 \tag{6.84}$$

Metal nitrates readily undergo bond homolysis to form metal oxides. A survey of the CID reactions of metal nitrate anions, $[\text{Metal}^x(\text{NO}_3)_{x+1}]^-$, with different oxidation states (x = 1–3) revealed the following competing fragmentation pathways: NO_2 loss to form a metal oxide (Eq. (6.85)), NO_3 loss with reduction of the metal center (Eq. (6.86)), formation of the nitrate anion, NO_3^- (Eq. (6.87)), and loss of two NO_2 neutrals (Eq. (6.88)). An examination of Table 6.2 reveals that the types of reactions observed depend on both the metal and its oxidation state. For the two metals (Cu and Fe) that were studied in two different oxidation states, CID of the higher oxidation state results in NO_3 loss with reduction of the metal to the lower oxidation state (Eq. (6.86)). This reaction does not occur for the lower oxidation state, which instead gives N–O bond homolysis via loss of NO_2 (Eq. (6.85)).

$$[\text{Metal}^x(\text{NO}_3)_{x+1}]^- \rightarrow [\text{Metal(O)}(\text{NO}_3)_x]^- + NO_2 \tag{6.85}$$

$$\rightarrow [\text{Metal}^{x-1}(\text{NO}_3)_x]^- + NO_3 \tag{6.86}$$

$$\rightarrow NO_3^- + [\text{Metal}^x(\text{NO}_3)_x] \tag{6.87}$$

$$\rightarrow [\text{Metal(O)}_2(\text{NO}_3)_{x-1}]^- + 2NO_2 \tag{6.88}$$

Table 6.2 Competing fragmentation reactions of metal nitrate anions $[\text{Metal}^x(\text{NO}_3)_{x+1}]^-$.

Metal =	x =	NO_2 loss observed (Eq. (6.85))?	NO_3 loss observed (Eq. (6.86))?	NO_3^- formed (Eq. (6.87))?	Eq. (6.88) observed?
Cu	1	yes	no	yes	no
Cu	2	no	yes	yes	no
Mn	2	yes	no	yes	yes
Co	2	yes	no	yes	no
Ni	2	yes	no	yes	no
Zn	2	yes	no	yes	no
Fe	2	yes	no	yes	yes
Fe	3	yes	yes	yes	yes
Cr	3	yes	no	yes	yes
Sc	3	yes	no	yes	yes
Y	3	yes	no	yes	yes
Al	3	yes	no	yes	yes
Ga	3	yes	yes	yes	no
In	3	yes	yes	yes	no

Although the ion-molecule reactions of the metal-oxo anions were not examined, the CID reactions of $[Fe(O)(NO_3)_3]^-$ and $[Fe_2(O)(NO_3)_5]^-$ were examined. Both of these anions undergo multiple losses of NO_2 (Eqs. (6.89)–(6.93)), representing access to anions with multiple oxo ligands.

$$[Fe(O)(NO_3)_3]^- \rightarrow [Fe(O)_2(NO_3)_2]^- + NO_2 \quad (6.89)$$

$$\rightarrow [Fe(O)_2(NO_3)]^- + 2NO_2 \quad (6.90)$$

$$[Fe_2(O)(NO_3)_5]^- \rightarrow [Fe_2(O)_2(NO_3)_4]^- + NO_2 \quad (6.91)$$

$$\rightarrow [Fe_2(O)_3(NO_3)_3]^- + 2NO_2 \quad (6.92)$$

$$\rightarrow [Fe_2(O)_4(NO_3)_2]^- + 3NO_2 \quad (6.93)$$

ESI/MS of uranium nitrate produces a wide range of cations and anions [95]. CID of anions with coordinated nitrates proceeds via sequential NO_2 loss (Eqs. (6.94) and (6.95)). The only CID product that reacts with D_2O is $[UO_2(O)(NO_3)_2]^-$ (Eq. (6.96)).

$$[UO_2(NO_3)_3]^- \rightarrow [UO_2(O)(NO_3)_2]^- + NO_2 \quad (6.94)$$

$$[UO_2(O)(NO_3)_2]^- \rightarrow [UO_2(O)_2(NO_3)]^- + NO_2 \quad (6.95)$$

$$[UO_2(O)(NO_3)_2]^- + D_2O \rightarrow [UO_2(OD)_2(NO_3)]^- \quad (6.96)$$

CID of $[Cu(L)NO_3]^+$, where L = phen, **15** (Scheme 6.9) results in ligand fragmentation via loss of NO_2 (Eq. (6.97)) in competition with ligand loss and metal reduction (Eq. (6.98)) [96]. The structure and reactivity of $[Cu(phen)O]^+$ was probed using a combination of DFT calculations, CID experiments, and ion-molecule reactions. The CID experiments give O atom loss (Eq. (6.99)) at lower CID energies and CO loss at higher CID energies. These results suggest that $[Cu(phen)O]^+$ is formed under low-energy conditions, but rearranges via O atom transfer to oxidize the phen ligand, most likely to phenanthrolinone, phenO (Eq. (6.100)). The resultant new complex can undergo decarbonylation (Eq. (6.101)).

$$[Cu(phen)NO_3]^+ \rightarrow [Cu(phen)O]^+ + NO_2 \quad (6.97)$$

$$\rightarrow [Cu(phen)]^+ + NO_3 \quad (6.98)$$

$$[Cu(phen)O]^+ \rightarrow [Cu(phen)]^+ + O \quad (6.99)$$

$$[Cu(phen)O]^+ \rightarrow [Cu(phenO)]^+ \quad (6.100)$$

$$[Cu(phenO)]^+ \rightarrow [Cu(phen\text{-}C)]^+ + CO \quad (6.101)$$

$[Cu(phen)O]^+$ is unreactive with methane and ethane, but undergoes H atom abstraction reactions (Eq. (6.102)) and O atom transfer reactions with larger alkanes such as propane, butane, and isobutene (Eq. (6.103)). Use of deuterated propanes ($CH_3CD_2CH_3$ and $CD_3CH_2CD_3$) reveals a slight preference for the activation of secondary C—H bonds. This resembles the selectivity for C—H bond activation of propane by chlorine atoms, suggesting radical-like reactivity of $[Cu(phen)O]^+$. A two-state reactivity model was used to rationalize the reactions of $[Cu(phen)O]^+$ with alkanes. Finally $[Cu(phen)O]^+$ reacts with the unsaturated hydrocarbons ethene, propene, and benzene, almost exclusivey via oxygen atom transfer.

$$[Cu(phen)O]^+ + RH \rightarrow [Cu(phen)OH]^+ + R \qquad (6.102)$$

$$\rightarrow [Cu(phen)]^+ + ROH \qquad (6.103)$$

ESI/MS of aqueous silver nitrate solutions not only produces the solvated silver(I) cation, but also gives rise to small polynuclear clusters of the type $[Ag(AgNO_3)_n(H_2O)_2]^+$ [97]. The simplest cluster (n = 2) fragments via sequential water loss to give the bare cluster $[Ag(AgNO_3)]^+$, which then fragments via $AgNO_3$ loss (Eq. (6.104)), NO_2 loss (Eq. (6.105)), and NO_3 loss (Eq. (6.106)). Although $AgNO_3$ loss is the dominant pathway, sufficient yields of Ag_2O^+ are formed to study its gas-phase chemistry. The ion-molecule reactions of Ag_2O^+ with a range of alkane and alkene substrates was studied. While Ag_2O^+ does not react with methane, it reacts with larger alkanes (ethane, propane, and butane) via H atom abstraction (Eq. (6.107)). When CD_3CH_3 is used as the substrate, the ratio of the signals of Ag_2OH^+ to Ag_2OD^+ is a direct measure of the kinetic isotope effect (KIE), which was determined to be 2.20 ± 0.06. Reaction of Ag_2O^+ with the deuterium-labeled propanes ($CH_3CD_2CH_3$ and $CD_3CH_2CD_3$) leads to the formation of Ag_2OH^+ as the major product in both cases, which highlights the interplay between regioselectivity, statistical effects, and the kinetic isotope effects associated with H atom abstraction. Using a kinetic model, both the reaction efficiencies and the branching ratios of Ag_2OH^+/Ag_2OD^+ were successfully reproduced using KIEs for the primary and secondary C—H of 1.68 and 2.59 respectively. These are moderate KIEs compared to other gas-phase C—H bond activation reactions mediated by gas-phase metal oxide ions. Arguably from the perspective of modeling catalysis by silver surfaces, the most interesting reactions studied were those of Ag_2O^+ with ethene (Eqs. (6.108)–(6.110)) and propene (Eqs. (6.111)–(6.114)). For both substrates, O atom transfer is observed to give rise to Ag_2^+ (Eqs. (6.108) and (6.111)) as well as a species arising from Ag atom loss (Eqs. (6.109) and (6.112)). The latter reacts with a second molecule of alkene to give rise to a ligand switching reactions (Eqs. (6.110) and (6.114)). In the case of propene, C—H bond activation occurs, which is driven by the formation of the resonance-stabilized allyl radical, as confirmed by deuterium-labeling studies. DFT calculations were carried out to get a better understanding of the nature of the O atom transfer products for Ag_2O^+ reacting with ethene. While acetaldehyde and coordinated acetaldehyde are predicted to be the thermodynamically favored products for Eqs. (6.108) and (6.109) respectively, ethylene oxide and coordinated ethylene oxide are predicted to be the kinetic products. The oxometallocycle, **19**, was proposed to be a key intermediate.

$$[Ag(AgNO_3)]^+ \rightarrow Ag^+ + AgNO_3 \tag{6.104}$$

$$\rightarrow Ag_2O^+ + NO_2 \tag{6.105}$$

$$\rightarrow Ag_2^+ + NO_3 \tag{6.106}$$

$$Ag_2O^+ + RH \rightarrow Ag_2OH^+ + R \tag{6.107}$$

$$Ag_2O^+ + CH_2=CH_2 \rightarrow Ag_2^+ + [C_2, H_4, O] \tag{6.108}$$

$$\rightarrow Ag(C_2, H_4, O)^+ + Ag \tag{6.109}$$

$$Ag(C_2, H_4, O)^+ + CH_2=CH_2 \rightarrow Ag(C_2H_4)^+ + [C_2, H_4, O] \tag{6.110}$$

$$Ag_2O + CH_3CH=CH_2 \rightarrow Ag_2^+ + [C_3, H_6, O] \tag{6.111}$$

$$\rightarrow Ag(C_3, H_6, O)^+ + Ag \tag{6.112}$$

$$\rightarrow Ag_2OH^+ + C_3H_5 \tag{6.113}$$

$$Ag(C_3, H_6, O)^+ + CH_3CH=CH_2 \rightarrow Ag(C_3H_6)^+ + [C_3, H_6, O] \tag{6.114}$$

19

Finally, the route to metal oxide cations from metal nitrate complexes can involve multistep reactions. For example, under harsh electrospray conditions (raised cone voltages) of aqueous iron nitrate solutions, FeO^+ can be formed [98]. CID reactions of various precursor ions suggest that the likely reactions that lead to this ion involve NO_2 loss to form the dihydroxide (Eq. (6.115)), which then undergoes water loss (Eq. (6.116)).

$$[(H_2O)Fe(NO_3)]^+ \rightarrow [Fe(OH)_2]^+ + NO_2 \tag{6.115}$$

$$[Fe(OH)_2]^+ \rightarrow FeO^+ + H_2O \tag{6.116}$$

6.2.4
Formation and Reactivity of Metal Nitrides and Related Species (Eq. (6.11), X = Y = Z = N)

Fragmentation of coordinated azide ligands can occur under the redox conditions associated with ESI (Scheme 6.9) [99] or via collisional activation of metal azide complexes (Scheme 6.10) [94, 100–102].

One of the first investigations to have employed ESI and tandem mass spectrometry to examine the gas-phase chemistry of a metal azide involved a study of the CID reactions of $[Pt(dien)N_3]^+$ [94]. Ligand fragmentation via loss of N_2 (Eq. (6.117), L = dien, **14**) was observed to dominate over loss of the protonated ligand (Eq. (6.118), L = dien), consistent with DFT calculations at a modest level of theory (B3LYP/LAV1S), which predicted that the reaction endothermicities favor ligand fragmen-

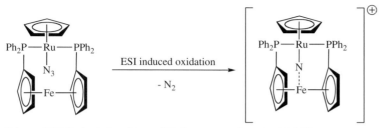

Scheme 6.10 Redox-induced loss of N_2 from a coordinated azide [99].

tation by over 6 kcal mol^{-1}.

$$[Pt(L)N_3]^+ \rightarrow [Pt(L)N]^+ + N_2 \qquad (6.117)$$

$$\rightarrow [Pt(L)\text{-}H]^+ + HN_3 \qquad (6.118)$$

Given that platinum nitrenes are unusual species, two follow-up studies examined the fragmentation of platinum azides in further detail. In the first, CID of [Pt(terpy)N$_3$]$^+$ was used to generate [Pt(terpy)N]$^+$ (Eq. (6.117), L = terpy, **16**), whose ion-molecule reactions were examined with a range of neutrals (acetonitrile, acetone, primary and secondary alcohols, diethylether) [103]. In all cases, formation of the adduct [Pt(terpy)N + L]$^+$ was observed (Eq. (6.119)). The resultant adducts were also subjected to CID. The adducts of the primary and secondary alcohols fragmented via oxidation of the alcohol to presumably convert the nitrene into an amide ligand, [Pt(terpy)NH$_2$]$^+$ (Eq. (6.120)). The diethyl ether adduct fragmented via the sequential loss of C$_2$H$_4$ and CH$_3$CHO to give the same amide, [Pt(terpy)NH$_2$]$^+$. In the second study, the fragmentation of [Pt(dien)N$_3$]$^+$ was re-examined using isotopic labeling experiments together with CID and IMR of [Pt(L)N]$^+$ [102]. CID of [Pt(L)N]$^+$ results in loss of ammonia (Eq. (6.121)). CID of the ^{15}N-labeled [Pt(L)N]$^+$ (where L = ^{15}N, N′,^{15}N″-dien) resulted in loss of ^{14}NH$_3$, indicating that the N of the NH$_3$ lost in Eq. (6.121) is the residual nitrogen of the azido ligand and does not originate from the amino group of the dien ligand. To establish the origin of the hydrogen atoms of the expelled ammonia, CID was carried out on [Pt(D$_5$-dien)N]$^+$, which was formed via CID of the azide precursor that had been allowed to undergo H/D exchange of the NH protons with D$_2$O. Both losses of ND$_3$ and ND$_2$H were observed, revealing that the nitrene activates the N–H and C–H bonds of the dien ligand. It appears as though the nitrene is transformed into an amide ligand (Eq. (6.122)), as slow ligand substitution is observed in ion-molecule reactions with methanol (Eq. (6.123)).

$$[Pt(terpy)N]^+ + L \rightarrow [Pt(terpy)N + L]^+ \qquad (6.119)$$

$$[Pt(terpy)N + L]^+ \rightarrow [Pt(terpy)NH_2]^+ + [L\text{-}2H] \qquad (6.120)$$

$$[Pt(L)N]^+ \rightarrow [Pt(L\text{-}3H)]^+ + NH_3 \qquad (6.121)$$

$$[Pt(L)N]^+ \rightarrow [Pt(L\text{-}2H)NH_2]^+ \qquad (6.122)$$

$$[Pt(L\text{-}2H)NH_2]^+ + CH_3OH \rightarrow [Pt(L\text{-}2H)OCH_3]^+ + NH_3 \qquad (6.123)$$

Schroeder and Schwarz have recently examined the formation of high-valent iron nitrides via CID of iron(III) azide ternary complexes (Eq. (6.124)). To date they have utilized two different auxiliary ligands: the cyclam-acetato ligand, **17**, which yields a complex with a formal charge of $+1$ (i.e., n = 1) [101]; and the 2,6-bis(1,1-di (aminomethyl)ethyl)pyridine ligand, **18**, which yields a complex with a formal charge of $+2$ (i.e., n = 2) [100]. In the first study, they noted the competition between N_2 loss (Eq. (6.124)) and HN_3 loss (Eq. (6.125)) and also examined the subsequent fragmentation reactions of the nitride, $[Fe(L)N]^{n+}$, using isotopic labeling experiments [101]. The low-energy CID spectrum of $[Fe(L)N]^{n+}$ is rich, and includes the losses of NH_2, NH_2, CO_2 and $[C,H_4,N]$. The losses of the aminyl radical and ammonia involve both C–H and N–H bond activation, as demonstrated by deuterium labeling.

$$[Fe(L)N_3]^{n+} \rightarrow [Fe(L)N]^{n+} + N_2 \qquad (6.124)$$

$$\rightarrow [Fe(L\text{-}H)]^{n+} + HN_3 \qquad (6.125)$$

$$\rightarrow [Fe(L)]^{n+} + N_3 \qquad (6.126)$$

In a second study, they noted that changing the auxiliary ligand to **18** resulted in competition between N_2 loss (Eq. (6.124)) and N_3 loss with reduction of the metal center (Eq. (6.126)). Experimental and theoretical estimates of the endothermicities of these two competing pathway were made. The appearance energy for N_2 loss was estimated to be $2.3\,\text{kcal mol}^{-1}$ and that for N_3 loss $27.7\,\text{kcal mol}^{-1}$. While the theoretical estimates for these fragmentation reactions were somewhat higher ($10.8\,\text{kcal mol}^{-1}$ for Eq. (6.124) and $38.7\,\text{kcal mol}^{-1}$ for Eq. (6.126)), they reproduce the experimental orders. Calculations on various isomers of $[Fe(L)N]^{n+}$ revealed that the nitrene is not the most stable isomer, with species resulting from C–H, N–H and C–C bond activation often being more stable. Indeed, CID of $[Fe(L)N]^{n+}$ gave rise to a range of losses that must arise via bond activation of the auxiliary ligand. In addition, the reactions of $[Fe(L)N]^{n+}$ with nine neutral reagents were examined [101]. While ethane, propene, 1-butene and *n*-butane were unreactive, cyclohexene underwent a formal NH_2^+ transfer reaction (Eq. (6.127)), and adduct formation was observed for methanol. $[Fe(L)N]^{2+}$ also underwent a number of interesting reactions with dienes. 1,3- and 1,4-cyclohexadiene react to give adduct formation and the ions $[Fe(L\text{-}2H)]^+$ (Eq. (6.128)) and $[Fe(L\text{-}H)]^+$ (Eq. (6.129)). The latter reaction is interesting since the complex remains doubly charged. Although the $[C_6,H_9,N]$ neutrals are not detected, based on thermochemical grounds, it seems reasonable to assume they arise from formation of NH_3 and benzene. Finally, a minor channel in the reaction of $[Fe(L)N]^{2+}$ with butadiene involves formal N atom transfer (Eq. (6.130)).

$$[Fe(L)N]^{2+} + c\text{-}C_6H_{10} \rightarrow [Fe(L\text{-}2H)]^+ + C_6H_{12}N^+ \qquad (6.127)$$

$$[Fe(L)N]^{2+} + c\text{-}C_6H_8 \to [Fe(L\text{-}2H)]^+ + C_6H_{10}N^+ \tag{6.128}$$

$$\to [Fe(L\text{-}H)]^{2+} + [C_6, H_9, N] \tag{6.129}$$

$$[Fe(L)N]^{2+} + C_4H_6 \to [Fe(L)]^{2+} + [C_4, H_6, N] \tag{6.130}$$

In the most recent work, the alkyne-nitrile metathesis reaction was examined by allowing $[Fe(L)N]^{2+}$ (where L = 4) to undergo ion-molecule reactions with a series of nine different alkynes (Table 6.3) [104]. When unsymmetrical alkynes are used, two different metathesis products are observed (Eqs. (6.131) and (6.132)).

$$[Fe(L)N]^{2+} + R^1 - C \equiv C - R^2 \to [Fe(L)CR^1]^{2+} + R^2 - C \equiv N \tag{6.131}$$

$$[Fe(L)N]^{2+} + R^1 - C \equiv C - R^2 \to [Fe(L)CR^2]^{2+} + R^1 - C \equiv N \tag{6.132}$$

The data shown in Table 6.3 show no obvious trends that may shed light on the mechanism(s) of the metathesis reactions. In terms of overall relative reactivity (k_{rel} of Table 6.3), phenylacetylene reacts fastest and pent-1-yne is the most sluggish alkyne, while ethyne and the parasubstituted phenylacetylenes are unreactive under the experimental conditions used. When the branching ratios and the relative reaction rates are combined, the alkyne–nitrile channel is the most productive for propargyl alcohol and phenylacetylene and least productive for pent-2-yne and pent-1-yne. The sterically congested alkyne exhibits modest reactivity for the metathesis reaction. No obvious relationship exists between the structure of the alkyne substrate and the propensity for a metathesis reaction. With regard to which metathesis reaction is favored for unsymmetrical alkynes, while no regioselectivity operates for pent-2-yne and phenylacetylene other terminal acetylenes favor the loss of the more substituted nitrile (Eq. (6.131)).

Table 6.3 Relative rates (k_{rel}) and branching ratios (BR, %) of metathesis products.

$R^1 =$	$R^2 =$	$k_{rel}{}^a$	BR for R^2CN loss (Eq. (6.131))	BR for R^2CN loss (Eq. (6.132))	BR of all other products[b]
H	H	No reaction	—	—	—
H	CH_2OH	57	68	—	32
H	Pr	4	72	—	28
H	tBu	23	25	8	67
H	Ph	100	15	16	69
H	$p\text{MeOC}_6H_4$	No reaction	—	—	—
H	$p\text{CF}_3OC_6H_4$	No reaction	—	—	—
Me	Et	10	14	14	72
Et	Et	69	34	—	66

$^a k_{rel}$ is expressed as a % of the fastest reaction.
[b]This represents the sum of the products of the other reaction channels, which include adduct formations, addition with elimination of CH_2NH, addition with elimination of NH_3, and N atom transfer.

biomolecules, Laskin, J. and Lifshitz, C. (eds.), Wiley-Interscience, Hoboken, 301.

55 Barlow, C.K., McFadyen, W.D. and O'Hair, R.A.J. (2005) *J. Am. Chem. Soc.*, **127**, 6109.

56 Lam, C.N.W. and Chu, I.K. (2006) *J. Am. Soc. Mass Spectrom.*, **17**, 1249.

57 Lam, A.K.Y., Abrahams, B.F., Grannas, M.J., McFadyen, W.D. and O'Hair, R.A.J. (2006) *Dalton Trans.*, 5051.

58 Combariza, M.Y., Fahey, A.M., Milshteyn, A. and Vachet, R.W. (2005) *Int. J. Mass Spectrom.*, **244**, 109.

59 Deacon, G.B., Faulks, S.J. and Pain, G.N. (1986) *Adv. Organomet. Chem.*, **25**, 237.

60 Mehrotra, R.C. and Bohra, R. (1983) *Metal Carboxylates*, Academic Press, London and New York.

61 Richey, H.G. Jr (ed.) (2000) *Grignard Reagents: New Developments*, Wiley, Chichester.

62 Krause, N. (ed.) (2002) *Modern Organocopper Chemistry*, Wiley-VCH, Weinheim.

63 Jacob, A.P., James, P.F. and O'Hair, R.A.J. (2006) *Int. J. Mass Spectrom.*, **255–256**, 45.

64 O'Hair, R.A.J., Vrkic, A.K. and James, P.F. (2004) *J. Am. Chem. Soc.*, **126**, 12173.

65 Waters, T., O'Hair, R.A.J. and Wedd, A.G. (2003) *Int. J. Mass Spectrom.*, **228**, 599.

66 James, P.F. and O'Hair, R.A.J. (2004) *Org. Lett.*, **6**, 2761.

67 O'Hair, R.A.J. (2002) *Chem. Commun.*, 20.

68 Rijs, N., Khairallah, G.N., Waters, T. and O'Hair, R.A.J. (2008) *J. Am. Chem. Soc.*, **130**, 1069.

69 Bachrach, S.M., Hare, M. and Kass, S.R. (1998) *J. Am. Chem. Soc.*, **120**, 12646.

70 O'Hair, R.A.J., Waters, T. and Cao, B. (2007) *Angew. Chem., Int. Ed.*, **46**, 7048.

71 Whitesides, G.M., Stedronsky, E.R., Casey, C.P. and San Filippo, J. Jr (1970) *J. Am. Chem. Soc.*, **92**, 1426.

72 Julian, R.R., May, J.A., Stoltz, B.M. and Beauchamp, J.L. (2003) *J. Am. Chem. Soc.*, **125**, 4478.

73 Armentrout, P.B. and Sunderlin, L.S. (1992) *Transition Metal Hydrides*, Dedien, A. (ed.), Wiley, New York, 1.

74 Khairallah, G.N. and O'Hair, R.A.J. (2006) *Int. J. Mass Spectrom.*, **254**, 145.

75 Novara, F.R., Gruene, P., Schroder, D. and Schwarz, H. (2008) *Chem.–Eur. J.*, **14**, 5957.

76 Novara, F.R., Schwarz, H. and Schroeder, D. (2007) *Helv. Chim. Acta*, **90**, 2274.

77 Waters, T., O'Hair, R.A.J. and Wedd, A.G. (2003) *J. Am. Chem. Soc.*, **125**, 3384.

78 Schlangen, M., Schroeder, D. and Schwarz, H. (2007) *Angew. Chem., Int. Ed.*, **46**, 1641.

79 Schlangen, M. and Schwarz, H. (2007) *Angew. Chem., Int. Ed.*, **46**, 5614.

80 Schlangen, M. and Schwarz, H. (2008) *Helv. Chim. Acta*, **91**, 379.

81 Khairallah, G.N. and O'Hair, R.A.J. (2005) *Dalton Trans.*, 2702.

82 Khairallah, G.N. and O'Hair, R.A.J. (2005) *Angew. Chem.*, **44**, 728.

83 Celio, H. and White, J.M. (2001) *J. Phys. Chem. B*, **105**, 3908.

84 Tamura, M. and Kochi, J.K. (1972) *Bull. Chem. Soc. Jap.*, **45**, 1120.

85 Khairallah, G.N. and O'Hair, R.A.J. (2007) *Dalton Trans.*, 3149.

86 Khairallah, G.N. and O'Hair, R.A.J. (2008) *Dalton Trans.*, 2956.

87 Khairallah, G.N., O'Hair, R.A.J. and Bruce, M.I. (2006) *Dalton Trans.*, 3699.

88 Schroeder, D. and Schwarz, H. (1995) *Angew. Chem., Int. Ed. Engl.*, **34**, 1973.

89 Feichtinger, D. and Plattner, D.A. (2000) *Perkin*, **2**, 1023.

90 Feichtinger, D. and Plattner, D.A. (2001) *Chem.-Eur. J.*, **7**, 591.

91 Plattner, D.A., Feichtinger, D., El-Bahraoui, J. and Wiest, O. (2000) *Int. J. Mass Spectrom.*, **195/196**, 351.

92 Feichtinger, D. and Plattner, D.A. (1997) *Angew. Chem., Int. Ed. Engl.*, **36**, 1718.

93 Tian, Z., Chan, B., Sullivan, M.B., Radom, L. and Kass, S.R. (2008) *Proc. Natl. Acad. Sci. U. S. A.*, **105**, 7647.

94 Styles, M.L., O'Hair, R.A.J., McFadyen, W.D., Tannous, L., Holmes, R.J. and Gable, R.W. (2000) *Dalton Trans.*, 93.

$$[Fe(L)N]^{2+} + c\text{-}C_6H_8 \rightarrow [Fe(L\text{-}2H)]^+ + C_6H_{10}N^+ \quad (6.128)$$

$$\rightarrow [Fe(L\text{-}H)]^{2+} + [C_6, H_9, N] \quad (6.129)$$

$$[Fe(L)N]^{2+} + C_4H_6 \rightarrow [Fe(L)]^{2+} + [C_4, H_6, N] \quad (6.130)$$

In the most recent work, the alkyne-nitrile metathesis reaction was examined by allowing $[Fe(L)N]^{2+}$ (where L = 4) to undergo ion-molecule reactions with a series of nine different alkynes (Table 6.3) [104]. When unsymmetrical alkynes are used, two different metathesis products are observed (Eqs. (6.131) and (6.132)).

$$[Fe(L)N]^{2+} + R^1 - C \equiv C - R^2 \rightarrow [Fe(L)CR^1]^{2+} + R^2 - C \equiv N \quad (6.131)$$

$$[Fe(L)N]^{2+} + R^1 - C \equiv C - R^2 \rightarrow [Fe(L)CR^2]^{2+} + R^1 - C \equiv N \quad (6.132)$$

The data shown in Table 6.3 show no obvious trends that may shed light on the mechanism(s) of the metathesis reactions. In terms of overall relative reactivity (k_{rel} of Table 6.3), phenylacetylene reacts fastest and pent-1-yne is the most sluggish alkyne, while ethyne and the parasubstituted phenylacetylenes are unreactive under the experimental conditions used. When the branching ratios and the relative reaction rates are combined, the alkyne – nitrile channel is the most productive for propargyl alcohol and phenylacetylene and least productive for pent-2-yne and pent-1-yne. The sterically congested alkyne exhibits modest reactivity for the metathesis reaction. No obvious relationship exists between the structure of the alkyne substrate and the propensity for a metathesis reaction. With regard to which metathesis reaction is favored for unsymmetrical alkynes, while no regioselectivity operates for pent-2-yne and phenylacetylene other terminal acetylenes favor the loss of the more substituted nitrile (Eq. (6.131)).

Table 6.3 Relative rates (k_{rel}) and branching ratios (BR, %) of metathesis products.

$R^1 =$	$R^2 =$	k_{rel}[a]	BR for R^2CN loss (Eq. (6.131))	BR for R^2CN loss (Eq. (6.132))	BR of all other products[b]
H	H	No reaction	—	—	—
H	CH$_2$OH	57	68	—	32
H	Pr	4	72	—	28
H	tBu	23	25	8	67
H	Ph	100	15	16	69
H	pMeOC$_6$H$_4$	No reaction	—	—	—
H	pCF$_3$OC$_6$H$_4$	No reaction	—	—	—
Me	Et	10	14	14	72
Et	Et	69	34	—	66

[a] k_{rel} is expressed as a % of the fastest reaction.
[b] This represents the sum of the products of the other reaction channels, which include adduct formations, addition with elimination of CH$_2$NH, addition with elimination of NH$_3$, and N atom transfer.

6.3
Conclusions

This chapter has highlighted the potential of using ESI in conjunction with CID to generate reactive metallic species via ligand cleavage in the gas phase. Most work to date has focused on generating organometallics, metal hydrides, metal oxides, and metal nitrenes for subsequent reactivity studies. A key challenge for the gas-phase 'synthesis' of the often highly reactive metal oxides and metal nitrenes is to avoid intramolecular attack of the auxiliary ligand(s). Given the wealth of knowledge in ligand design in coordination chemistry, this problem should be readily overcome. Thus, the prospects for studying a wide range of reactive intermediates via the combined use of ESI and CID appear to be excellent.

Acknowledgments

I would like to thank the Australian Research Council for financial support for our studies on metal-mediated chemistry over the past 8 years (Grants: A00103008; DP0558430) and most recently via the ARC Centers of Excellence program (ARC Centre of Excellence for Free Radical Chemistry and Biotechnology). In addition, the intellectual input of my collaborators and students listed in the cited papers is gratefully acknowledged.

Note: Since this article was written, a number of other studies have appeared on the generation and reactions of gas phase reactive metallic species, including: the use of decarboxylation reactions to produce organometallic ions [105–107] and the lithium acetate enolate anion [108]; silver and silver hydride cluster ions [109]; gold carbenes [110, 111] and metal-oxo cations [112].

References

1 Yamaguchi, K. (2003) *J. Mass Spectrom.*, **38**, 473.
2 Evans, W.J., Miller, K.A., Ziller, J.W. and Greaves, J. (2007) *Inorg. Chem.*, **46**, 8008.
3 Eelman, M.D., Blacquiere, J.M., Moriarty, M.M. and Fogg, D.E. (2008) *Angew. Chem., Int. Ed.*, **47**, 303.
4 Henderson, W. and McIndoe, J.S. (2005) *Mass Spectrometry of Inorganic, Coordination and Organometallic Compounds: Tools – Techniques – Tips*, John Wiley & Sons, Ltd, Chichester.
5 Lubben, A.T., McIndoe, J.S. and Weller, A.S. (2008) *Organometallics*, **27**, 3303.
6 Wilson, S.R., Perez, J. and Pasternak, A. (1993) *J. Am. Chem. Soc.*, **115**, 1994.
7 Wilson, S.R. and Wu, Y. (1993) *Organometallics*, **12**, 1478.
8 Santos, L.S., Knaack, L. and Metzger, J.O. (2005) *Int. J. Mass Spectrom.*, **246**, 84.
9 Leonardo, S.S. (2008) *Eur. J. Org. Chem.*, **2008**, 235.
10 O'Hair, R.A.J. (2006) *Chem. Commun.*, 1469.
11 Poznyak, A.L. and Pavlovskii, V.I. (1988) *Angew. Chem.*, **100**, 812.
12 Fischer, R.A. (ed.) (2005) Precursor Chemistry of Advanced Materials CVD, ALD and Nanoparticles, Springer, Berlin.

13 Boehme, D.K. and Schwarz, H. (2005) *Angew. Chem., Int. Ed.*, **44**, 2336.
14 Eller, K. and Schwarz, H. (1991) *Chem. Rev.*, **91**, 1121.
15 Schroeder, D., Schwarz, H. and Shaik, S. (2000) *Struct. Bonding (Berlin)*, **97**, 91.
16 Schroeder, D. and Schwarz, H. (2001) *Essays in Contemporary Chemistry*, Quinkert, G. and Kisakürek, M.V. (eds.) Wiley-VCH, Weinheim; Chichester, 131.
17 Schroeder, D. and Schwarz, H. (2007) *Top. Organomet. Chem.*, **22**, 1.
18 Schroeder, D., Shaik, S. and Schwarz, H. (2000) *Acc. Chem. Res.*, **33**, 139.
19 Eller, K. (1993) *Coord. Chem. Rev.*, **126**, 93.
20 Armentrout, P.B. (1992) *Gas-Phase Metal Reactions*, Fontijn, A., (ed.), Elsevier: Netherlands, 301.
21 Armentrout, P.B. (1992) *Adv. Gas Phase Ion Chem.*, **1**, 83.
22 Armentrout, P.B. (1999) *Top. Organomet. Chem.*, **4**, 1.
23 Armentrout, P.B. (2000) *Int. J. Mass Spectrom.*, **200**, 219.
24 Armentrout, P.B. (2001) *Annu. Rev. Phys. Chem.*, **52**, 423.
25 Armentrout, P.B. (2003) *Eur. J. Mass Spectrom.*, **9**, 531.
26 Armentrout, P.B. (2003) *Top. Curr. Chem.*, **225**, 233.
27 Armentrout, P.B. (2003) *Int. J. Mass Spectrom.*, **227**, 289.
28 Armentrout, P.B. and Beauchamp, J.L. (1989) *Acc. Chem. Res.*, **22**, 315.
29 Graul, S.T. and Squires, R.R. (1988) *J. Am. Chem. Soc.*, **110**, 607.
30 Graul, S.T. and Squires, R.R. (1989) *J. Am. Chem. Soc.*, **111**, 892.
31 Graul, S.T. and Squires, R.R. (1990) *J. Am. Chem. Soc.*, **112**, 2506.
32 Squires, R.R. (1992) *Acc. Chem. Res.*, **25**, 461.
33 Damrauer, R. and Hankin, J.A. (1995) *Chem. Rev.*, **95**, 1137.
34 Froelicher, S.W., Freiser, B.S. and Squires, R.R. (1984) *J. Am. Chem. Soc.*, **106**, 6863.
35 Hankin, J.A., Krempp, M. and Damrauer, R. (1995) *Organometallics*, **14**, 2652.
36 Gronert, S., O'Hair, R.A.J., Prodnuk, S., Suelzle, D., Damrauer, R. and DePuy, C.H. (1990) *J. Am. Chem. Soc.*, **112**, 997.
37 Damrauer, R., Krempp, M. and O'Hair, R.A.J. (1993) *J. Am. Chem. Soc.*, **115**, 1998.
38 Jacobson, D.B. and Freiser, B.S. (1984) *J. Am. Chem. Soc.*, **106**, 4623.
39 Carlin, T.J., Sallans, L., Cassady, C.J., Jacobson, D.B. and Freiser, B.S. (1983) *J. Am. Chem. Soc.*, **105**, 6320.
40 Colton, R., D'Agostino, A. and Traeger, J.C. (1995) *Mass Spectrom. Rev.*, **14**, 79.
41 Traeger, J.C. (2000) *Int. J. Mass Spectrom.*, **200**, 387.
42 Traeger, J.C. and Colton, R. (1998) *Adv. Mass Spectrom.*, **14**, 637.
43 Gatlin, C.L. and Turecek, F. (1997) *Electrospray ionization mass spectrometry: fundamentals, instrumentation and applications*, Cole, R.B. (ed.), Wiley, New York.
44 Henderson, W. and McIndoe, J.S. (2004) *Compr. Coord. Chem. II*, **2**, 387.
45 Henderson, W., Nicholson, B.K. and McCaffrey, L.J. (1998) *Polyhedron*, **17**, 4291.
46 Tsipis, C. (2004) *Comments Inorg. Chem.*, **25**, 19.
47 Damrauer, R. (2004) *Organometallics*, **23**, 1462.
48 Farrar, J.M. and Saunders, W.H. Jr (eds) (1988) *Techniques for the Study of Ion-Molecule Reactions*, Wiley-Interscience, New York.
49 Freiser, B.S. (1994) *Acc. Chem. Res.*, **27**, 353.
50 Freiser, B.S. (1996) *J. Mass Spectrom.*, **31**, 703.
51 Asamoto, B. (ed.) (1991) *FT-ICR/MS: Analytical Applications of Fourier Transform Ion Cyclotron Resonance Mass Spectrometry*, Wiley-VCH, Weinheim.
52 Chisholm, D.M. and McIndoe, J.S. (2008) *Dalton Trans.*, 3933.
53 Chu, I.K., Rodriquez, C.F., Lau, T.-C., Hopkinson, A.C. and Siu, K.W.M. (2000) *J. Phys. Chem. B*, **104**, 3393.
54 Hopkinson, A.C. and Siu, K.W.M. (2006) *Principles of mass spectrometry applied to*

biomolecules, Laskin, J. and Lifshitz, C. (eds.), Wiley-Interscience, Hoboken, 301.

55 Barlow, C.K., McFadyen, W.D. and O'Hair, R.A.J. (2005) *J. Am. Chem. Soc.*, **127**, 6109.

56 Lam, C.N.W. and Chu, I.K. (2006) *J. Am. Soc. Mass Spectrom.*, **17**, 1249.

57 Lam, A.K.Y., Abrahams, B.F., Grannas, M.J., McFadyen, W.D. and O'Hair, R.A.J. (2006) *Dalton Trans.*, 5051.

58 Combariza, M.Y., Fahey, A.M., Milshteyn, A. and Vachet, R.W. (2005) *Int. J. Mass Spectrom.*, **244**, 109.

59 Deacon, G.B., Faulks, S.J. and Pain, G.N. (1986) *Adv. Organomet. Chem.*, **25**, 237.

60 Mehrotra, R.C. and Bohra, R. (1983) *Metal Carboxylates*, Academic Press, London and New York.

61 Richey, H.G. Jr (ed.) (2000) *Grignard Reagents: New Developments*, Wiley, Chichester.

62 Krause, N. (ed.) (2002) *Modern Organocopper Chemistry*, Wiley-VCH, Weinheim.

63 Jacob, A.P., James, P.F. and O'Hair, R.A.J. (2006) *Int. J. Mass Spectrom.*, **255–256**, 45.

64 O'Hair, R.A.J., Vrkic, A.K. and James, P.F. (2004) *J. Am. Chem. Soc.*, **126**, 12173.

65 Waters, T., O'Hair, R.A.J. and Wedd, A.G. (2003) *Int. J. Mass Spectrom.*, **228**, 599.

66 James, P.F. and O'Hair, R.A.J. (2004) *Org. Lett.*, **6**, 2761.

67 O'Hair, R.A.J. (2002) *Chem. Commun.*, 20.

68 Rijs, N., Khairallah, G.N., Waters, T. and O'Hair, R.A.J. (2008) *J. Am. Chem. Soc.*, **130**, 1069.

69 Bachrach, S.M., Hare, M. and Kass, S.R. (1998) *J. Am. Chem. Soc.*, **120**, 12646.

70 O'Hair, R.A.J., Waters, T. and Cao, B. (2007) *Angew. Chem., Int. Ed.*, **46**, 7048.

71 Whitesides, G.M., Stedronsky, E.R., Casey, C.P. and San Filippo, J. Jr (1970) *J. Am. Chem. Soc.*, **92**, 1426.

72 Julian, R.R., May, J.A., Stoltz, B.M. and Beauchamp, J.L. (2003) *J. Am. Chem. Soc.*, **125**, 4478.

73 Armentrout, P.B. and Sunderlin, L.S. (1992) *Transition Metal Hydrides*, Dedien, A. (ed.), Wiley, New York, 1.

74 Khairallah, G.N. and O'Hair, R.A.J. (2006) *Int. J. Mass Spectrom.*, **254**, 145.

75 Novara, F.R., Gruene, P., Schroder, D. and Schwarz, H. (2008) *Chem.–Eur. J.*, **14**, 5957.

76 Novara, F.R., Schwarz, H. and Schroeder, D. (2007) *Helv. Chim. Acta*, **90**, 2274.

77 Waters, T., O'Hair, R.A.J. and Wedd, A.G. (2003) *J. Am. Chem. Soc.*, **125**, 3384.

78 Schlangen, M., Schroeder, D. and Schwarz, H. (2007) *Angew. Chem., Int. Ed.*, **46**, 1641.

79 Schlangen, M. and Schwarz, H. (2007) *Angew. Chem., Int. Ed.*, **46**, 5614.

80 Schlangen, M. and Schwarz, H. (2008) *Helv. Chim. Acta*, **91**, 379.

81 Khairallah, G.N. and O'Hair, R.A.J. (2005) *Dalton Trans.*, 2702.

82 Khairallah, G.N. and O'Hair, R.A.J. (2005) *Angew. Chem.*, **44**, 728.

83 Celio, H. and White, J.M. (2001) *J. Phys. Chem. B*, **105**, 3908.

84 Tamura, M. and Kochi, J.K. (1972) *Bull. Chem. Soc. Jap.*, **45**, 1120.

85 Khairallah, G.N. and O'Hair, R.A.J. (2007) *Dalton Trans.*, 3149.

86 Khairallah, G.N. and O'Hair, R.A.J. (2008) *Dalton Trans.*, 2956.

87 Khairallah, G.N., O'Hair, R.A.J. and Bruce, M.I. (2006) *Dalton Trans.*, 3699.

88 Schroeder, D. and Schwarz, H. (1995) *Angew. Chem., Int. Ed. Engl.*, **34**, 1973.

89 Feichtinger, D. and Plattner, D.A. (2000) *Perkin*, **2**, 1023.

90 Feichtinger, D. and Plattner, D.A. (2001) *Chem.-Eur. J.*, **7**, 591.

91 Plattner, D.A., Feichtinger, D., El-Bahraoui, J. and Wiest, O. (2000) *Int. J. Mass Spectrom.*, **195/196**, 351.

92 Feichtinger, D. and Plattner, D.A. (1997) *Angew. Chem., Int. Ed. Engl.*, **36**, 1718.

93 Tian, Z., Chan, B., Sullivan, M.B., Radom, L. and Kass, S.R. (2008) *Proc. Natl. Acad. Sci. U. S. A.*, **105**, 7647.

94 Styles, M.L., O'Hair, R.A.J., McFadyen, W.D., Tannous, L., Holmes, R.J. and Gable, R.W. (2000) *Dalton Trans.*, 93.

95 Pasilis, S., Somogyi, A., Herrmann, K. and Pemberton, J.E. (2006) *J. Am. Soc. Mass Spectrom.*, **17**, 230.
96 Schroeder, D., Holthausen, M.C. and Schwarz, H. (2004) *J. Phys. Chem. B*, **108**, 14407.
97 Roithova, J. and Schroeder, D. (2007) *J. Am. Chem. Soc.*, **129**, 15311.
98 Schroeder, D., Roithova, J. and Schwarz, H. (2006) *Int. J. Mass Spectrom.*, **254**, 197.
99 Paim, L.A., Augusti, D.V., Dalmazio, I., Alves, T.M.d.A., Augusti, R. and Siebald, H.G.L. (2005) *Polyhedron*, **24**, 1153.
100 Schlangen, M., Neugebauer, J., Reiher, M., Schroeder, D., Lopez, J.P., Haryono, M., Heinemann, F.W., Grohmann, A. and Schwarz, H. (2008) *J. Am. Chem. Soc.*, **130**, 4285.
101 Schroeder, D., Schwarz, H., Aliaga-Alcalde, N. and Neese, F. (2007) *Eur. J. Inorg. Chem.*, 816.
102 Wee, S., White, J.M., McFadyen, W.D. and O'Hair, R.A.J. (2003) *Aust. J. Chem.*, **56**, 1201.
103 Wee, S., Grannas, M.J., McFadyen, W.D. and O'Hair, R.A.J. (2001) *Aust. J. Chem.*, **54**, 245.
104 Boyd, J.P., Schlangen, M., Grohmann, A. and Schwarz, H. (2008) *Helv. Chim. Acta*, **91**, 1430.
105 Rijs, N. and O'Hair, R.A.J. (2009) *Organometallics*, **28**, 2684.
106 Khairallah, G.N., Thum, C. and O'Hair, R.A.J. (2009) *Organometallics*, **28**, 5002.
107 Khairallah, G.N., Waters, T. and O'Hair, R.A.J. (2009) *Dalton Trans.*, 2832.
108 Meyer, M.M., Khairallah, G.N., Kass, S.R. and O'Hair, R.A.J. (2009) *Angew. Chem. Int. Ed.*, **48**, 2934.
109 Wang, F.Q., Khairallah, G.N. and O'Hair, R.A.J. (2009) *Int. J. Mass Spectrom.*, **283**, 17.
110 Fedorov, A., Moret, M.-E. and Chen, P. (2008) *J. Am. Chem. Soc.*, **130**, 8880.
111 Fedorov, A. and Chen, P. (2009) *Organometallics*, **28**, 1278.
112 Schroeder, D., de Jong, K.P. and Roithová, J. (2009) *Eur. J. Inorg. Chem.*, 2121.

7
Palladium Intermediates in Solution[1]
Anna Roglans and Anna Pla-Quintana

7.1
Introduction

Carbon-carbon bond formation reactions are the most important processes in organic chemistry. The discovery in the early 1970s of metal-catalyzed cross-coupling reactions permitted a significant step forward resulting in these reactions being considered as useful and powerful synthetic tools [1]. Among metal-mediated cross-coupling reactions, the use of palladium catalysts has radically changed the way organic molecules are made [1, 2]. However, in order to improve these catalytic processes it is necessary to understand the reaction mechanisms. Most chemical reactions occur through a complex sequence of steps via reactive intermediates. An attempt has been made to detect these intermediates indirectly using chemical and physical methods and in a direct and detailed way using spectroscopic methods. Most of the spectroscopic methods used do not allow the reactive intermediates to be studied and detected directly from the reaction mixtures. Furthermore, it is often not possible to monitor the substrates, intermediates, and final products by the same method as that used for the detection of the intermediates.

Mass spectrometry with electrospray ionization (ESI-MS) [3] has become a useful alternative for the determination of the reactive intermediates given that it enables the simple direct detection of the components of the reaction mixture and, at the same time, allows starting compounds, intermediates and final products to be monitored. The technique has recently been applied to the determination of mechanisms in highly diverse reactions [4]. The main function of the ESI process is to transfer analyte species which are normally ionized in the condensed phase into the gas phase as isolated entities. The species must be charged in order to allow the detection of intermediates and to have a sufficiently long lifetime. For samples which already contain ions, no further ionization is needed. However, the common sample

---

1) Dedicated to the memory of our friend Prof. Marcial Moreno-Mañas.

Reactive Intermediates: MS Investigations in Solution. Edited by Leonardo S. Santos.
Copyright © 2010 WILEY-VCH Verlag GmbH & Co. KGaA, Weinheim
ISBN: 978-3-527-32351-7

ionization process for neutral species is to abstract a proton from the solvent to give an $[M + H]^+$ cation. For these cases the sample needs to have a functional group susceptible to being protonated. The $[M + H]^+$ cations can be detected in positive-ion mode ESI(+). Similarly, species containing acidic groups may lose a proton, generating the corresponding negative ion observed as $[M - H]^-$. In this case the $[M - H]^-$ anions can be detected in negative-ion mode ESI(−). Furthermore, electrochemical reactions affecting the analyte solution may occur during the electrospray process [5]. This is particularly important when examining organometallic and coordination compounds, as the oxidation state of the metal may be affected [6]. However, neutral species can also be oxidized or reduced at the probe tip to produce $[M]^{\bullet+}$ or $[M]^{\bullet-}$, permitting their characterization by ESI-MS. The tandem version, ESI-MS/MS, is also a valuable tool for the identification of charged complexes and their structural assignment, since it enables the first mass analyzer to fish out species with a certain m/z ratio so that they can be fragmented in the collision cell, usually by Collision-Induced Dissociation (CID). The generated fragments can then be analyzed with the second mass analyzer.

In this chapter we will concentrate on the mechanistic studies that have been conducted to date on palladium-catalyzed reactions, including not only those ESI-MS studies which are highly specific and complete but also ones that merely touch on this technique. In these mechanistic studies it is crucial to take into account the need to analyze the detected species carefully by ESI-MS to make absolutely certain that the ones which prevail in the solution are in fact intermediates on the reaction path.

A series of steps must be taken to ensure that the mechanistic study is methodologically correct. (i) In a first set of experiments the individual behavior of each of the reactants, final products, catalysts and additives needs to be studied by ESI-MS. (ii) Binary and ternary mixtures should be made and control experiments conducted. (iii) The reaction is studied by sampling at constant time intervals during the whole course of the reaction. There are certain tricks that can be used to find the intermediates: (a) the stoichiometric ratio of the reagents can be changed starting from the actual synthetic conditions and going up to, for example, stoichiometric catalyst quantities, (b) the aliquot can be cooled down to freeze the intermediates prior to injection, (c) a quenching additive, such as an acid/base quencher or silver triflate which abstracts halogens from the metallic complexes, can be used to help in ionizing the intermediates. The studies in this chapter cover all of these possibilities.

One of the advantages of these mechanistic studies involving palladium species is that the peaks in the mass spectra due to intermediates containing this metal are immediately identified by their characteristic isotope distribution [isotope distribution of palladium: ^{102}Pd (1.02%), ^{104}Pd (11.14%), ^{105}Pd (22.33%), ^{106}Pd (27.33%), ^{108}Pd (26.46%) and ^{110}Pd (11.72%), see Figure 7.1] and are easily distinguished from other non-palladium-containing species. In this chapter all m/z values of observed palladium species will be reported for ^{106}Pd unless other isotopically rich atoms make up part of the cluster, as in the case of Cl, Br, Sn, Te, in which cases the largest peak will be given.

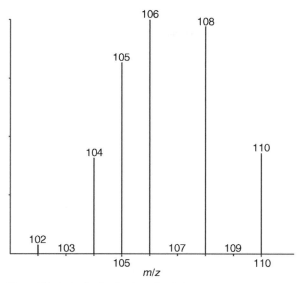
Figure 7.1 Isotopic distribution of palladium.

7.2
ESI-MS Studies in Suzuki-Miyaura Cross-Coupling and Related Reactions

One of the first ESI-MS mechanistic studies was undertaken by Canary et al. in 1994 [7]. Although the electrospray source coupled to a mass analyzer was first used by Fenn in the 1980s [8], its original application was based on the study of biomolecules such as peptides and proteins for which he won the Nobel Prize in 2002. A decade later, the first studies were published taking advantage of the soft ionization of the electrospray which made it possible to observe directly the transient catalytic intermediates under real reaction conditions by injecting the reaction mixture into the spectrometer. These studies have given rise to an advance in mechanistic hypotheses. Canary was a pioneer in publishing the first complete study of palladium-catalyzed Suzuki-Miyaura cross-coupling reactions between several arylboronic acids and bromopyridines using the ESI-MS technique. The Suzuki-Miyaura reaction involves the cross-coupling between organoboron compounds and organic halides or triflates and is one of the most convenient synthetic methods for the formation of carbon-carbon bonds [9]. The Suzuki-Miyaura reaction and its most commonly accepted catalytic cycle is shown in Scheme 7.1.

Initially, there is an oxidative addition of the electrophile R^1-X to a Pd(0) complex to afford intermediate **I1**. In the case of using a Pd(II) complex as precatalyst, a previous reduction step of Pd(II) to Pd(0) occurs, generating the true catalytic species. Transmetalation between R^1-Pd-X and the boronic acid $R^2B(OH)_2$ with the aid of bases leads to the formation of intermediate **I2**. Reductive elimination then gives the cross-coupled compound R^1-R^2 and the Pd(0) species is regenerated, restarting the

R1—X + R2—B(OH)2
[Pd]
base
R1—R2 + X—B(OH)2
Pd(II)
Pd(0)
R1—R2
R1—X
reductive elimination
oxidative addition
R1—Pd(II)—R2
I2
R1—Pd(II)—X
I1
transmetallation
B(OH)3
R2—B(OH)2 + base

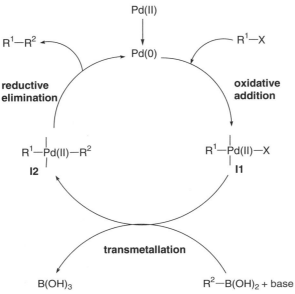

Scheme 7.1 The generally accepted catalytic cycle for the Suzuki-Miyaura cross-coupling reaction.

catalytic cycle. Various ligands, especially phosphines, are effective in stabilizing the palladium species during the cross-coupling reaction.

Canary used pyridyl halides (**1** and **2**) instead of the typically used phenyl derivatives to introduce a protonable group into the complexes, which otherwise might not be detected by the ESI-MS (Scheme 7.2).

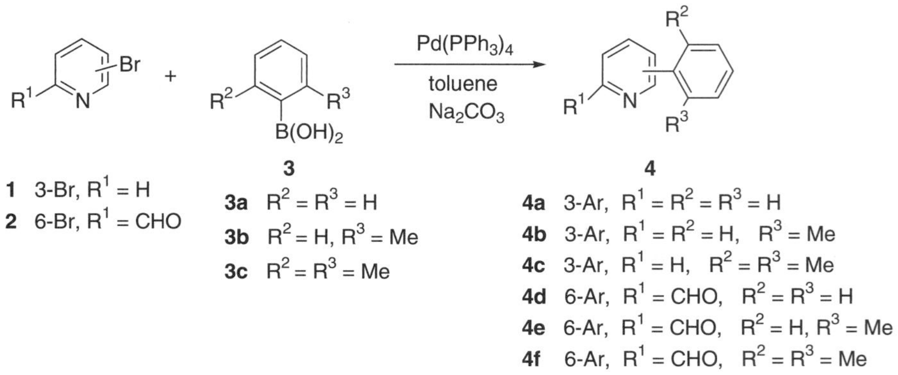

Br
R1
N
+
R2
R3
B(OH)2
3
Pd(PPh3)4
toluene
Na2CO3
R2
R1
N
R3
4
1 3-Br, R1 = H
2 6-Br, R1 = CHO
3a R2 = R3 = H
3b R2 = H, R3 = Me
3c R2 = R3 = Me
4a 3-Ar, R1 = R2 = R3 = H
4b 3-Ar, R1 = R2 = H, R3 = Me
4c 3-Ar, R1 = H, R2 = R3 = Me
4d 6-Ar, R1 = CHO, R2 = R3 = H
4e 6-Ar, R1 = CHO, R2 = H, R3 = Me
4f 6-Ar, R1 = CHO, R2 = R3 = Me

Scheme 7.2 Suzuki-Miyaura cross-coupling reactions of pyridyl halides **1** and **2** with arylboronic acids **3** studied by ESI-MS.

Table 7.1 Electrospray mass spectral data for reactions shown in Scheme 7.2.

Entry	Pyridyl halide	Arylboronic acid	R^1, R^2, R^3	Species observed by ESI-MS
1	1	3a	$R^1 = H$ $R^2 = R^3 = H$	I3 ($m/z = 788$), I4 ($m/z = 708$) I5 ($m/z = 786$), I7 ($m/z = 578$)
2	1	3b	$R^1 = H$ $R^2 = H$, $R^3 = Me$	I3 ($m/z = 788$), I4 ($m/z = 708$) I5 ($m/z = 800$), I7 ($m/z = 578$)
3	1	3c	$R^1 = H$ $R^2 = R^3 = Me$	I4 ($m/z = 708$) I5 ($m/z = 814$), I6 ($m/z = 735$)
4[a]	2	3a	$R^1 = CH(OH)(OMe)$ $R^2 = R^3 = H$	I3 ($m/z = 848$), I4 ($m/z = 768$) I5 ($m/z = 846$), I6 ($m/z = 707$) I7 ($m/z = 578$)
5[a]	2	3b	$R^1 = CH(OH)(OMe)$ $R^2 = H$ $R^3 = Me$	I3 ($m/z = 848$), I4 ($m/z = 768$) I5 ($m/z = 860$)
6[a]	2	3c	$R^1 = CH(OH)(OMe)$ $R^2 = R^3 = Me$	I3 ($m/z = 848$), I4 ($m/z = 768$) I5 ($m/z = 874$), I6 ($m/z = 735$)

[a] Aldehydes observed as methanol hemiacetals.

The reaction was studied by taking aliquots from the organic layer of the reaction mixture at regular time intervals until the reaction was completed. The aliquot was quenched in cold methanol and then heated to room temperature before injection into the spectrometer. Scheme 7.2 shows the reactions studied, and Table 7.1 gives the intermediates detected by ESI-MS.

The key intermediates of the Suzuki-Miyaura reaction, species I3 and I4 corresponding to oxidative addition intermediates and species I5 corresponding to transmetalation intermediates, were detected by ESI-MS in all the reactions for the first time in authentic reaction mixtures. Species I6 and I7, detected in most cases, were assigned to $[ArPd(PPh_3)_2]^+$ and $[Pd(PPh_3)_4]^{2+}$, respectively, and were also observed in control experiments when the aryl halide was excluded from the reaction mixture. The authors postulate that intermediate I6 is a result of transmetalation with I7. This first study showed for the first time that the ESI-MS technique has great potential as a complementary analytical tool for the determination of reaction mechanisms.

Moreno-Mañas, Aramendia et al. [10], following on from Canary's study, investigated the mechanistic aspects of palladium-catalyzed self-coupling reactions of arylboronic acids using the same technique. Since the authors had experienced the formation of self-coupling products when attempting difficult cross-couplings they decided to study this process from a mechanistic viewpoint. This oxidative coupling was also observed previously as a side reaction in several Pd-catalyzed Suzuki-Miyaura cross-couplings [11], and later this oxidative homocoupling of arylboronic acids in the presence of an oxidant was described by other authors [12]. Moreno-Mañas et al. [13] initially studied the self-coupling of 4-(trifluoromethyl)phenylboronic acid **5a** under different experimental conditions and investigated the mechanism of the process by monitoring the reaction using ^{19}F NMR. Later [10], an ESI-MS study permitted the mechanistic scheme of the self-coupling reaction to be completed. More recently, the mechanism of this process in the presence of dioxygen has also been studied by Jutand et al. by means of ^{31}P and ^{1}H NMR spectroscopy, electrochemical methods, and ab initio calculations [14]. The two self-couplings shown in Scheme 7.3 were monitored by sampling and ESI-MS analysis at different time intervals.

The intermediates detected in this study [10], which allowed the authors to make a mechanistic proposal for the process shown in Scheme 7.4, are described in Table 7.2

Certain conclusions were drawn from this study. (i) In the case of arylboronic acid **5a**, oxidative addition intermediate **I9** was only detected when acid quenching was performed. However, species **I15** were observed in the two reactions (Entries 1 and 2) and were probably formed by cleavage of intermediates **I9**, which were ascribed as being an indication of the presence of the oxidative addition intermediate. (ii) Intermediates **I11** were not detected. However, species **I12** accumulated, suggesting that the rate-determining step was between **I12** and the beginning of the catalytic cycle. (iii) Intermediate **I13** was only detected in the case of arylboronic **5b** (Entry 2), and this was the most advanced intermediate observed in the catalytic cycle. (iv) Species **I16** were detected in both cases and were related to the formation of phenols in side reactions. An interesting point of this study is that it revealed the steps occurring after the formation of biaryl leading to the recovery of the catalytic species.

Two parallel studies were published in 2007 regarding the Suzuki-Miyaura reaction using diazonium salts as electrophilic partners and potassium organotrifluoroborates as the organoboron reagent (Scheme 7.5, Table 7.3). The coupling of these two partners is a modification of the more popular Suzuki-Miyaura cross-coupling reactions in an attempt to improve and broaden the scope of this process [15]. One of the advantages of analyzing reaction intermediates using

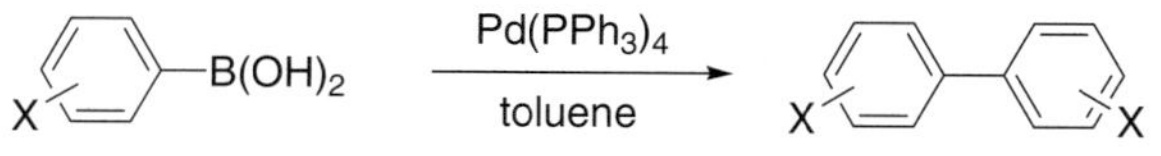

5a X = 4-CF$_3$
5b X = 3-NH$_2$

6a X = 4-CF$_3$
6b X = 3-NH$_2$

Scheme 7.3 Palladium-catalyzed self-coupling reactions of arylboronic acids **5** studied by ESI-MS.

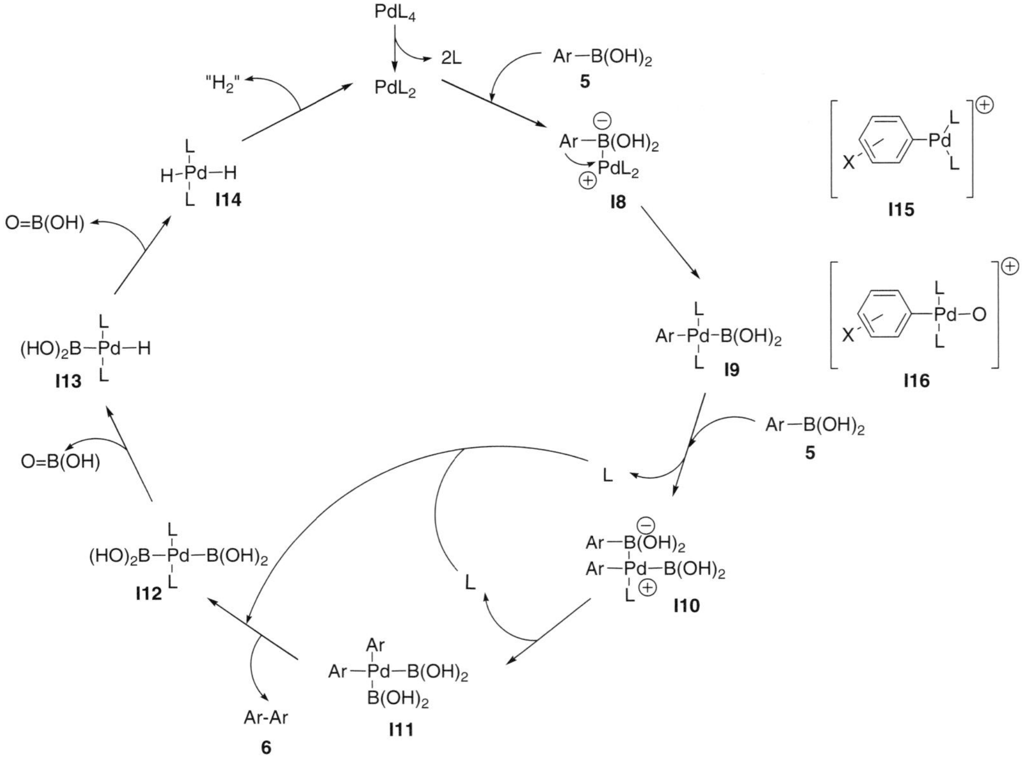

7.2 ESI-MS Studies in Suzuki-Miyaura Cross-Coupling and Related Reactions

Scheme 7.4 Proposed mechanistic cycle for palladium-catalyzed self-coupling of arylboronic acids.

diazonium salts is that in the oxidative addition step a molecular nitrogen extrusion takes place to give arylpalladium species, whose cationic nature facilitates the monitoring of the reaction by means of ESI-MS. One of these studies, by Roglans et al., [16] uses $Pd_2(dba)_3$ as a catalytic system (dba = dibenzylideneacetone) and the other, by Mastrorilli et al. [17], uses bis(μ-acetato)bis(4,4′-difluoroazobenzene-C^2,N) dipalladium(II) as a precatalyst.

In the case where $Pd_2(dba)_3$ was used as the catalyst, ESI-MS allowed oxidative addition intermediates **I18** and diaryl Pd(II) transmetalated intermediates **I19** to be

Table 7.2 Intermediates observed by ESI-MS in reactions of Scheme 7.3.

Entry	Arylboronic acid	Species observed by ESI-MS
1	5a	**I8** or **I9** ($m/z = 839$)[a], **I12** ($m/z = 743$)[b], **I15** ($m/z = 775$), **I16** ($m/z = 791$)
2	5b	**I12** ($m/z = 743$)[b], **I13** ($m/z = 677$), **I15** ($m/z = 722$), **I16** ($m/z = 738$)

[a] The sample was quenched with trifluoroacetic acid before injecting into the spectrometer.
[b] Intermediate **I12** adds Na^+, which was presumably leached from the glassware or reagent-grade solvents.

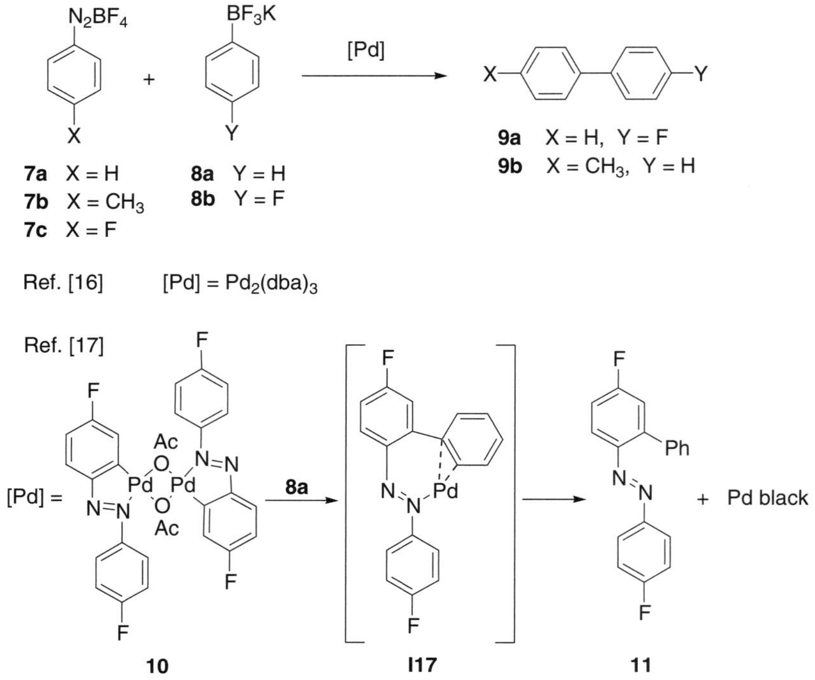

Scheme 7.5 Cross-coupling reaction between aryldiazonium tetrafluoroborates **7** and potassium aryltrifluoroborates **8** using two different catalytic species.

Table 7.3 Intermediate species detected by ESI mass spectrometry of reactions shown in Scheme 7.5.

Entry [Ref.]	ArN$_2$BF$_4$	ArBF$_3$K	[Pd]	Species observed by ESI-MS
1 [16]	7a X = H	8b Y = F	Pd$_2$(dba)$_3$	**I18** (n = 2, m = 0, m/z = 651) **I19** (n = 1, m = 0: m/z = 513; n = 1, m = 1: m/z = 554)
2 [16]	7b X = CH$_3$	8a Y = H	Pd$_2$(dba)$_3$	**I18** (n = 2, m = 0: m/z = 665; n = 1, m = 1: m/z = 472; n = 1, m = 0: m/z = 431) **I19** (n = 1, m = 0: m/z = 509; n = 1, m = 1: m/z = 550)
3[a] [17]	—	8a Y = H	10	**I17** (M + F$^-$, m/z = 419)
4 [17]	7c X = F	8a Y = H	10	**I20** (n = 1: m/z = 495; n = 2: m/z = 789)

[a]HRMS was performed in negative ion mode ESI(−).

Ref. [16]

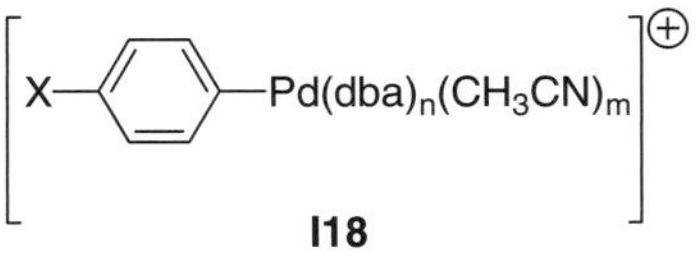

$$\left[X\text{-}\underset{\text{I18}}{\text{C}_6\text{H}_4}\text{-Pd(dba)}_n(\text{CH}_3\text{CN})_m \right]^{\oplus} \quad \left[X\text{-}\underset{\text{I19}}{\underset{Y}{\text{C}_6\text{H}_4}}\text{-Pd(dba)}_n(\text{CH}_3\text{CN})_m\text{-C}_6\text{H}_4 \right]^{\oplus} + \text{H}$$

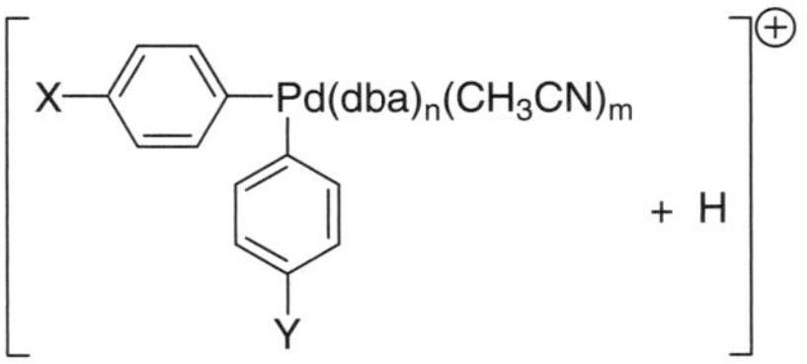

Ref. [17]

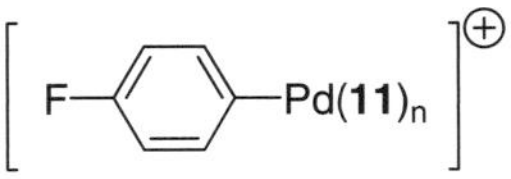

$$\left[F\text{-}\underset{\text{I20}}{\text{C}_6\text{H}_4}\text{-Pd(11)}_n \right]^{\oplus}$$

detected (Entries 1 and 2, Table 7.3). Two different reactions were studied (**7a** + **8b** and **7b** + **8a**) in order to confirm the nature of the species detected. The two species **I18** and **I19** added different ligands to the coordinating sphere of palladium, either the dba of the starting catalyst or CH_3CN molecules, the solvent used in the ESI-MS study. In addition, intermediates of type $[Ar_2Pd(dba)_n(CH_3CN)_m]^+$ giving rise to homocoupling products of the aryldiazonium counterpart were also detected. These homocoupled products, identified by GC-MS, were obtained at traces levels along with the expected biaryl derivatives.

Mastrorilli *et al.* described a combined ESI HRMS and ^{19}F NMR mechanistic study of this reaction with palladacycle complex **10** (Scheme 7.5) as the precatalyst [17], from which the authors postulated the formation of the true catalytic species in a first step. It is suggested that palladacycle **10** reacts with potassium trifluorophenylborate **8a** to give Pd(0) intermediate **I17** (Entry 3, Table 7.3), which starts the catalytic cycle. This species undergoes oxidative addition of the aryldiazonium salt to give the cationic aryl-palladium(II) complexes **I20**, which contain azobenzene **11** as a ligand (Entry 4, Table 7.3). In this study, no intermediate was detected corresponding to transmetalation species.

7.3
ESI-MS Studies in the Identification of Oxidative Addition Intermediates

A study combining X-ray crystallographic methods with ESI mass spectrometry was conducted by Hor *et al.* [18]. These authors investigated the behavior of several palladium complexes formed by oxidative addition of 2-bromopyridine to Pd(0), which were stabilized either by triphenylphosphine [Pd(PPh$_3$)$_4$] or several bidentate phosphines [Pd(L)$_2$, L = dppm, dppp, dppb, or dppf], and examined the stabilizing effects and structural influence of these different phosphines on a dinuclear skeleton (Scheme 7.6).

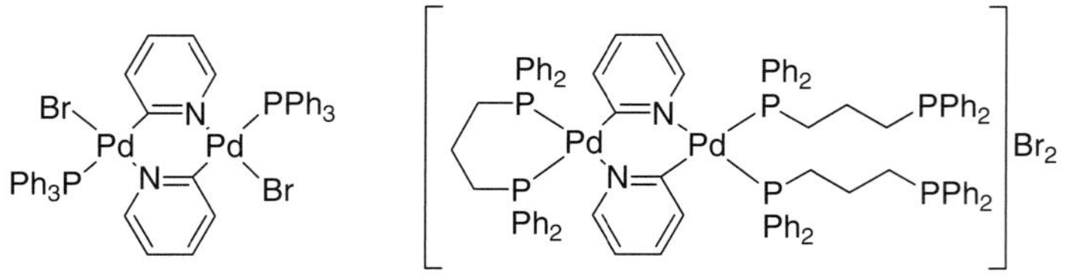

Scheme 7.6 Binuclear Pd-complexes **12** and **13**.

An array of structural possibilities was envisaged based on the different coordination modes of the pyridine, phosphines and bromide ligands. The study of the complex derived from the oxidative addition of 2-bromopyridine to Pd(PPh$_3$)$_4$, **12**, was carried out by dissolving the complex with CH$_3$CN/H$_2$O and injecting the sample into the spectrometer at different cone voltages (from 5 V to 40 V). In ESI-MS, fragmentation can often be induced by applying a higher cone voltage, which results in lower charged ions. Furthermore, increasing the cone voltage can minimize aggregates or adducts of ions with solvent molecules [3]. Table 7.4 (Entries 1–4) shows the species detected by ESI-MS of complex **12** at different cone voltages. At low voltages, only binuclear oxidative addition species of type **I21** were observed with different ligands (PPh$_3$, CH$_3$CN or Br$^-$) at the coordination sphere of the two palladium atoms proceeding from a ligand exchange (Entries 1–3). When the voltage was increased, fragmentation to monopalladium species **I22** occurred (Entry 4). These experiments give support to the hypothesis that dinuclear structures are maintained in solution. Another complex studied in ESI-MS was that resulting from the oxidative addition of 2-bromopyridine to Pd(dppp)$_2$ (dppp: 1,3-bis-(diphenylphosphino)propane), **13**. In the case of bidentate phosphines, the fact that these are bridging phosphines increases the possibilities of ligand coordination, ligand

Table 7.4 Species detected by ESI-MS of **12** and **13** at different cone voltages.

Entry	Palladium complex	Cone voltage (V)	Species observed by ESI-MS
1	12	5	[**I21a**]$^+$ (L = PPh$_3$, S = CH$_3$CN, X = Br: m/z = 1013) [**I21b**]$^{2+}$ (L = PPh$_3$, S = CH$_3$CN, X = CH$_3$CN: m/z = 487)
2	12	20	[**I21c**]$^{2+}$ (L = PPh$_3$, S = X = none: m/z = 446) [**I21d**]$^{2+}$ (L = PPh$_3$, S = CH$_3$CN, X = none: m/z = 466)
3	12	30	[**I21c**]$^{2+}$ (L = PPh$_3$, S = X = none: m/z = 446) [**I21d**]$^{2+}$ (L = PPh$_3$, S = CH$_3$CN, X = none: m/z = 466) [Ph$_3$PC$_5$H$_4$N]$^+$ (m/z = 340)
4	12	40	**I22a** (L = PPh$_3$, S = CH$_3$CN, X = none: m/z = 487) **I22b** (L = PPh$_3$, S = X = none: m/z = 446) [**I21e**]$^+$ (L = PPh$_3$, S = none, X = Br: m/z = 972)
5	13	5–40	**I23** (n = 1: m/z = 843; n = 2: m/z = 884) **I24** (m/z = 676), **125** (m/z = 1089), **126** (m/z = 1008), **I27** (m/z = 596)

7.3 ESI-MS Studies in the Identification of Oxidative Addition Intermediates

$$\left[\begin{array}{c}\text{L, N=N, X}\\ \text{Pd, Pd}\\ \text{S, N=N, L}\end{array}\right] \quad \left[\begin{array}{c}\text{S, L}\\ \text{Pd}\\ \text{X, N}\end{array}\right]^{\oplus} \quad \left[\begin{array}{c}\text{Ph}_2\text{P, N=N, P, Ph}_2\text{PPh}_2\\ \text{Pd, Pd}\\ \text{P, N=N, P, PPh}_2\\ \text{Ph}_2 \quad \text{Ph}_2\end{array}\right]^{2+} + n\text{HBr}$$

I21 **I22** **I23**

$$\left[\text{PdH(Br)(pyr)(dppp)}\right]^{\oplus} \quad \left[\text{PdH(Br)(pyr)(dppp)}_2\right]^{\oplus} \quad \left[\text{Pd(pyr)(dppp)}_2\right]^{\oplus} \quad \left[\text{Pd(pyr)(dppp)}\right]^{\oplus}$$

I24 **I25** **I26** **I27**

transformation and molecular fragmentation. This was clear in the ESI-MS, where a greater number of molecular peaks were seen in comparison with PPh$_3$ (Entry 5). Moreover, the fragmentation of the binuclear complex **13** was facilitated by the presence of HBr, which caused the breaking of the pyridine ligands bridging the two palladiums. This gave rise to the mononuclear species **I24–I27**.

As a result of this study, the authors concluded that the structure of the phosphine and the ligands present in solution influenced the formation of different structures of the palladium species in solution, which may in turn influence the catalytic capacity of the different compounds.

Jutand et al., who specialize in mechanistic studies of palladium-catalyzed cross-coupling processes, have occasionally employed ESI-MS in their studies. They particularly use amperometry and conductivity techniques. The intermediates postulated are also characterized by other spectroscopies such as IR, NMR, and mass spectrometry. The species detected by ESI-MS will be referred to here (Table 7.5).

Their first study was based on the rate of oxidative addition of aryl triflates to Pd(PPh$_3$)$_4$ in a coordinating solvent such as DMF [19]. In one case, the authors isolated a

Table 7.5 Oxidative addition species detected by ESI-MS of R-OTf to Pd(PPh$_3$)$_4$.

$$\text{Pd(PPh}_3)_4 + \text{R-OTf} \longrightarrow \left[\text{R-Pd(PPh}_3)_n(\text{DMF})_m\right]$$
$$\textbf{I28}$$

Entry	R	Species observed by ESI-MS	Reference
1	Cl–C$_6$H$_4$–	**I28** (n = 2, m = 0: m/z = 741)	[19]
2	CH$_2$=C(nC$_4$H$_9$)–	**I28** (n = 2, m = 0: m/z = 713)	[20]
3	(CH$_3$)$_2$C=C$_6$H$_4$–	**I28** (n = 2, m = 0: m/z = 767)	[20]

complex from the p-ClC$_6$H$_4$OTf with Pd(PPh$_3$)$_4$ in DMF. The ESI mass spectrum of this complex exhibited a peak at $m/z = 741$ corresponding to the mass of the oxidative addition cation [p-ClC$_6$H$_4$Pd(PPh$_3$)$_2$]$^+$ (Entry 1). A second example was the oxidative addition of vinyl triflates to Pd(PPh$_3$)$_4$ in DMF, giving in the ESI mass spectra the cationic complexes [(η^1-vinyl)Pd(PPh$_3$)$_2$]$^+$ when the vinyl radicals are prop-1-en-2-yl (Entry 2) and 4-*tert*-butyl-1-cyclohexen-1-yl (Entry 3) [20]. In contrast to the cationic complexes derived from aryl triflates, these vinyl intermediates were less stable since they decomposed to phosphonium salt [vinyl-PPh$_3$]$^+$TfO$^-$ and Pd(0) complexes. The two phosphonium salts were also detected by ESI-MS at $m/z = 345$ and $m/z = 399$ respectively.

7.4
ESI-MS Studies in Mizoroki-Heck and Related Reactions

Another of the standard palladium-catalyzed C—C bond formations is the Mizoroki-Heck reaction. This reaction is based on the palladium-catalyzed coupling of olefins with aryl or vinyl halides under basic conditions [21]. The catalytic cycle that is typically proposed for this reaction is outlined in Scheme 7.7. The first step of the reaction postulated in the mechanism is an oxidative addition of R^1-X to a Pd(0) complex. The next step is the insertion of the alkene to the Pd complex I1. In order for this to be possible, an uncharged ligand has to break away giving a neutral Pd(II) complex that will be coordinated by the olefin. The insertion of the alkene into the Pd—R^1 bond results in the C—C bond-forming step to give I30. Rotation around the C—C bond and β-hydride elimination yields the new substituted olefin and intermediate I31. Regeneration of the active catalytic species occurs by the addition of a base.

The first studies to detect intermediates in this reaction were performed by Brown *et al.* in 1996 [22] and 1997 [23]. However, this group did not undertake an exhaustive ESI-MS analysis but rather made a mechanistic study using ^{31}P and ^{13}C NMR spectroscopy especially. In these two studies the authors identified crucial intermediates in the Pd-catalyzed Heck arylation of methyl acrylate 15a and 2,3-dihydrofuran 18 (Scheme 7.8). The aryl halide complexes [P$_2$Pd(Ar)(X)] generated by oxidative addition of aryl halides to palladium complexes stabilized by bidentate phosphines were treated with AgOTf in order to obtain the respective cationic complexes [(P$_2$Pd)Ar]$^+$ 14. In the case of the oxidative addition complex generated by the reaction of 3,5-trifluoromethylphenyl bromide with a diphenylphosphinoferrocene-stabilized palladium complex [22], a solid was isolated at 0 °C giving a peak at $m/z = 873$ corresponding to the [(P$_2$Pd)Ar]$^+$ cation 14 in the ESI mass spectrum.

The addition of an excess of one of the two above-mentioned olefins to the previously prepared cationic complexes 14 allowed the authors to observe by ESI-MS intermediates I32 and I34 at $m/z = 875$, corresponding to the olefin insertion step, as well as compounds I33 at $m/z = 747$ and I35 at $m/z = 799$ derived from the addition of Pd—H to the olefins 15a and 18, respectively (Scheme 7.8).

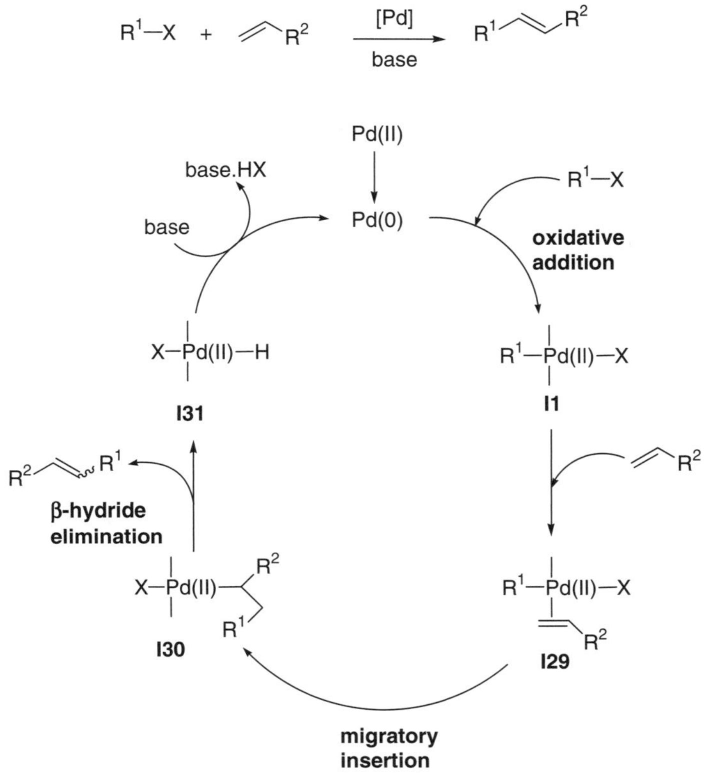

7.4 ESI-MS Studies in Mizoroki-Heck and Related Reactions | 241

$$R^1\text{--}X + \diagup\!\!\diagup R^2 \xrightarrow[\text{base}]{[Pd]} R^1\diagup\!\!\diagup\!\!\diagdown R^2$$

Scheme 7.7 General catalytic cycle for the Mizoroki-Heck reaction.

A more extensive use of the ESI-MS technique in Mizoroki-Heck reactions, also following on from Canary's work, was undertaken by Hallberg et al. [24] (Scheme 7.9). The synthetic study was based on the preparation of spiro compounds of type **21** constructed by palladium-catalyzed intramolecular arylation of cyclic enamidine **20**. The authors managed to synthesize the three possible isomers deriving from the migration of the double bond using different ligands. In order to investigate the mechanistic aspects further, an electrospray ionization mass spectrometric study was conducted to detect any catalytic intermediate and observe the role of the amidine nitrogen atom in this process. Given that the presence of saturated amidine **22** in the starting material resulting from its synthesis impeded the cyclization of **20**, the authors hypothesized that amidine **22** deactivated the catalyst. In order to advance the reaction and obtain the spiro derivative, it was necessary to add methyl acrylate to the mixture. Reactions of a 3:1 mixture of **20**:**22** with Pd(OAc)$_2$ and two different phosphines, PPh$_3$ and P(o-tol)$_3$, were studied by ESI-MS. In the case of both phosphines, the intermediates detected corresponded to the oxidative addition of amidine **22** (intermediates **I36**) but not of enamidine **20**. When amidine **22** was not added to the reaction mixture, the oxidative addition intermediate **I37** corresponding to the addition of iodoenamidine **20** was detected. The authors deduced that in the

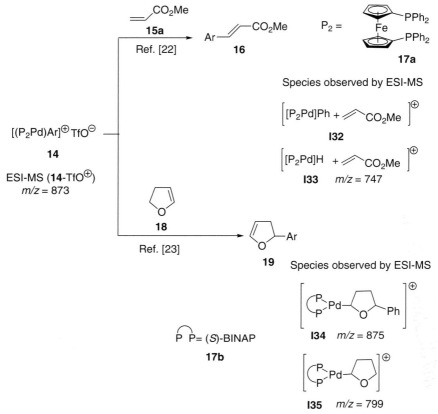

Scheme 7.8 Reactions studied and intermediates detected by ESI-MS in Mizoroki-Heck reactions by Brown.

intramolecular oxidative addition of the iodoenamidine or iodoamidine to Pd(0), the amidine nitrogen acted as a ligand displacing one of the phosphine ligands. The addition of methyl acrylate to the reaction mixture made intermediates **I36** disappear, forming final product **23**, even though no new peaks in the ESI-MS spectra corresponding to the intermolecular insertion of methyl acrylate **15a** to give **23** were observed. Scheme 7.9 shows the ESI mass spectrometry study and the Pd species postulated in this study.

Other interesting Heck reaction studies are those employing aryldiazonium salts as electrophilic agents. These are good electrophilic partners which furthermore give cationic intermediates making them ideal for ESI-MS study. In 2003, Roglans *et al.* [25] described the first Matsuda-Heck reaction of aryldiazonium salts with acrylates and styrene with a recoverable Pd-triolefinic azamacrocyclic ligand **24** as a catalyst (Scheme 7.10). In this synthetic study preliminary ESI mass spectrometry investigation was undertaken and a full study was published later [26]. 15-Membered azamacrocyclic palladium complex **24** and several aryldiazonium salts **7a–d** were mixed in methanol at room temperature until nitrogen gas bubbling was finished

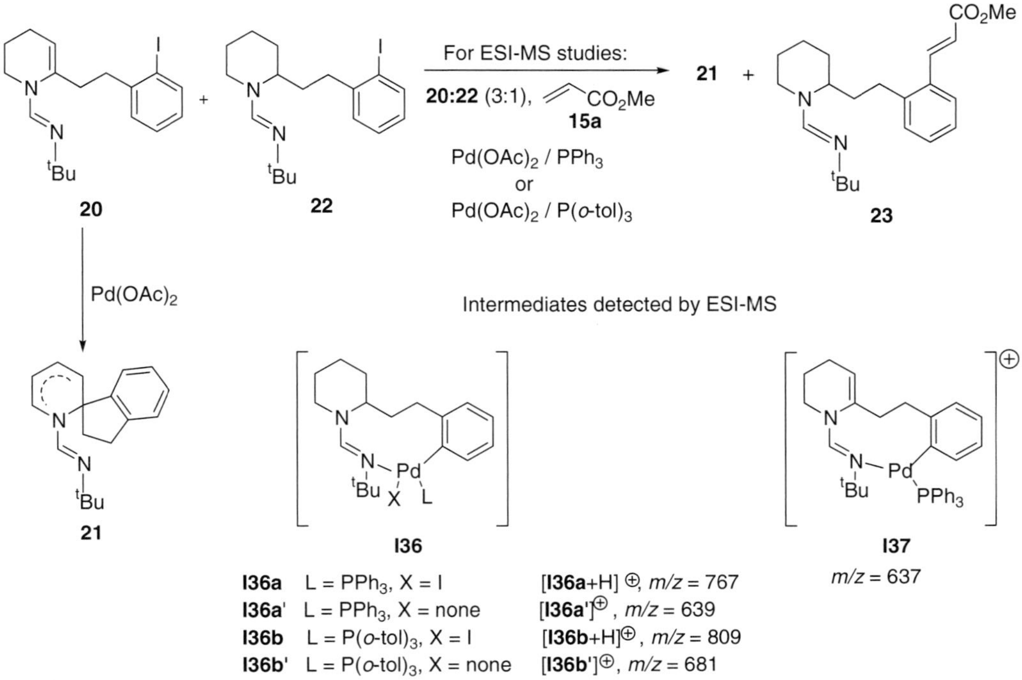
For ESI-MS studies:
20:22 (3:1), CO2Me
15a
Pd(OAc)2 / PPh3
or
Pd(OAc)2 / P(o-tol)3
21 +
CO2Me
20
22
23
Pd(OAc)2
21
Intermediates detected by ESI-MS
I36
I37
m/z = 637
I36a L = PPh3, X = I [I36a+H]⊕, m/z = 767
I36a' L = PPh3, X = none [I36a']⊕, m/z = 639
I36b L = P(o-tol)3, X = I [I36b+H]⊕, m/z = 809
I36b' L = P(o-tol)3, X = none [I36b']⊕, m/z = 681

Scheme 7.9 Intramolecular Mizoroki-Heck arylation of enamidine **20** and intermediates observed by ESI-MS.

(see Table 7.6). In all cases a peak corresponding to the oxidative addition species **I38** was observed (Entries 1–4). However, when the olefinic compound (either ethyl acrylate or *tert*-butyl acrylate) was added to the previous reaction mixture and monitored by ESI-MS, no new peaks containing the olefin could be detected. The disappearance of oxidative addition species and the reappearance of the Pd(0) catalyst **24** was observed in the ESI-MS, proving the recovery of the catalyst.

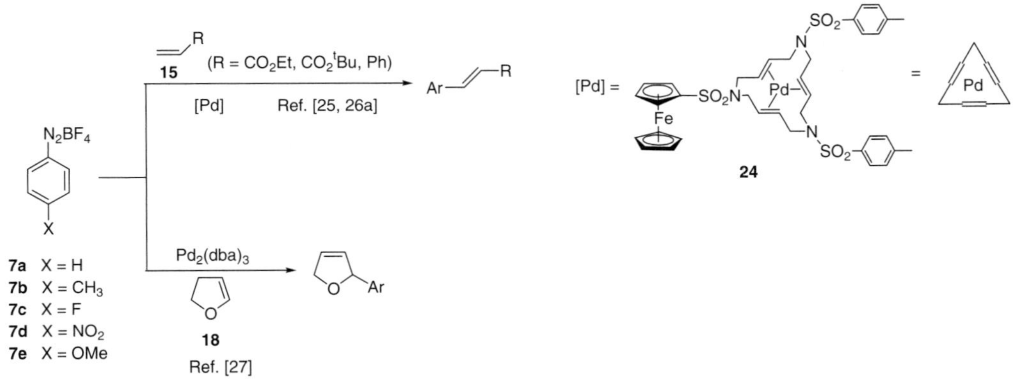
15
(R = CO2Et, CO2tBu, Ph)
[Pd]
Ref. [25, 26a]
Ar–CH=CH–R
N2BF4
X
7a X = H
7b X = CH3
7c X = F
7d X = NO2
7e X = OMe
Pd2(dba)3
18
Ref. [27]
Ar
[Pd] =
Fe
24
Pd

Scheme 7.10 Matsuda-Heck reactions with aryldiazonium salts studied by ESI-MS.

Table 7.6 Species detected by ESI mass spectrometry of reactions shown in Scheme 7.10.

Entry	Reaction mixtures	Species observed by ESI-MS	Ref.
1	7a + [Pd], 24	I38a (X = H, m/z = 946)	[25,26a]
2	7b + [Pd], 24	I38b (X = CH$_3$, m/z = 960)	[25,26a]
3	7c + [Pd], 24	I38c (X = F, m/z = 964)	[25,26a]
4	7d + [Pd], 24	I38d (X = NO$_2$, m/z = 991)	[25,26a]
5	7e + Pd$_2$(dba)$_3$	I39 (n = 2, m = 0: m/z = 295)	[27]
		I39′ (n = 3, m = 0: m/z = 336)	
		I40 (n = 1, m = 1: m/z = 488)	
6	7e + Pd$_2$(dba)$_3$ + 2,3-tetrahydrofuran 18	I41 (n = 1, m = 0: m/z = 324)	[27]
		I42 (n = 0, m = 1: m/z = 517)	
		I43 (n = 1, m = 1: m/z = 558)	

I38

I39 n = 2, m = 0
I39′ n = 3, m = 0
I40 n = 1, m = 1

I41 n = 1, m = 0
I42 n = 0, m = 1
I43 n = 1, m = 1

Correia, Eberlin et al. [27] conducted an analog study using Pd$_2$(dba)$_3$ as the catalyst. In this case, however, an ESI-MS/MS study is performed to characterize structurally the intermediates detected. They first studied the oxidative addition step using 4-MeOPhN$_2$BF$_4$, **7e**, as the diazonium salt and [Pd$_2$(dba)$_3$].dba in acetonitrile. In this first set of experiments, cationic Pd(II) species **I39–I40** containing CH$_3$CN, and in some cases dba, were observed (Entry 5, Table 7.6) corresponding to the oxidative addition of the aryldiazonium salt to palladium catalyst. Species **I39–I40** only differ from each other in the number and type of ligands coordinated to the palladium atom. Intermediate **I40** was found to be the most stable in solution after 90 minutes of mixing. On adding 2,3-dihydrofuran (**18**), the insertion of the olefin to the palladium species giving intermediates **I41–I43** was observed (Entry 6, Table 7.6). The authors repeated these experiments using two other olefins, 3,4-dihydro-2H-pyran and N-Boc-4,5-dihydro-1H-pyrrole, and observed analog intermediates. The structural assignments of species **I41–I43** were studied by means of tandem mass spectrometry MS/MS. For intermediate **I43**, which contained one acetonitrile molecule and one dba molecule, dissociation chemistry led to one fragment at m/z = 517 and another at m/z = 283, corresponding to the consecutive loss of these two ligands. Two other species were detected corresponding to steps occurring after the reductive elimination stage of this type of catalytic processes. These are [HPd(CH$_3$CN)(dba)]$^+$ (**I44**) at m/z = 382 and [HPd(dba)]$^+$ at m/z = 341.

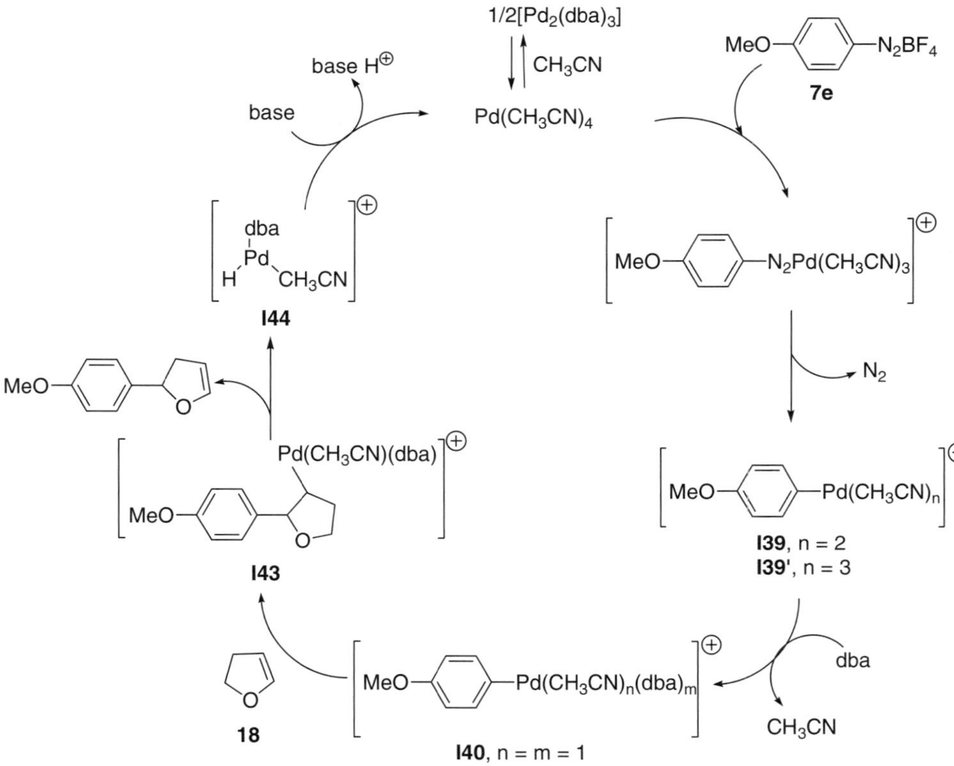

Scheme 7.11 Mechanism probed by ESI-MS of the Matsuda-Heck reaction with aryldiazonium salts.

This represents the first ESI-MS complete study of Matsuda-Heck arylation with aryldiazonium salts and allowed the authors to propose a detailed catalytic cycle for this reaction, outlined in Scheme 7.11. The main cationic intermediates of this catalytic cycle were detected and structurally characterized directly from the reaction mixture.

Another Mizoroki-Heck mechanistic study using ESI-MS was undertaken by Nilsson et al. [28]. In this case, the specific reaction is the microwave-assisted arylation of n-butyl vinyl ether (**25**), an electron-rich olefin, with aryl triflates in the presence of different palladium sources and with the addition of chelating ligands. Using aryl triflates as electrophilic partners also generates cationic intermediates, which may be more easily detected by ESI-MS. Scheme 7.12 shows the reactions studied by ESI-MS in this methodological work.

Two specific reactions were studied: the arylation of n-butyl vinyl ether (**25**) with two different triflates, 1-naphthyl triflate (**26**), and 4-cyanophenyl triflate (**27**). The four catalytic systems studied were based on palladacycles **30** and **31**, pincer complex **32**, Pd(OAc)$_2$ and polyurea encapsulated Pd EnCat30. The authors used dppp (1,3-bis(diphenylphosphino)propane) (**33**) and (2S,4S)-(-)-2,4-bis(diphenylpho-

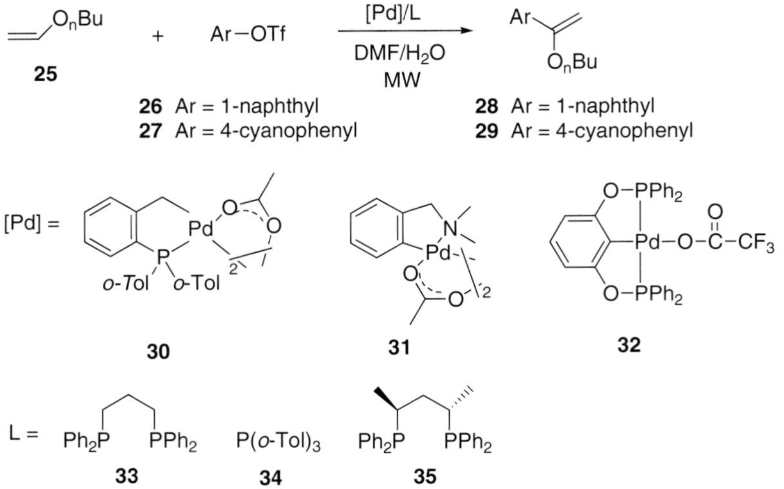

Scheme 7.12 Mizoroki-Heck reactions with n-butyl vinyl ether **25** studied by ESI-MS.

sphino)pentane (**35**) as the bidentated phosphine ligands, and monodentate phosphine **34**. The combination of two palladium catalysts and phosphines used in this study is shown in Table 7.7. ESI-MS analysis of the different reactions heated at 120 °C at predetermined time intervals showed two significant palladium species as detected cationic intermediates (Table 7.7): (i) species **I45** and **I46** corresponding to ligand exchange with additional bidentate phosphine from the initial palladium complex **30** and **31** (Entries 1 and 7), and (ii) species **I47**, **I48** and **I49** corresponding to the oxidative addition of the aryl triflate to the Pd(0)-bidentate phosphine complex and the subsequent loss of the initial palladacycle of the catalyst. These intermediates were observed as bidentate ligand-chelated cationic aryl palladium species. Species **I45** was confirmed by collision-induced MS/MS experiments demonstrating the elimination of tri-o-tolyl phosphine (**34**) and dppp (**33**).

Table 7.7 Palladium species detected by ESI-MS of reactions shown in Scheme 7.12.

Entry	Aryl triflate	Palladium source	Additive	Species observed by ESI-MS
1	26	30	33	**I45** ($m/z = 821$), **I47** ($m/z = 1057$), **I47'** ($m/z = 1073$)
2	26	Pd(OAc)$_2$	33 + 34	**I45** ($m/z = 821$), **I47'** ($m/z = 1073$)
3	27	Pd(OAc)$_2$	33	**I48'** ($m/z = 1048$)
4	27	Pd(OAc)$_2$	35	**I49** ($m/z = 1104$)
5	27	Pd EnCat30	33	**I48'** ($m/z = 1048$)
6[a]	27	30	33	**I45** ($m/z = 821$), **I48'** ($m/z = 1048$)
7	27	31	33	**I46** ($m/z = 652$), **I48'** ($m/z = 1048$)
8	27	32	33	**I48'** ($m/z = 1048$)

[a]When the reaction was run at 70 °C instead of 120 °C, only intermediate **I45** was observed.

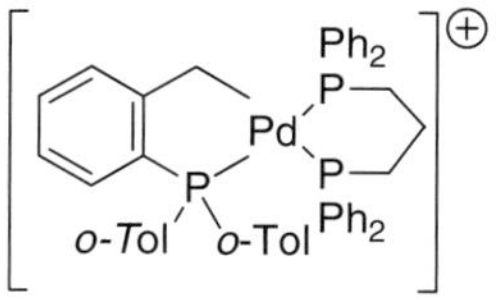

[Structures 145, 146, 147, 147', 148', 149 with:
Ar = 1-naphthyl { 147 R = PPh₂ ; 147' R = P(=O)Ph₂ }
Ar = 4-cyanophenyl { 148' R = P(=O)Ph₂ }]

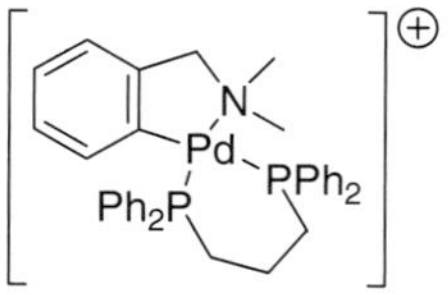

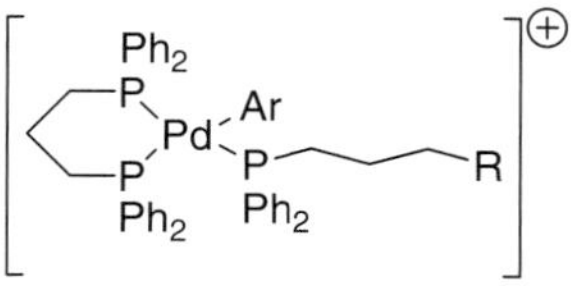

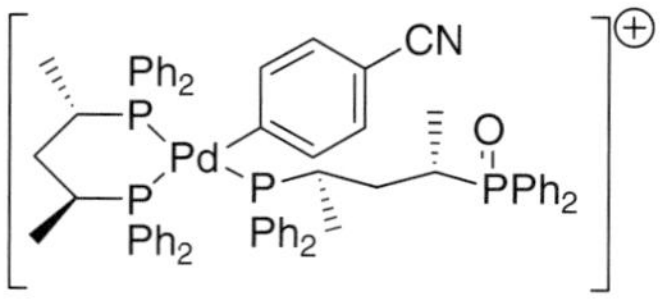

The insertion of *n*-butyl vinyl ether (**25**), which would be the next step in the mechanism postulated by the Mizoroki-Heck reaction, was not observed in any case. The authors suggest that such complexes were either present in very low concentrations or just too unstable to be detected by ESI-MS.

The same authors also used ESI-MS/MS to study an air-promoted oxidative Heck reaction between arylboronic acids and enamides as electron-rich olefins (Scheme 7.13) [29]. In addition to the conventional Mizoroki-Heck reaction of unsaturated compounds with organic halides or triflates as electrophiles, the use of nucleophilic organometallic reagents, such as organoboron compounds, has

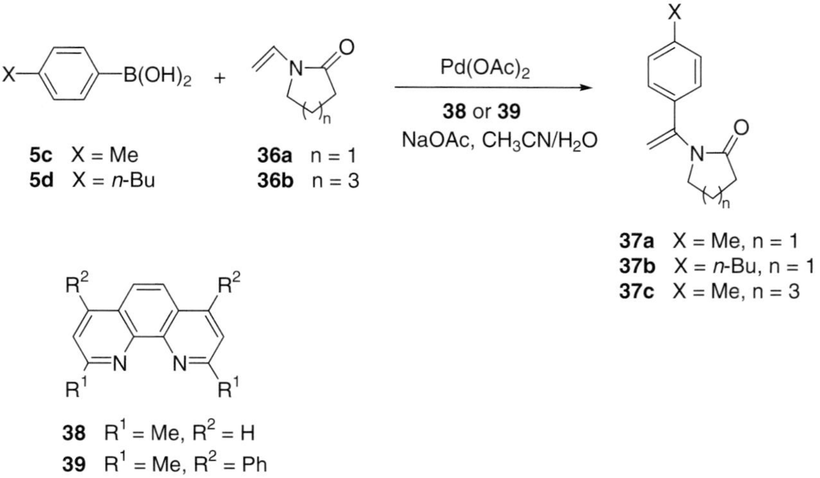

5c X = Me
5d X = *n*-Bu

36a n = 1
36b n = 3

Pd(OAc)$_2$
38 or **39**
NaOAc, CH$_3$CN/H$_2$O

37a X = Me, n = 1
37b X = *n*-Bu, n = 1
37c X = Me, n = 3

38 R^1 = Me, R^2 = H
39 R^1 = Me, R^2 = Ph

Scheme 7.13 Oxidative Heck reactions investigated by ESI-MS.

also been studied [30]. A proposed catalytic cycle for this oxidative coupling suggests an aryl transfer from the organoboron derivative to palladium(II) complex followed by olefin insertion and β-hydride elimination to afford the desired product and recover Pd(0) species, which are oxidized to Pd(II) reinitiating the catalytic cycle [30c,31]. The work of Sjöberg, Larhed *et al.* was based on the oxidative Heck reaction between arylboronic acids **5c** and **5d** and enamides **36a** and **36b** using Pd(OAc)$_2$ and bidentate phenanthroline ligands **38** and **39** as the catalytic system.

The study started by selecting an oxidative Heck coupling between arylboronic acid **5c**, enamide **36a**, and 2,9-dimethyl-1,10-phenanthroline (**38**) as the ligand. The aliquots were taken from the reaction mixture after 3 h at room temperature and diluted with acetonitrile in a 1 : 10 ratio before being injected into the mass spectrometer. Table 7.8, showing the detected organopalladium intermediates found in this study, supports the typically catalytic cycle of an oxidative Heck reaction. Entry 1 shows all the intermediates observed during the whole process.

Table 7.8 Palladium species detected by ESI mass spectrometry of reactions shown in Scheme 7.13.

Entry	Arylboronic acid	Enamidine	Catalytic system	Species observed by ESI-MS
1	5c	36a	Pd(OAc)$_2$/38	**I50A** (X = OAc, L = none: m/z = 373) **I50A'** (X = OAc, L = 38: m/z = 581) **I51A** (X = OAc, L = CH$_3$CN: m/z = 455) **I52A** (L = none: m/z = 405) **I52A'** (L = CH$_3$CN: m/z = 446) **I52A''** (L = 38: mz = 613) **I53A** (m/z = 516)
2	5c	36a	Pd(OAc)$_2$/39	Experiment to confirm species **I50** **I50B** (X = OAc, L = none: m/z = 525) **I50B'** (X = OAc, L = 39: m/z = 885)
3	5c	36a	Pd(OAc)$_2$/39	Experiment to confirm species **I52** **I52B** (L = none: m/z = 557) **I52B'** (L = CH$_3$CN: m/z = 598) **I52B''** (L = 39: m/z = 917)
4	5d	36a	Pd(OAc)$_2$/38	Experiment to confirm species **I52** **I52C** (L = none: m/z = 447) **I52C'** (L = CH$_3$CN: m/z = 488) **I52C''** (L = 38: m/z = 655)
5	5c	36a	Pd(OAc)$_2$/39	Experiment to confirm species **I53** **I53B** (m/z = 668)
6	5c	36b	Pd(OAc)$_2$/38	Experiment to confirm species **I53** **I53C** (m/z = 544)
7	5d	36a	Pd(OAc)$_2$/38	Experiment to confirm species **I53** **I53D** (m/z = 558)

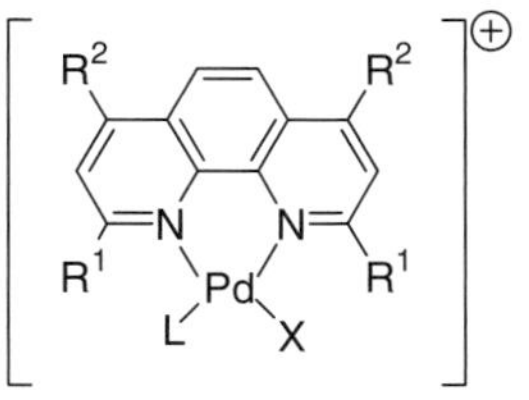

I50 X = OAc, L = none
I50' X = OAc, L = **38** or **39**

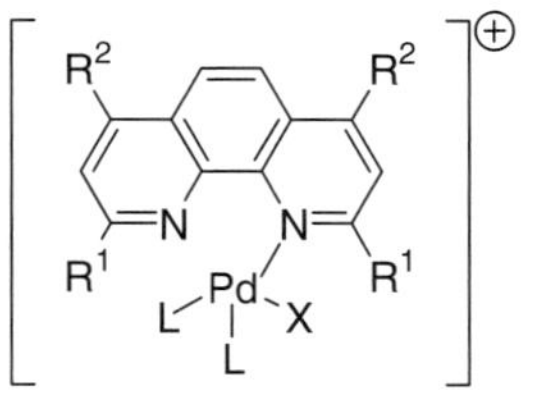

I51 X = OAc, L = CH$_3$CN

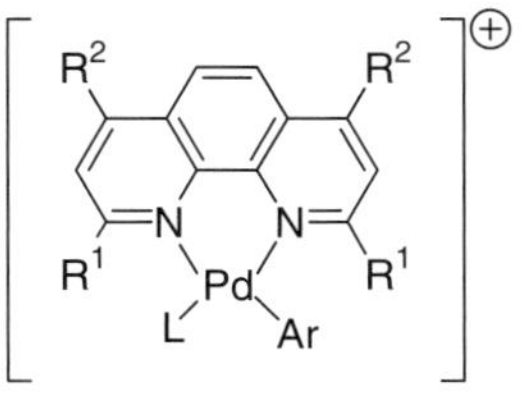

I52 L = none
I52' L = CH$_3$CN
I52" L = **38** or **39**

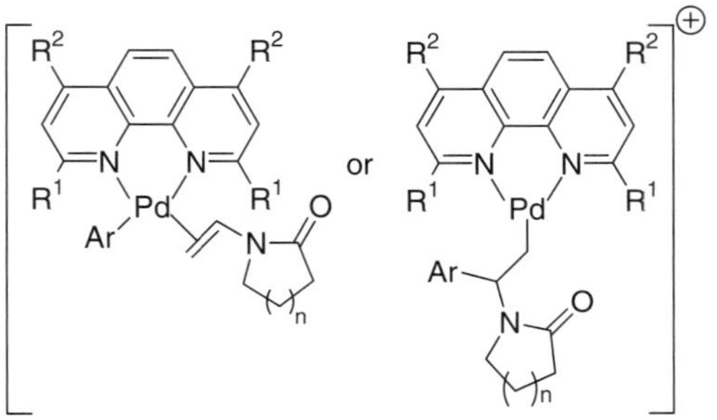

I53

In order to confirm the species identified in the standard reaction (Entry 1), the components were separately substituted by homologous derivatives. An ESI mass spectrum was recorded for each of these specific reactions (Entries 2–7). Further study of palladium-containing cations was performed by MS/MS. The same type of intermediates was detected in all cases. The first intermediates observed were the palladium species coordinated to ligands **38** and **39**. These ligands behaved either as bidentates (**I50, I50'**) or monodentates (**I51**). Oxidative addition species of type **I52** were also observed with different ligands in the coordination sphere of the palladium atom. The structure of these ions was validated by MS/MS experiments. Finally, species **I53**, which correspond to the insertion of the olefin, could not clearly be assigned even after performing MS/MS experiments. This study permitted the authors to draw up a plausible catalytic cycle of these oxidative Heck reactions as shown in Scheme 7.14. It was suggested that bidentate ligands **38** or **39** were attached to the metal center during the entire catalytic cycle.

Apart from the catalytic systems based on Pd/phosphines typically used in Mizoroki-Heck reactions, many other types of new palladium catalysts have been developed over the last decade. Avoiding the use of the phosphine ligands is a great advantage as they usually cannot be recovered and they frequently hamper the isolation and purification of the final product. One viable alternative is the use of ligand-free palladium catalysts usually in the form of Pd(OAc)$_2$. At the high temperatures required for Mizoroki-Heck reactions, most ligand-free palladium complexes are unstable and have a tendency to form soluble Pd(0) nanoparticles [32]. The question arises as to the role played by the Pd nanoparticles formed and whether

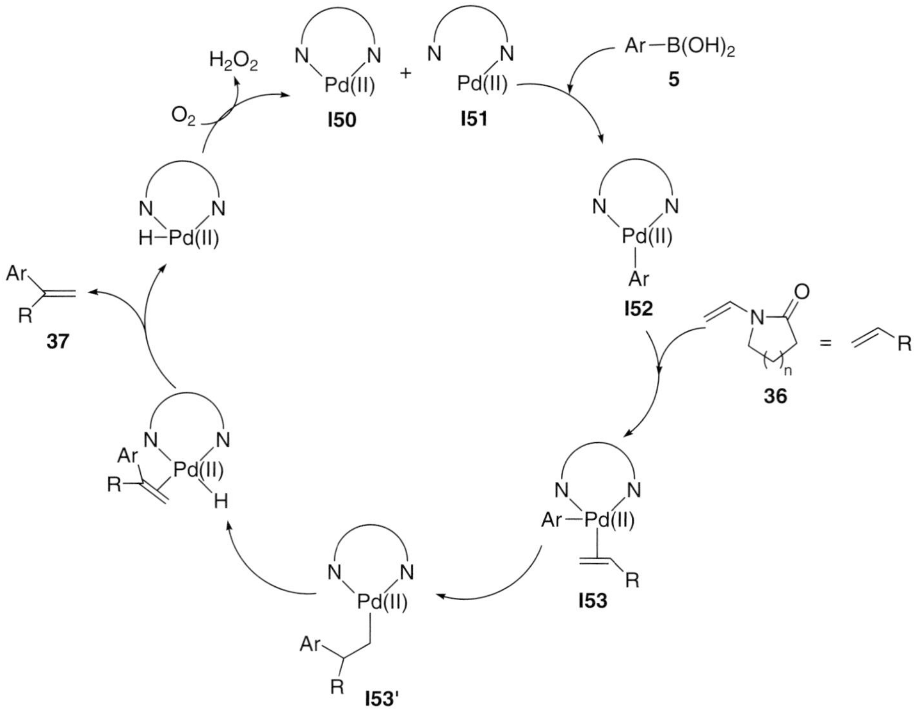

Scheme 7.14 Proposed catalytic cycle for the oxidative Heck reaction shown in Scheme 7.13.

or not they are the true catalytic species. In order to gain more mechanistic information about the ligand-free systems, spectroscopic techniques such as EXAFS and TEM together with a mass spectrometry study were conducted by de Vries *et al.* [32b–e]. The mass spectrometry analysis was performed with negative-ion electrospray-mode ESI(−) in order to observe anionic intermediates, since mechanisms based on anionic species had been proposed by Jutand *et al.* [33] for palladium phosphine catalysts in the Mizoroki-Heck reaction. Two arylation reactions were studied by de Vries *et al.* using Pd(OAc)$_2$ as the ligand-free palladium catalyst: the reaction of iodo- or bromobenzene (**40**) with butyl acrylate (**15b**) (Scheme 7.15). In the case of iodobenzene, the authors developed a practical method for the recycling and reactivation of Pd(OAc)$_2$. The precipitated palladium sediments were easily recovered by filtration and reactivated by the addition of an oxidizing agent such as I$_2$.

The species detected by negative ESI-MS are shown in Scheme 7.15. In the case of iodobenzene, a palladium(0) intermediate (**I54**) was observed at the beginning of the reaction and two palladium(II) species (**I55** and **I56a**) were detected when the conversion of the reaction was higher than 5%. Species **I55** corresponded to the oxidative addition of the iodobenzene **40a** to Pd(0). In the case of bromobenzene, the only palladium species found in the ESI(−) MS studies was [PdBr$_3$]$^-$ (**I56b**). The absence of oxidative addition species of type **I55** prompted the authors to postulate that in the case of phenyl bromide this oxidative addition step was the

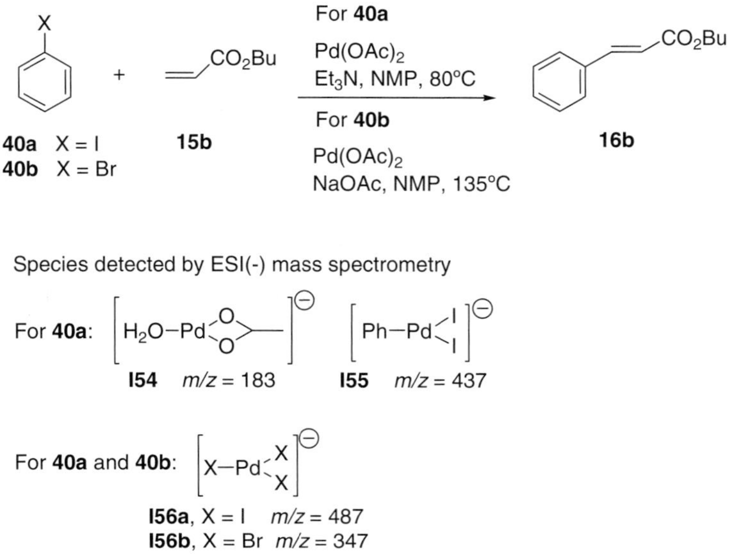

Scheme 7.15

For **40a**:
Pd(OAc)$_2$, Et$_3$N, NMP, 80°C

For **40b**:
Pd(OAc)$_2$, NaOAc, NMP, 135°C

40a X = I
40b X = Br

15b: =CO$_2$Bu

16b: product

Species detected by ESI(-) mass spectrometry

For **40a**:
$[\text{H}_2\text{O}-\text{Pd}(\text{O}_2\text{CO})]^{-}$
I54 m/z = 183

$[\text{Ph}-\text{PdI}_2]^{-}$
I55 m/z = 437

For **40a** and **40b**:
$[\text{X}-\text{PdX}_2]^{-}$
I56a, X = I m/z = 487
I56b, X = Br m/z = 347

Scheme 7.15 Ligand-free Heck reactions on halobenzenes.

rate-determining step and that most of the palladium in a zero oxidation state must be locked within the nanoparticles. These ESI-MS studies suggested that although there was a large amount of palladium in the form of soluble nanoparticles, the catalytic process really took place via monomeric or dimeric anionic species. The Pd(II) complex was rapidly reduced to Pd(0), which had a strong tendency to form soluble nanoparticles. The arylating agent attacked the Pd atoms at the rim of the nanoparticles leading to the formation of soluble anionic complexes. Scheme 7.16 shows a mechanistic proposal for a ligand-free Mizoroki-Heck reaction on iodobenzene. In a first step, Pd(II) has to be reduced to Pd(0). At the beginning of the reaction, intermediate **I54** was formed, which then underwent oxidative addition of iodobenzene to give intermediate **I55**. The authors postulated that either the underligated palladium species had a solvent molecule which was coordinated to the metal center or that the species was a dimer. The following steps in the mechanism are those which are typically proposed for the Mizoroki-Heck reaction: olefin insertion and β-hydride elimination to afford olefin **16b**. After hydride elimination, the authors postulate three possibilities: (i) a reaction with I$_2$ to form intermediate **I56a**; (ii) the formation of soluble nanoparticles, or (iii) a reaction with Ar-I.

7.5
ESI-MS Studies in Stille Cross-Coupling Reactions

The Stille reaction is the palladium-catalyzed coupling of unsaturated halides or triflates with organostannanes [34]. The accepted mechanism for this reaction is based on the three steps which typically work for this type of cross-coupling:

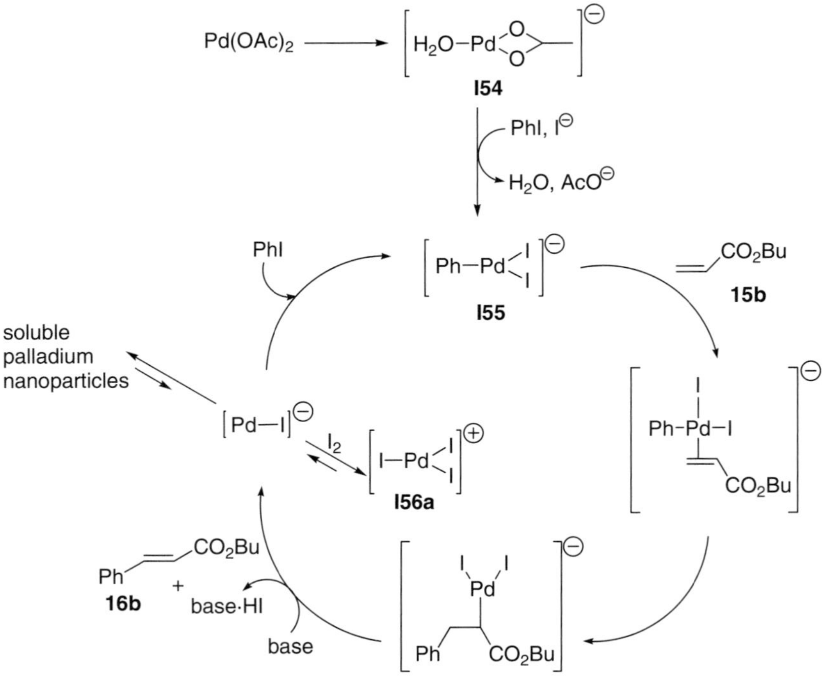

Scheme 7.16 Mechanistic proposal for a ligand-free Mizoroki-Heck reaction.

(i) an oxidative addition of the electrophilic component to Pd(0), (ii) a transmetalation step, and (iii) a reductive elimination (Scheme 7.17).

Several mechanistic studies of the Stille reaction have been undertaken based on the characterization of the main intermediates by NMR spectroscopy [35] and using electrochemical techniques [36]. However, Santos, Eberlin et al. [37] undertook an ESI-MS/MS investigation of a Stille cross-coupling in their study of 3,4-dichloro-iodobenzene (**40c**) and vinyltributyltin (**41**) using Pd(PPh$_3$)$_4$ as the catalyst (Scheme 7.18 and Table 7.9).

In a first step, the authors analyzed the species of the catalyst in the working conditions of the spectrometer. Both when starting with Pd(PPh$_3$)$_4$ and when starting with a mixture of Pd(OAc)$_2$ and PPh$_3$, species corresponding to PdL$_2$ (**I57**) were observed (Entry 1, Table 7.9). The authors postulate that species **I57** were radical cations [M]$^{•+}$ derived from oxidation at the probe tip during the ESI process of the corresponding neutral species. In a second step, oxidative addition mixing of Pd(PPh$_3$)$_4$ and the corresponding iodoarene **40c** was studied. Two species of type **I58** were detected by ESI-MS corresponding to the oxidative addition step (Entry 2, Table 7.9). ESI-MS/MS characterization of species **I58** was performed by selecting the peak at 777 for CID. The loss of a phosphine ligand PPh$_3$ was observed together with cation [PPh$_4$]$^+$ at the $m/z = 339$. In the case of **I58'**, the loss of CH$_3$CN was observed. After this, a mixture of the whole reaction was made in order to inject it into the spectrometer at regular intervals over a period of two hours (Entry 3, Table 7.9).

7.5 ESI-MS Studies in Stille Cross-Coupling Reactions

$$R^1-X + R^2SnR_3^3 \xrightarrow{[Pd]} R^1-R^2 + R_3^3SnX$$

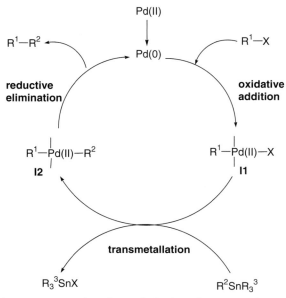

Scheme 7.17 General catalytic cycle for the Stille cross-coupling reaction.

Several new species involved in the transmetalation step were now detected: the radical cation **I59** derived from a neutral species and the cationic homolog **I60**. Tin-containing species **I61** was also observed and assigned as another transmetalation product. At $m/z = 657$, species **I62** might proceed from neutral species [ArPd

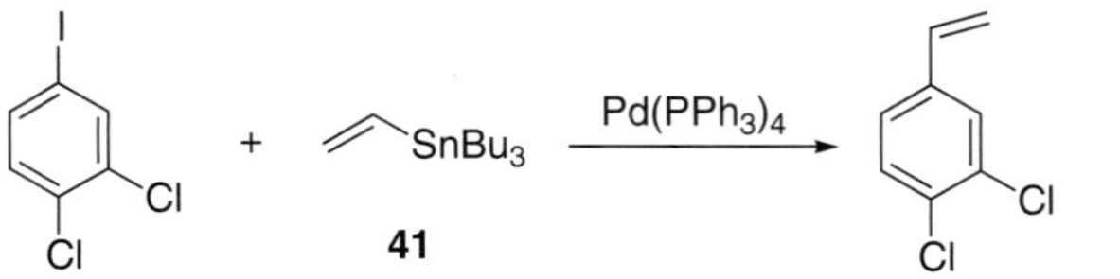

Scheme 7.18 Stille cross-coupling reaction studied by ESI-MS/MS.

Table 7.9 Species detected by ESI mass spectrometry of the reaction shown in Scheme 7.18.

Entry	Reaction mixtures	Species observed by ESI-MS
1	Pd(PPh$_3$)$_4$ or Pd(OAc)$_2$/PPh$_3$	**I57** ($m/z = 630$), **I57'** ($m/z = 689$), **I57''** ($m/z = 707$)
2	Pd(PPh$_3$)$_4$ + **40c**	**I58** ($m/z = 777$), **I58'** ($m/z = 818$)
3	Pd(PPh$_3$)$_4$ + **40c** + **41**	**I59** ($m/z = 958$), **I60** ($m/z = 569$), **I61** ($m/z = 291$), **I62** ($m/z = 657$)

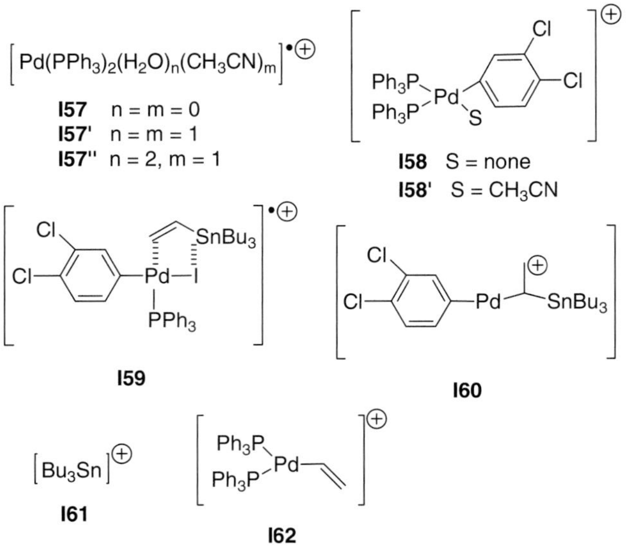

$\left[Pd(PPh_3)_2(H_2O)_n(CH_3CN)_m \right]^{\bullet \oplus}$

I57 n = m = 0
I57' n = m = 1
I57'' n = 2, m = 1

I58 S = none
I58' S = CH$_3$CN

I59

I60

$[Bu_3Sn]^{\oplus}$
I61

I62

(CH=CH$_2$)(PPh$_3$)$_2$], where on losing the aryl an easily detected cationic species remained. The transmetalated species were also subjected to CID for further characterization. Species **I59** dissociated to form a fragment ion at $m/z = 831$ corresponding to the loss of iodo and a fragment at $m/z = 569$ corresponding to the loss of PPh$_3$. Fragmentation of intermediate **I60** afforded the final product **42** and [Bu$_3$Sn-Pd]$^+$. The anionic complexes postulated by Jutand et al. [33] were not detected in the ionization negative mode ESI(−). I$^-$ was observed as the only anionic species. Scheme 7.19 shows the proposed catalytic cycle of the Stille reaction based on ESI-MS experiments (Scheme 7.17).

7.6
ESI-MS Studies in Palladium-Catalyzed Reactions Involving Allenes

Palladium-catalyzed reactions of allenes have become an important area of study in synthetic organic chemistry [38]. Efficient regio- and stereoselective palladium-catalyzed addition of organoboronic acids to allenes have been described [39]. In continuing studies of the chemistry of allenes [40], Guo, Ma et al. [41] conducted an ESI-MS/MS study of the arylation of allene **43** with arylboronic acids **5e** and **5f** using Pd(PPh$_3$)$_4$ as the catalyst and in the presence of acetic acid (Scheme 7.20).

Aliquots of the reaction mixture were diluted with methanol and transferred to the mass spectrometer at different intervals of time. The species detected and characterized with high resolution mass measurements allowed the authors to postulate a catalytic cycle, which is shown in Scheme 7.21.

At the beginning of the reaction, the first species observed was intermediate **I63**, which may proceed from the intermediate **I64**. It was suggested that **I64** was the result of an oxidative addition of acetic acid to a Pd(0) complex. In a later injection,

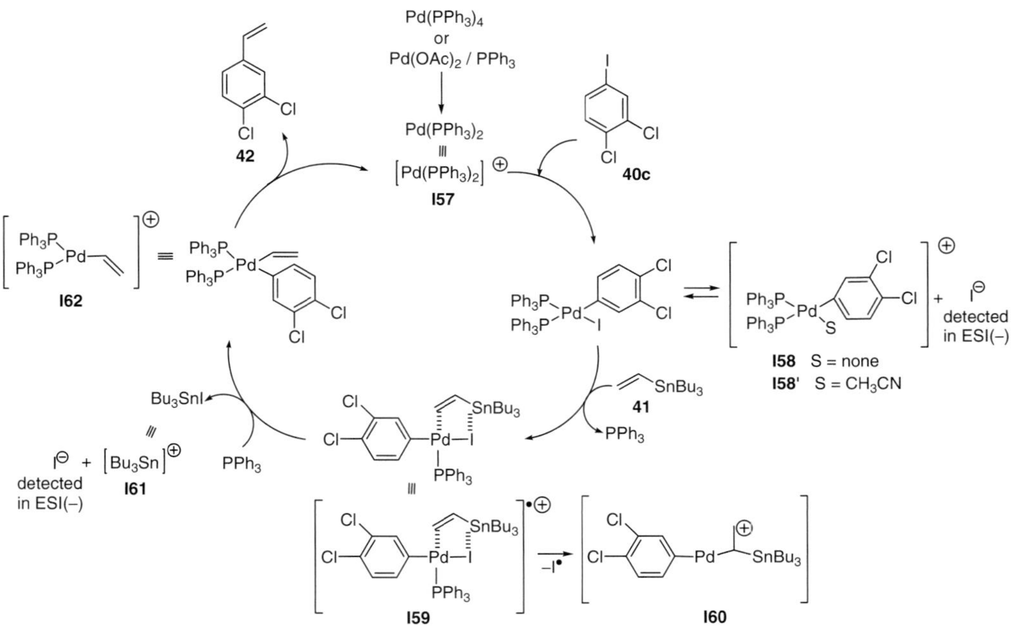

7.6 ESI-MS Studies in Palladium-Catalyzed Reactions Involving Allenes | 255

Scheme 7.19 Proposed mechanistic cycle for the Stille cross-coupling based on ESI-MS/MS studies.

intermediate **I65** was observed resulting from a hydropalladation reaction of allene **43** to intermediate **I64**. In this step of the mechanism the regioselectivity of the process was determined given that the hydropalladation took place on the less sterically hindered terminal C=C bond in the allene moiety. Another detected species was the radical cation **I66**, which was postulated as proceeding from oxidation in the mass spectrometer of the neutral HPd(PPh$_3$)$_2$-vinyl species. The transmetalation of intermediate **I65** with phenylboronic acid **5e** generated species **I68**, which was also intercepted by ESI-MS as a cationic π-allyl intermediate **I67**. The authors justified its formation as being a sequential oxidation during the ESI process and the elimination of H• from an (η^3-allyl)palladium complex. Reductive elimination of intermediate **I68** afforded the final product **44e** and Pd(0) catalytic species. When intermediates **I65**, **I66**, and **I67** were subjected to MS/MS for further structural characterization, the loss of one PPh$_3$ as a ligand was seen in all three cases, giving

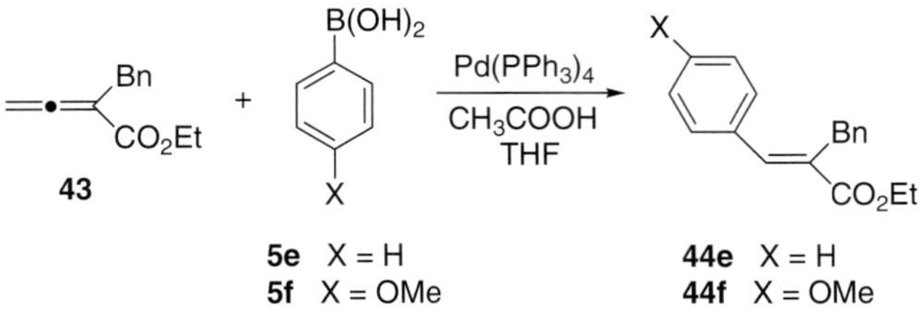

Scheme 7.20 Arylation of allene **43** with arylboronic acids **5e** and **5f** studied by ESI-MS/MS.

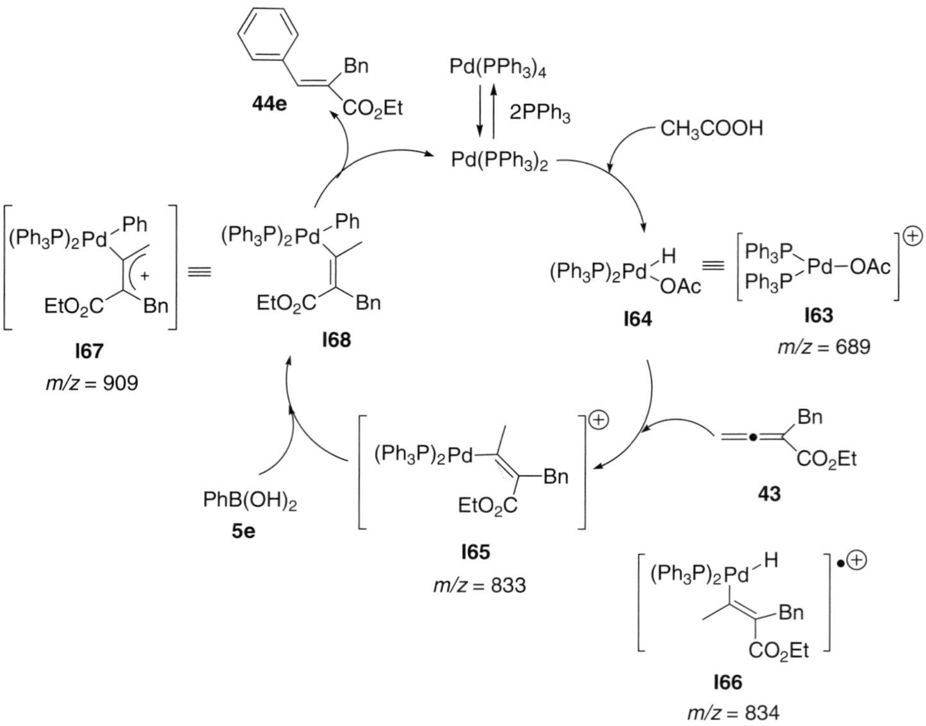

Scheme 7.21 Mechanistic proposal of Pd-catalyzed arylation of allenes probed by ESI-MS/MS.

peaks at $m/z = 571$ for [**165**-PPh$_3$]$^+$, $m/z = 572$ for [**166**- PPh$_3$]$^+$ and $m/z = 647$ for [**167**-PPh$_3$]$^+$. The same mechanistic study was performed with arylboronic acid **5f**, giving similar results.

In another study of palladium-catalyzed reactions involving allenes [38], Guo et al. described a three-component tandem double addition-cyclization reaction of allenyl malonates, aryl halides and imines to give pyrrolidine derivatives [42] (Scheme 7.22). Apart from the methodological synthetic study, the authors conducted a mechanistic study of the process using the ESI-MS technique [43]. The particular reaction studied is shown in Scheme 7.22.

The ESI-MS mechanistic study consisted of injecting samples into the mass spectrometer to which the different reactants were gradually added. The species detected in this study are shown in Table 7.10.

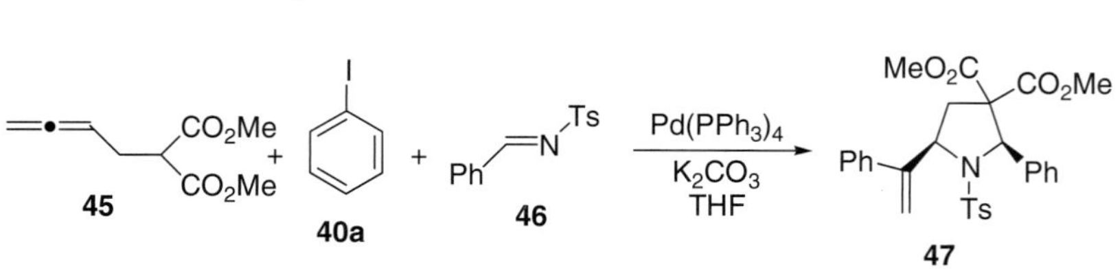

Scheme 7.22 Tandem double addition-cyclization reaction studied by ESI-MS.

Table 7.10 Species detected by ESI mass spectrometry of reaction shown in Scheme 7.22.

Entry	Mixtures of components[a]	Species observed by ESI-MS
1	45 + 40a + Pd(PPh$_3$)$_4$	**I69** ($m/z = 707$), **I70** ($m/z = 891$)
2	45 + 40a + 46 + Pd(PPh$_3$)$_4$	**I69** ($m/z = 707$), **I70** ($m/z = 891$)
3	45 + 40a + 46 + Pd(PPh$_3$)$_4$ + K$_2$CO$_3$	5 min: **I69** ($m/z = 707$), **I69'** ($m/z = 748$)
		20 min: **I69** ($m/z = 707$), **I69'** ($m/z = 748$)
		I70 ($m/z = 891$)
		40 min: **I69** ($m/z = 707$), **I69'** ($m/z = 748$)
		I70 ($m/z = 891$), **I71** ($m/z = 1150$)
		[**45** + K]$^+$ ($m/z = 223$), [**46** + K]$^+$ ($m/z = 298$), [**47** + K]$^+$ ($m/z = 558$)

[a]The mixtures were dissolved in THF and stirred at 85 °C in a nitrogen atmosphere and diluted with CH$_3$CN prior to injection into the spectrometer.

$$\left[(PPh_3)_2Pd\underset{S}{\overset{Ph}{\lessgtr}}\right]^{\oplus}$$

I69 S = none
I69' S = CH$_3$CN

I70

I71

Intermediate **I69**, corresponding to the oxidative addition of iodobenzene (**40a**) to Pd(0) species, and **I70**, proceeding from the carbopalladation of **I69** with allenylmalonate **45**, were detected in a first mixture of 2-(2,3-allenyl)malonate **45**, iodobenzene **40a** and Pd(PPh$_3$)$_4$ (Entry 1). A new mixture of all the reactants and the Pd catalyst, but excluding the base (Entry 2), was injected into the mass spectrometer after 30 min and 2 h and only **I69** and **I70** were observed. A third mixture with all the components, including the base K$_2$CO$_3$, was injected at different time intervals (Entry 3). After a 40 min reaction time, a new intermediate **I71** involving imine **46** was observed. The authors postulated that the base deprotonated the malonate moiety of the intermediate **I70** with the subsequent addition of imine **46** to afford **I71**. This intermediate suffered an intramolecular allylic amination to afford the pyrrolidine derivative **47** and regenerate the catalytic species. Other species observed in the mass spectra were the potassium adducts of neutral starting and final products. The reaction was injected into the mass spectrometer after 5 h, 24 h and 36 h. At the end, all the intermediates had disappeared and only the potassium adduct corresponding to the final product was observed. Further structural characterization of all intermediates detected in this study was undertaken with MS/MS experiments and accurate mass determination. Scheme 7.23 shows the proposed catalytic cycle of the tandem cyclization reaction given in Scheme 7.22.

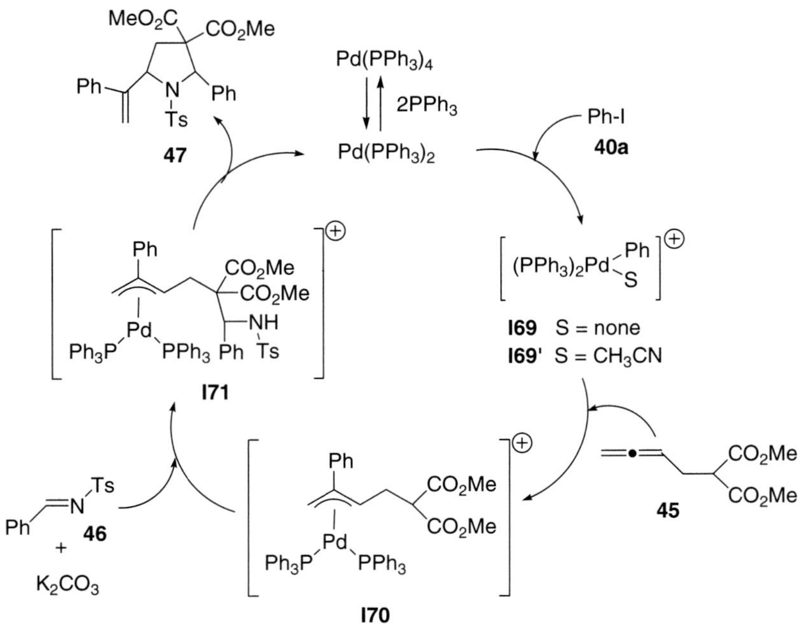

Scheme 7.23 Proposed mechanism for the Pd(0)-catalyzed three-component tandem double addition-cyclization reaction based on ESI-MS studies.

7.7
ESI-MS Studies in Palladium-Catalyzed Alkynylation Reactions

Palladium-catalyzed alkynylation is an important method for the synthesis of alkynes [44]. Comasseto et al. described the efficient coupling of vinylic tellurides with alkynes using $PdCl_2/CuCl_2$ as the catalytic system to afford enyne derivatives. In order to further understand the mechanism of this reaction, a study using the ESI-MS/MS technique was performed by the authors [45]. The particular reaction studied by mass spectrometry is shown in Scheme 7.24. The coupling retains the double-bond geometry.

The reaction shown in Scheme 7.24 was stirred for 1 h and injected into the mass spectrometer for analysis. Apart from several species proceeding from the starting vinylic telluride **48**, which were also detected in control experiments of a methanolic solution of **48**, several cationic Pd-Te intermediates were observed and structurally characterized by means of tandem mass spectrometric experiments. These are

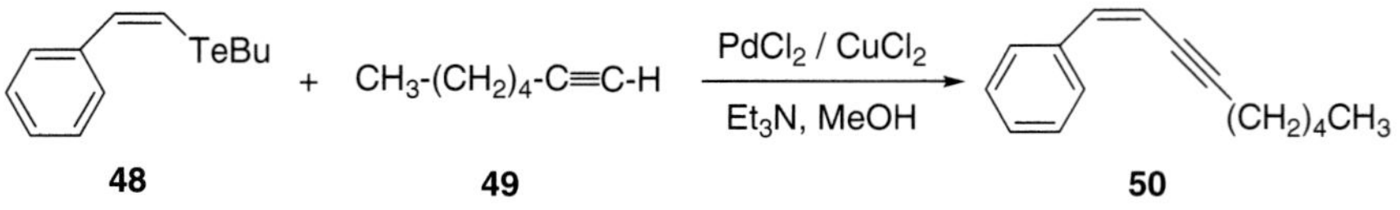

Scheme 7.24 $PdCl_2/CuCl_2$-catalyzed coupling reaction of vinylic telluride **48** with acetylene **49**.

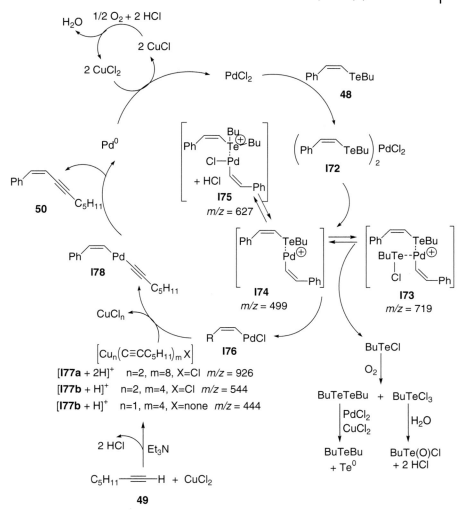

Scheme 7.25 Complete proposed catalytic cycle for the reaction shown in Scheme 7.24 based on ESI-MS/MS experiments.

essential intermediates for the validation of the proposed mechanism. Organocopper intermediates were also detected by ESI-MS and further characterized by MS/MS analysis. Given these results, the authors proposed a catalytic cycle for this reaction, which is shown in Scheme 7.25.

Palladium-tellurium cationic intermediates **I73**, **I74**, and **I75** were observed in the mass spectra. It has been suggested that these were formed by ionization of the neutral species L_nPdCl_2. Several alkynylcopper species of type **I77** were detected with the characteristic Cu isotopic pattern. Negative ion-mode ESI-MS(−) monitoring of the reaction failed to reveal any metal anions.

Sonogashira cross-coupling is another typical palladium-catalyzed process consisting of the alkynylation of aryl or alkenyl halides in the presence of Cu(I) salts [46].

In a methodological synthetic study of a Sonogashira cross-coupling of halogenated 2-pyrones with terminal acetylenes using $Pd(PPh_3)_2Cl_2/CuI$ as the catalytic system performed by Fairlamb et al. [47], the authors attempted to determine the nature of the soluble species of Pd/Cu. To this end, they performed a limited ESI-MS study. They found a species at $m/z = 636$ in ESI negative mode, which they tentatively assigned as $[PPh_3PdCu(PhC=C)_2H]^-$. None of the anionic species such as $[PdBr_3]^-$ detected in de Vries' studies, nor other related phosphine-ligated anionic palladium species, were observed. In positive mode ESI(+), two peaks were observed, which the authors postulate as being bimetallic species containing Pd and Cu. However, no further structural analysis was performed.

7.8
ESI-MS Studies in Palladium-Catalyzed Allylic Substitution Reactions

Palladium-catalyzed allylic substitution reactions, known as Tsuji-Trost reactions, are a well-established method for carbon-carbon bond forming processes [48]. The generally accepted mechanism for this reaction involves the oxidative addition of the allylic substrate to Pd(0) to provide a π-allylpalladium complex. The subsequent reaction of the electrophilic π-allylpalladium complex with the nucleophile affords the substituted product and Pd(0), which is regenerated to start the catalytic cycle (Scheme 7.26).

In an ongoing study using water as a solvent in palladium chemistry, Muzart et al. [49, 50] described the substitution of allylic acetates **51** with different nucleophiles such as acetylacetone (**52**) and sodium para-toluenesulfinate (**54**) in aqueous media using a palladium complex and the hydrophilic ligand $[(HOCH_2CH_2NH\text{-}COCH_2)_2NCH_2]_2$ (**56**). Scheme 7.27 shows the specific reactions studied by ESI-MS in order to further investigate mechanistic aspects of these reactions. Binary and ternary mixtures, as well as mixtures of the whole reaction, in 1:1 $MeOH/H_2O$ solvent media were injected into the mass spectrometer for analysis. The hydrophilic ligand **56** has some groups that are susceptible to being protonated, giving positively charged species, but it also has other groups, that are acidic, which can lose a proton to generate negatively charged species. Therefore, the authors conducted the study both in ESI(+) and ESI(−) ionization modes. The most significant data from these studies are given in Table 7.11.

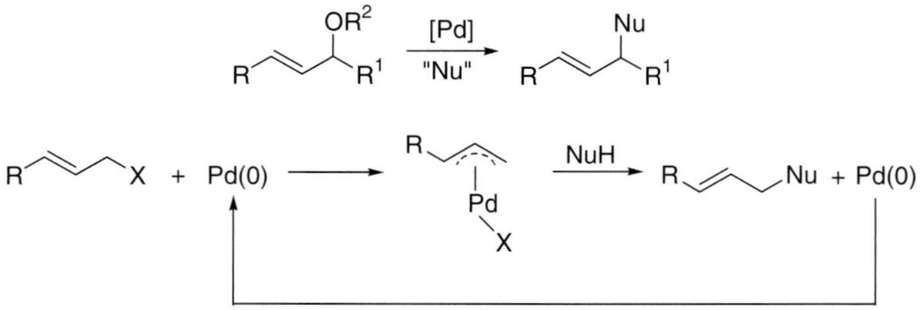

Scheme 7.26 General scheme for Tsuji-Trost reaction.

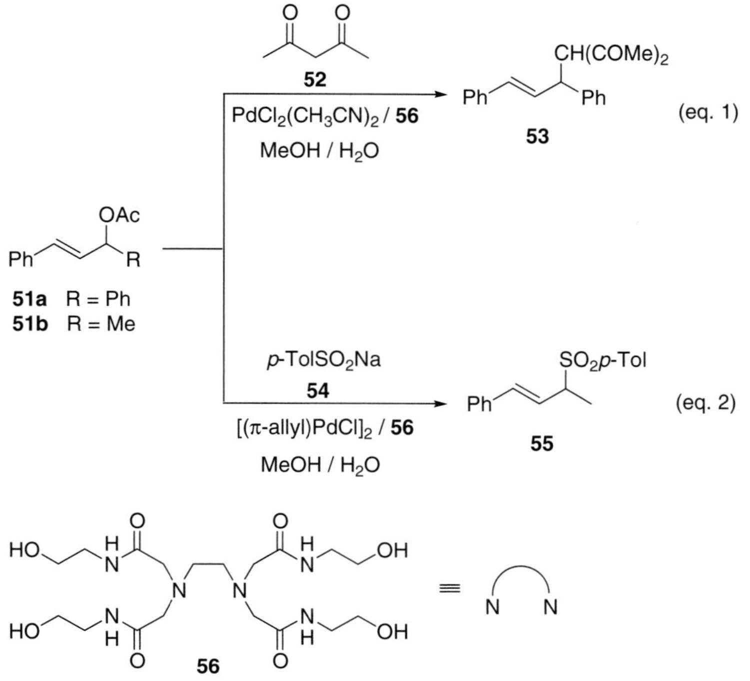

7.8 ESI-MS Studies in Palladium-Catalyzed Allylic Substitution Reactions

Scheme 7.27 Palladium-catalyzed allylic substitution reactions studied by ESI-MS.

Table 7.11 Most important palladium species detected by ESI-MS of the reactions shown in Scheme 7.27.

Entry (ref)	Mixtures of components	Species observed by ESI-MS
1 [49]	$PdCl_2(CH_3CN)_2 + 56$	$[56+H]^+$ ($m/z = 465$) $[I79a - 2Cl - H]^+$ ($m/z = 569$), $[I79a - Cl]^+$ ($m/z = 607$)
2 [49]	$PdCl_2(CH_3CN)_2 + 56 + 51a$	$[56+H]^+$ ($m/z = 465$) $[I79a - 2Cl - H]^+$ ($m/z = 569$), $[I79a - Cl]^+$ ($m/z = 607$)
3 [49]	$PdCl_2(CH_3CN)_2 + 56 + 51a + 52 + K_2CO_3$	$[56+K]^+$ ($m/z = 503$) $[I79a - 2Cl - 2H + K]^+$ ($m/z = 607$), $[I79a - Cl - H - K]^+$ ($m/z = 645$), $[I80 - H - X + K]^+$ ($m/z = 707$)
4 [49]	0.2 $PdCl_2(CH_3CN)_2$ + 0.2 **56** + **51a** + 2 **52** + Na_2CO_3	$[56+Na]^+$ ($m/z = 487$) $[I79a - 2Cl - 2H + Na]^+$ ($m/z = 591$), $[I80 - H - X - Na]^+$ ($m/z = 691$)
5 [49]	**51a** + $Pd(PPh_3)_4$	**I81a** ($m/z = 823$), **I81a'** ($m/z = 561$)
6 [50]	$[(\pi\text{-allyl})PdCl]_2 + 56$	**I79b** ($m/z = 611$)
7 [50]	$[(\pi\text{-allyl})PdCl]_2 + 56 + 51b + 54$	**I79b** ($m/z = 611$), **I81b** ($m/z = 701$), $[I82 + Na]^+$ or $[I82' + Na]^+$ ($m/z = 879$)
8 [50]	Aqueous phase of the extracted reaction mixture of entry 7	**I79b** ($m/z = 611$), $[I83 - H + Na]^+$ ($m/z = 747$), $[I83 - 2H + 2Na]^+$ ($m/z = 769$)

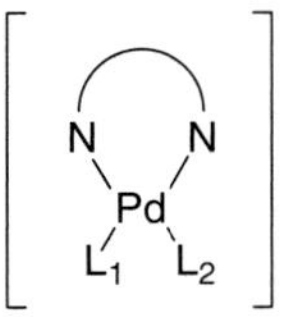

179a $L_1 = L_2 = Cl$

179b $\begin{cases} L_1 = \eta^3\text{-allyl} & L_2 = \text{none} \\ L_1 = \eta^1\text{-allyl} & L_2 = \text{none} \end{cases}$

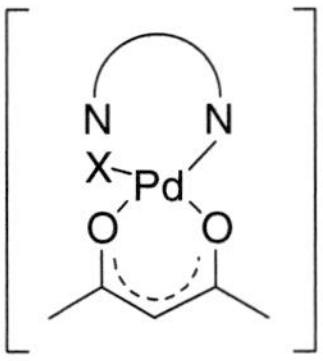

180 X = OAc or Cl

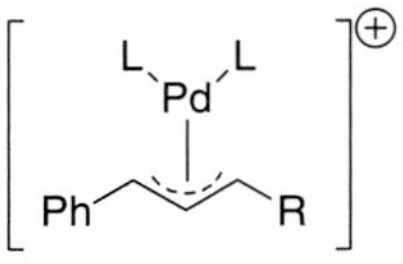

181a R = Ph   L = PPh$_3$
181'a R = Ph   L = none
181b R = Me   L = 56

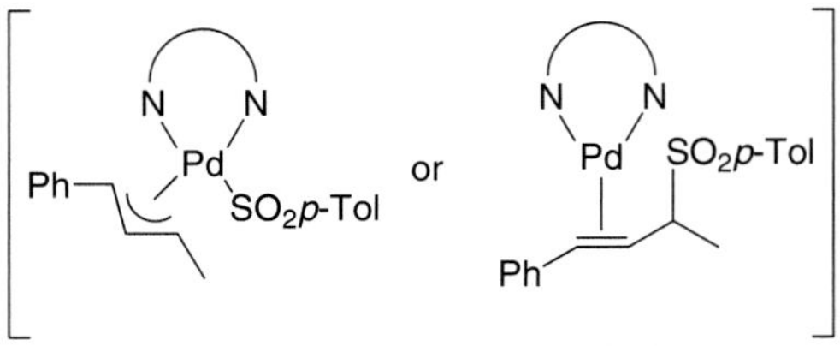

182   **182'**

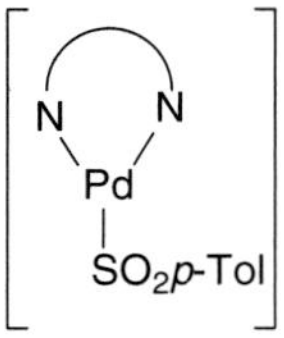

183

Initially, the first reaction of Scheme 7.27 (Eq. 1) was studied by ESI-MS. An equimolar mixture of PdCl$_2$(CH$_3$CN)$_2$ and ligand **56** at room temperature led the authors to assume that complex **I79a** was formed (Entry 1). A stoichiometric mixture of PdCl$_2$(CH$_3$CN)$_2$, ligand **56**, and allylic acetate **51a** gave the same peaks as those observed in Entry 1 even when the mixture was heated at 50 °C for 90 min (Entry 2). The same equimolar mixture as in Entry 2, but now also with 2 equivalents of nucleophile **52** and K$_2$CO$_3$, was heated at 50 °C and monitored. The same three intermediates observed in Entry 2 were seen here but now as potassium adducts (Entry 3). In addition, a new peak was detected corresponding to the formation of the palladium enolate complex **I80**. This enolate complex was also found in two other mixtures: firstly, the same mixture of Entry 3 but excluding allylic acetate **51a**, and secondly, excluding both **51a** and K$_2$CO$_3$. It was suggested that the formation of enolate **I80** could be explained by the direct reaction of PdCl$_2$(CH$_3$CN)$_2$ with **52** or by a transmetalation reaction with the potassium enolate of **52**. The same mixtures of Entries 1 and 3 were injected into the mass spectrometer in negative-ion mode ESI (−), and the same kinds of species were observed but now as anions. Another mixture was based on all the components under catalytic conditions and using Na$_2$CO$_3$ instead of K$_2$CO$_3$ to test the postulated intermediates (Entry 4). The same two intermediates **I79a** and **I80** were observed as sodium adducts. Under the above

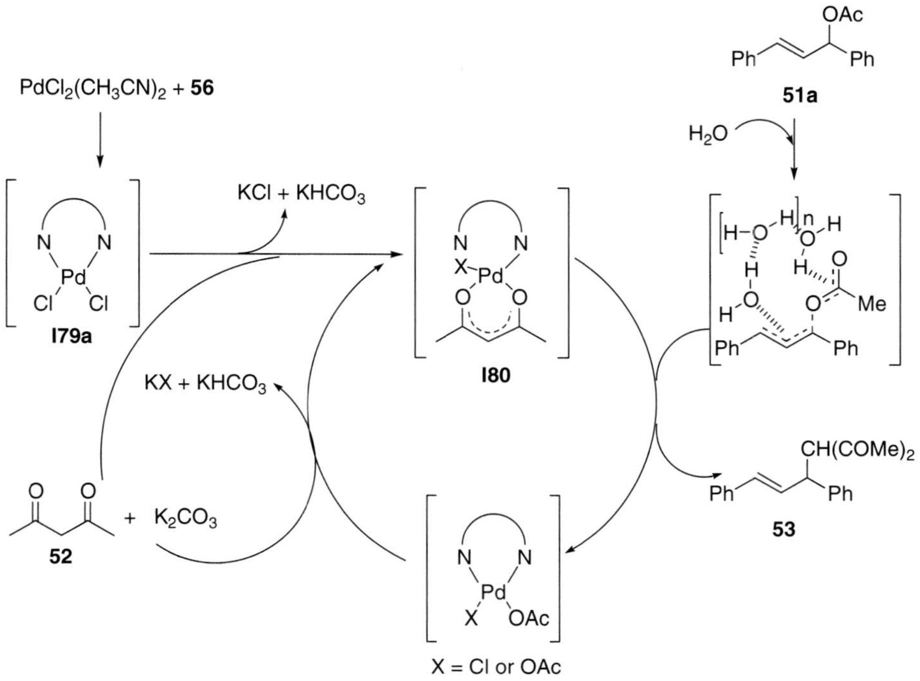

Scheme 7.28 Proposed catalytic cycle for the reaction shown in Scheme 7.27.

experimental conditions, no peak corresponding to any π-allylpalladium intermediate was observable by ESI-MS. In a final experiment, stoichiometric quantities of allylic acetate **51a** were mixed with a Pd(0) source such as Pd(PPh$_3$)$_4$. Two intermediates corresponding to the *in situ* formation of π-allyl complexes **I81** were detected (Entry 5). These ESI-MS studies revealed that in the allylic substitution of Eq. (1) in Scheme 7.27, the formation of palladium acetylacetonate complex **I80** rather that the typically postulated π-allylpalladium intermediate of type **I81** was the key intermediate in this process. Scheme 7.28 shows the mechanistic proposal for this reaction in which the palladium acetylacetonate complex **I80** is involved in the formation of **53**.

The second related ESI-MS mechanistic study [50] examined the substitution of allylic acetate **51b** with *para*-toluenesulfinate **54** using [(π-allyl)PdCl]$_2$ and the same water-soluble ligand **56** as the catalytic system (Eq. (2) in Scheme 7.27). Entries 6–8 of Table 7.11 give the most significant palladium species detected in these experiments.

In a first stage, a 1:2 mixture of [(π-allyl)PdCl]$_2$ and **56** revealed a peak corresponding to the intermediate **I79b**, which again showed the coordination of the palladium atom with the hydrophilic ligand. This complex had an allyl moiety coordinated to the metal in an η^3 or η^1 coordination mode (Entry 6). No new peaks were observed when the allylic acetate **51b** was added to the above mixture. However, when nucleophile **54** was added, two new peaks were detected after 1 h. at 50 °C: (i) a π-allyl palladium complex (**I81b**) and (ii) an intermediate consistent with two possible species **182** or **182'** (Entry 7). Twenty-four hours later, an ESI-MS analysis of the aqueous phase after extraction of the crude mixture with CH$_2$Cl$_2$ revealed the intermediate **179b** together with two new clusters corresponding to species **183** (Entry 8). Given these results, the authors suggested a possible catalytic cycle involving a Pd(IV) intermediate. Scheme 7.29 shows the postulated mechanistic cycle.

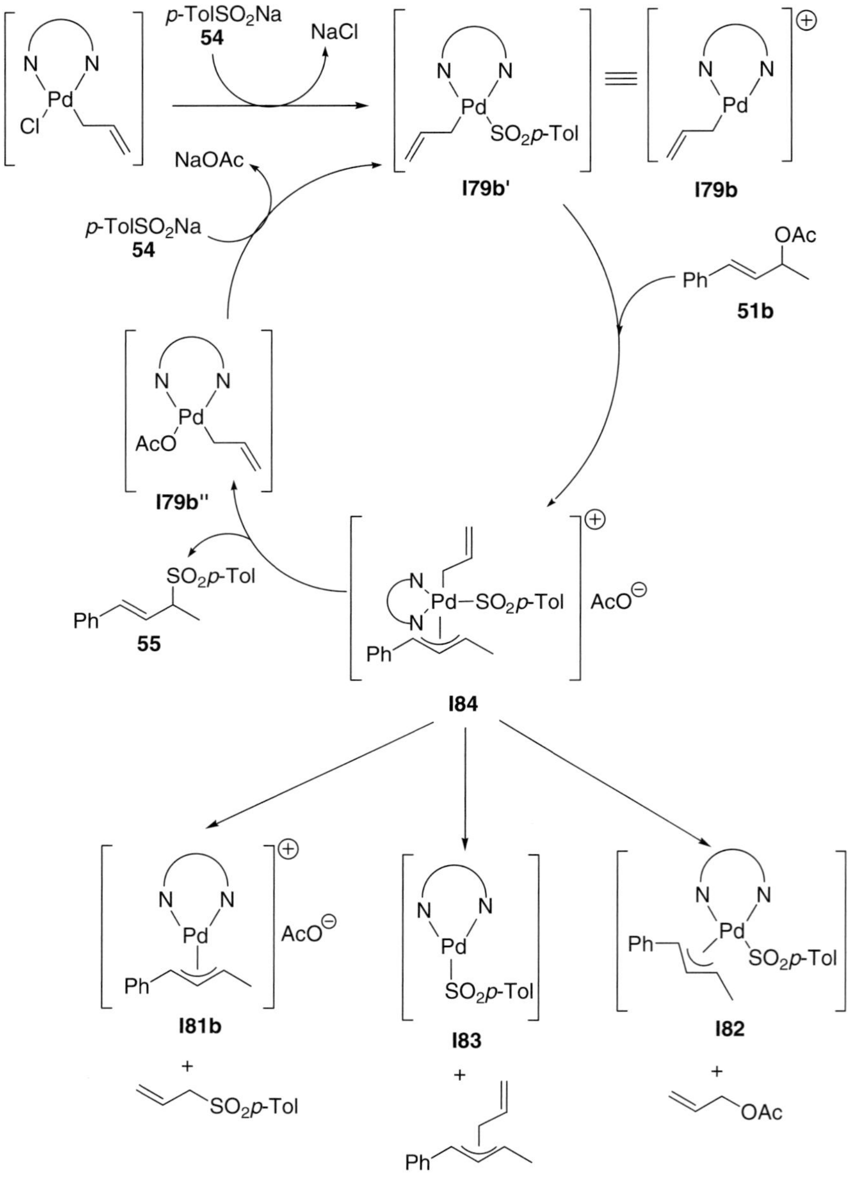

Scheme 7.29 Proposed catalytic cycle for the reaction shown in Eq. (2) of Scheme 7.27.

The authors suggested that the real catalyst was the neutral intermediate **I79b′**, detected as a cluster at $m/z = 611$, corresponding to **I79b**. The oxidative addition of allylic acetate **51b** to the above intermediate afforded the formation of a Pd(IV) π-allyl complex **I84**, which was not detected directly by ESI-MS. However, the authors postulate that this complex may be related to intermediates **I81b,I83** and **I82**, which were all observed in the ESI mass spectra. Reductive elimination from **I84** and interaction with **54** close the catalytic cycle.

Pfaltz et al. [51, 52] applied ESI-MS to the study of the kinetic resolution of palladium-catalyzed enantioselective allylation of diethyl ethylmalonate with allylic esters (Scheme 7.30). However, in this particular case the authors used the ESI-MS technique for the high-throughput parallel screening of the chiral catalyst [53], suspecting that it might be possible to determine the catalyst's capacity for enantio-discrimination by examining catalyst-bound reactant complexes rather than by analyzing the final product. This study was inspired by the screening method for homogeneous polymerization catalysts developed by Chen [53], which we will look at in Section 7.10.

In their initial study [51a], the authors used pseudoenantiomers **57a** and **57b** with different arylic units (Ar = phenyl for **57a** and Ar = 4-methylphenyl for **57b**), which did not greatly affect the reactivity of the allylic ester but which allowed the complexes formed to be distinguished by mass spectrometry. The mechanistic scheme which was proposed for the reaction in question is shown in Scheme 7.30. The first step of the reaction generated a cationic π-allyl palladium complex **I85/I86** which was easily detectable by ESI-MS. Then, the nucleophile attack afforded the

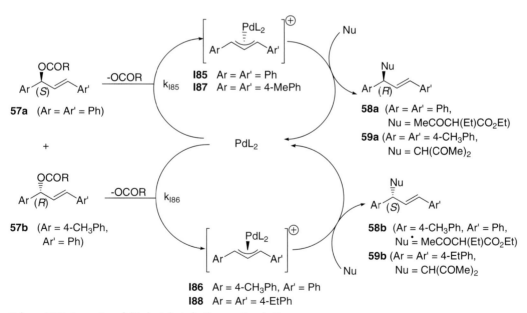

Scheme 7.30 Screening of chiral catalysts for the enantioselective allylations of diethyl ethylmalonate studied by ESI-MS.

final product. The fact that the two π-allyl palladium complexes had different masses allowed the ESI-MS to quantify the relative ratio in which they were formed. The authors treated equimolar mixtures of **57a** and **57b** with a catalytic system consisting of [Pd(C$_3$H$_5$)(CH$_3$CN)$_2$]OTf and several chiral phosphines (**60–64**, Scheme 7.31) and diethyl ethylmalonate as the nucleophile and injected the mixtures into the spectrometer [51a]. The **I85/I86** ratio revealed the intrinsic ability of the chiral catalyst (k_{I85}/k_{I86}) to discriminate between the two pseudoenantiomers. In order to broaden the analytical method, the authors tested whether the method could be used for screening mixtures of several palladium catalysts in one single reaction but found that this was only possible when working at low temperatures where ligand exchange was avoided. A qualitative reactivity order of the different chiral catalysts was established from the spectrum. The structure of the different phosphines and the relation of intermediate Pd-complexes **I85/I86** are given in Scheme 7.31. The order of selectivity between the different ligands was found to be **63 > 62 > 61 ≈ 64 ≈ 60**. In a later study, the authors extended the methodology to introduce a double mass-labeling strategy in the context of a catalyst optimization study [51b]. In this case, a library of ligands was prepared in a single batch by condensation of chiral diamines and chiral diols and further *in situ* complexation with [Pd(C$_3$H$_5$)(CH$_3$CN)$_2$]OTf in order to screen the corresponding catalyst mixture. Scheme 7.31 shows a series of ligands with S configurations which were tested for screening. The π-allyl palladium complex ratios generated by each catalytic system are given next to each ligand. Here as well it can be concluded that the most selective ligand in the kinetic resolution step, which is the sterically most demanding one, is **70** with naphthyl groups in the sulfonamide moieties. Another application of this screening method designed by the same authors [52] was to monitor the back reaction of quasienantiomeric final products in order to determine the enantioselectivity in the nucleophilic addition step. The authors explained the fact that the enantioselectivity determined from back-reaction screening was identical to that of the forward reaction in terms of the principle of microscopic reversibility, which states that transition states of forward and back reactions are identical. In this case, an equimolar mixture of the quasienantiomeric allylation products **59a** and **59b** (Scheme 7.30) was mixed with the catalytic system formed by [Pd(C$_3$H$_5$)(CH$_3$CN)$_2$]OTf and several chiral phoshines (**71–75** in Scheme 7.31) and injected into the mass spectrometer. The ratio between the two palladium allyl complexes **I87/I88**, formed by the elimination of acetyl acetonate from **59a/59b**, which was dependent on the nature of the phosphine ligand, is shown in Scheme 7.31. The ratios of the signal heights of the two clusters were very similar to the enantiomeric ratios determined by HPLC analysis of the previously obtained products. When the phosphines had an *o*-tolyl substituent (**74** and **75**), only moderate enantioselectivity was induced. In this case, the screening of mixtures of palladium complexes was also undertaken, working at −20 °C to avoid ligand exchange. Despite the lower ratios in the multicatalyst screening, the order of selectivity was unchanged.

This ESI mass spectrometry method of chiral palladium-catalyst screening, which requires only small amounts of substrates, is fast, reliable, and operationally simple,

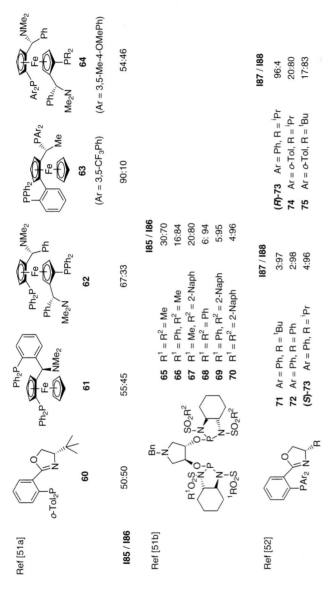

Scheme 7.31 Structure of chiral ligands used in reactions of Scheme 7.30.

and samples can be injected directly from the reaction mixtures, avoiding isolation and purification steps.

7.9
ESI-MS Studies in Palladium-Catalyzed Oxidation of 2-Allylphenols

Muzart *et al.* [54] applied the analytical tool to study mechanistically the hydroxyalkoxylation of *o*-allylphenol **76** catalyzed by a mixture of $Pd(OCOCF_3)_2$ and the water-soluble ligand **56** (also used in the palladium-catalyzed substitution of allylic acetates in aqueous media [49, 50]) with H_2O_2 in H_2O/MeOH to afford 2-(1,2-dihydroxypropyl)phenol derivatives **77** and **78** (Scheme 7.32). It was suggested that the reaction went through a cascade process in which, in a first step, isomerization of the double bond took place followed by epoxidaton of the olefin and the opening of the resulting oxirane (Scheme 7.32). The ESI-MS/MS study performed by the authors supported the proposed mechanism.

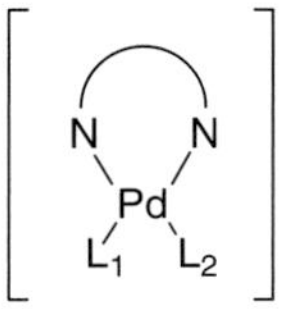

I79c $L_1 = L_2$ = none
I79c' $L_1 = OCOCF_3$, L_2 = none
I79c'' $L_1 = L_2 = OCOCF_3$

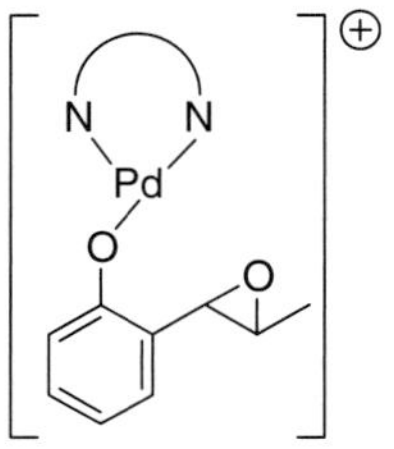

I89

The palladium species observed by ESI-MS are described in Table 7.12.

The equimolar mixture of $Pd(OCOCF_3)_2$ with the hydrophilic ligand **56** dissolved in a 1 : 1 $MeOH/H_2O$ mixture was injected into the mass spectrometer, and three

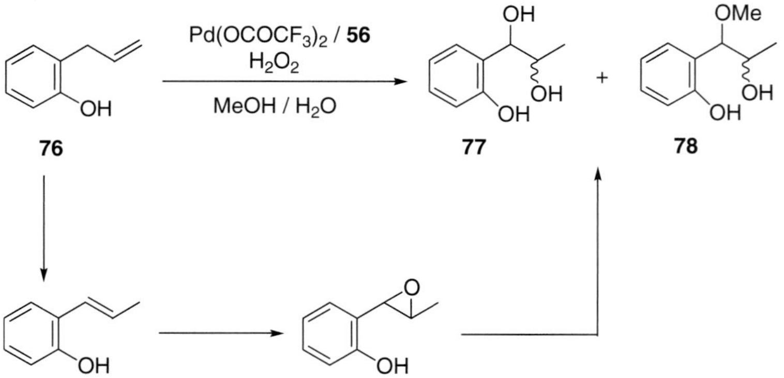

Scheme 7.32 Palladium-catalyzed hydroxyalkoxylation of *o*-allylphenol **76**.

Table 7.12 Palladium species detected by ESI-MS in the reaction of Scheme 7.32.

Entry	Mixtures of components	Species observed by ESI-MS
1	Pd(OCOCF$_3$)$_2$ + 56	ESI(+): [**I79c** − H]$^+$ (m/z = 569), **I79c'** (m/z = 683), [**I79c'** − H + Na]$^+$ (m/z = 705)
2	Pd(OCOCF$_3$)$_2$ + 56	ESI(−): [**I79c** − 3H]$^-$ (m/z = 567), [**I79c'** − 2H]$^-$ (m/z = 681), [**I79c''** − H]$^-$ (m/z = 795)
3	Pd(OCOCF$_3$)$_2$ + 56 + 76 + H$_2$O$_2$	ESI(+): [**I79c** − 2H + Na]$^+$ (m/z = 590), **I89** (m/z = 719)

peaks were observed corresponding to intermediate **I79c**, in which the palladium atom was coordinated to ligand **56** and had zero, one, or two trifluoroacetate anions ligated to the metal (Entry 1). Since **56** had functional groups susceptible to be deprotonated, the same mixture was analyzed in ESI(−) mode. The same kind of intermediate **I79c** was detected, although here as an anionic species (Entry 2). When allylphenol **76** and the oxidant were added to the former mixture and heated at 50 °C for 4 h, intermediate **I79c** together with a new peak at m/z = 719 were observed. The new cluster corresponded to intermediate **I89**, permitting the authors to propose an epoxide intermediate in the catalytic cycle.

7.10
ESI-MS Studies in Palladium-Catalyzed Polymerization Reactions

As mentioned above, Chen was a pioneer in using electrospray ionization tandem mass spectrometry as a fast and efficient method for high-throughput catalyst screening [53]. His group used Brookhart-type Pd(II) olefin polymerization catalysts [55] to show that ESI-MS/MS could screen the different catalysts by their propensity for polymer formation [56]. Polymer chains continued to grow as they remained bound to the catalyst during polymerization, resulting in the most effective catalysts being revealed by the longest chains. The experimental work started with an equimolar mixture of eight palladium complexes **79a–h** (Scheme 7.33) being synthesized simultaneously. They were then analyzed by ESI-MS and were all observed in the mass spectrum. Afterwards, an excess of ethylene was bubbled into the catalyst solution at −10 °C for 1 h and quenched with DMSO. Electrospraying the mixture into the mass spectrometer, selecting only ions above a 2200 m/z ratio cutoff, and then inducing a β-hydride elimination (Scheme 7.33) produced a spectrum in which the highest peak corresponded to intermediate **I91c**, which was in effect catalyst **79c**, as the experimental results had anticipated [55]. The response given by the different catalysts in the ESI-MS permitted the authors to establish an order of efficiency of the different ligands matching steric considerations (**79c** ≫ **79b** ≈ **79d** ≈ **79e** > others). The authors used ESI-MS to rapidly screen a series of polymerization catalysts using a competition experiment.

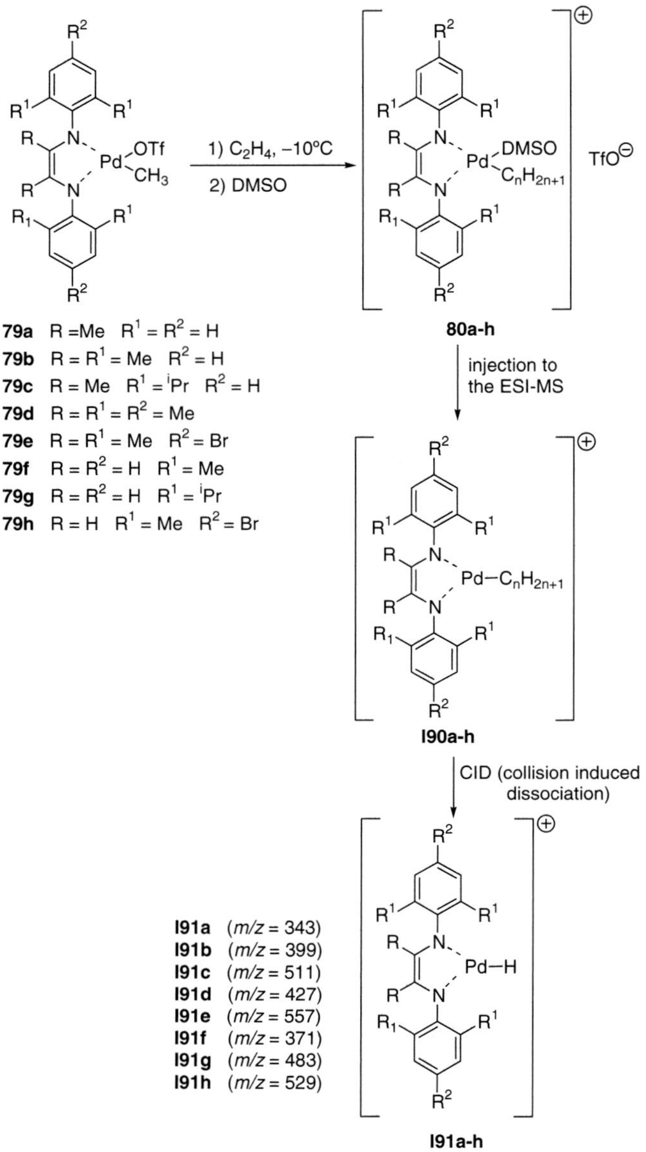
1) C_2H_4, −10°C
2) DMSO
TfO⊖
80a-h
injection to the ESI-MS
I90a-h
CID (collision induced dissociation)
I91a-h
79a R = Me R¹ = R² = H
79b R = R¹ = Me R² = H
79c R = Me R¹ = ⁱPr R² = H
79d R = R¹ = R² = Me
79e R = R¹ = Me R² = Br
79f R = R² = H R¹ = Me
79g R = R² = H R¹ = ⁱPr
79h R = H R¹ = Me R² = Br
I91a (m/z = 343)
I91b (m/z = 399)
I91c (m/z = 511)
I91d (m/z = 427)
I91e (m/z = 557)
I91f (m/z = 371)
I91g (m/z = 483)
I91h (m/z = 529)

Scheme 7.33 Simultaneous preparation and pool screening of eight palladium catalysts in ethylene polymerization.

The ESI-MS technique, together with NMR and IR spectroscopy, has also been used in order to understand the mechanism operative for the synthesis of acrylonitrile polymers by insertion polymerization using palladium catalysts [57]. Six types of $[L_2PdCH_3]^+$ palladium catalysts, where L were bidentate N-donor ligands **79c** and **83–87**, were used in this study. The nitrogen ligands were chosen to evaluate the

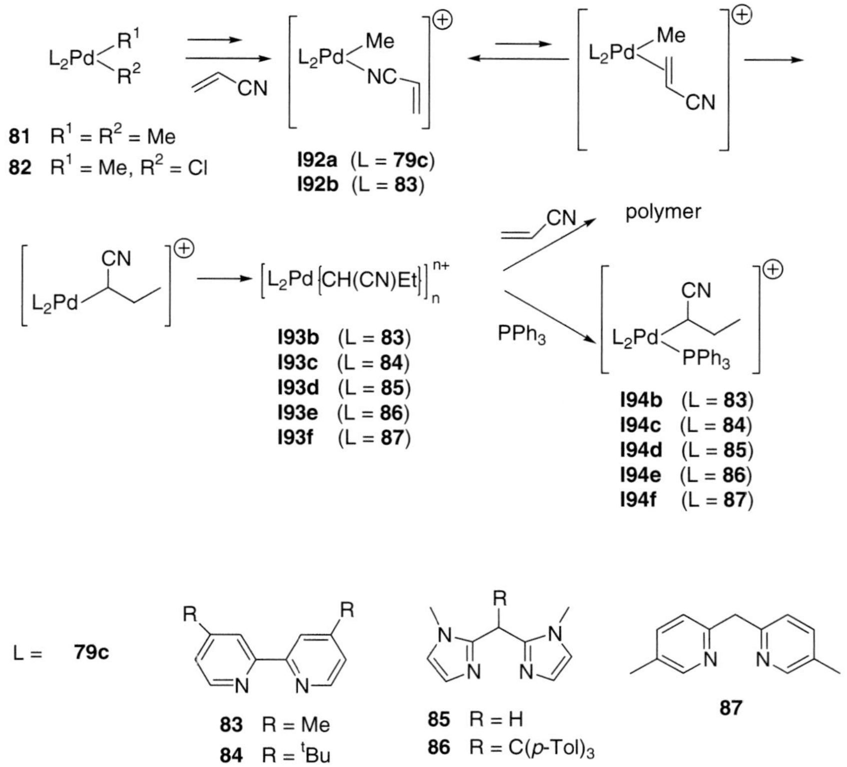

7.10 ESI-MS Studies in Palladium-Catalyzed Polymerization Reactions | 271

Scheme 7.34 Palladium-catalyzed acrylonitrile polymerization reactions.

electronic and steric effects of the palladium complexes. The different spectroscopic techniques were used to investigate whether acrylonitrile insertion occurred by 1,2- or by 2,1- regioselectivity.

Scheme 7.34 shows the general path for the insertion polymerization of acrylonitrile. The species observed by ESI-MS are set out in Table 7.13.

The reactions of palladium complexes **81** and **82** with acrylonitrile afforded intermediates **I92**, which have been demonstrated by NMR and IR spectrometry techniques to have N-coordination of the acrylonitrile to the palladium center rather than a C=C π–bond coordination. ESI mass experiments were performed for complexes containing ligands **79c** and **86** (Entries 1 and 2) to confirm the presence of acrylonitrile in their structure. Intermediates **I92** undergo a 2,1-acrylonitrile insertion to give aggregated species **I93**. The authors suggest that this occurs through isomerization of **I92** to the C=C π complexes (even though this is not observed by spectroscopy) followed by migratory insertion. In the case of **I92a**, it was suggested that, because of the steric hindrance of the ligand **79c**, no insertion took place and therefore the polymerization did not proceed. Intermediates **I93b–f**, observed as aggregates with n going from 1 to 3, were characterized by ESI-MS in all the palladium complexes tested (Entries 2–6). It has been suggested that the monomer

Table 7.13 Palladium species observed by ESI-MS in the polymerization of acrylonitrile.

Entry	Ligand (L)	Species observed by ESI-MS
1	79c	I92a ($m/z = 578$)
2	83	I92b ($m/z = 358$)
		I93b (n = 1, $m/z = 358$)
		I94b ($m/z = 620$)
3	84	I93c (n = 1, $m/z = 442$)
		I94c ($m/z = 704$)
4	85	I93d (n = 1 to 3, $m/z = 350$)
		I94d ($m/z = 612$)
5	86	I93e (n = 1, $m/z = 634$)
		I94e ($m/z = 896$)
6	87	I93f (n = 1 to 3, $m/z = 372$)
		I94f ($m/z = 634$)

units are linked by PdCHEtCN...Pd bridges. In order to support the formulation of intermediates **I93** as 2,1-insertion products, a chemical derivatization experiment was performed. Intermediates **I93** were treated with 1 equiv. of PPh$_3$ to form species **I94**, which enabled the definitive characterization by NMR of the 2,1-insertion regiochemistry and the confirmation of the mass by ESI-MS.

7.11
Conclusions

In summary, since ESI-MS is a soft ionization mass spectrometry, its ability to detect ionic species in solution provides a convenient technique for the direct observation and identification of short-lived palladium intermediates and, as a result, has become increasingly popular as a mechanistic tool in palladium-catalyzed reactions. The fact that the ions are formed in ESI-MS in such a gentle way is an important feature of its application in palladium chemistry. Furthermore, ESI-MS also makes it possible to work directly from diluted solutions, which is an important advantage in studying catalytic species that exist only under these conditions. The application of this technique in several palladium-catalyzed reactions has led to a great advance in mechanistic hypotheses of these processes. The analytical potential of this technique is still being actively investigated, and we can look forward to interesting new findings.

Acknowledgments

We gratefully acknowledge the financial support received for our own research from the Ministry of Education and Science of Spain (Project no. CTQ2008-05409-C02), 'Generalitat de Catalunya' (2009SGR637), and the contribution made by our doctoral students, particularly Sandra Brun (grant from the University of Girona).

References

1. de Meijere, A. and Diederich, F. (eds) (2004) *Metal-Catalyzed Cross-Coupling Reactions*, 2nd edn, Wiley-VCH, Weinheim.
2. Negishi, E. (ed.) (2002) *Handbook of Organopalladium Chemistry for Organic Synthesis*, John Wiley & Sons, Inc., New York.
3. Cole, R.B. (ed.) (1997) *Electrospray Ionization Mass Spectrometry. Fundamentals, Instrumentation and Applications*, John Wiley & Sons, Inc., New York.
4. Santos, L.S., Knaack, L., and Metzger, J.O. (2005) *Int. J. Mass Spectrom.*, **246**, 84–104; (b) Santos, L.S. (2008) *Eur. J. Org. Chem.*, 235–253; (c) Plattner, D.A. (2003) *Top Curr. Chem.*, **225**, 153–203.
5. (a) Van Berkel, G.J., McLuckey, S.A., and Glish, G.L. (1992) *Anal. Chem.*, **64**, 1586–1593; (b) Van Berkel, G.J. (2000) *J. Mass Spectrom.*, **35**, 773–783.
6. (a) Henderson, W. and McIndoe, J.S. (2005) *Mass Spectrometry of Inorganic, Coordination and Organometallic Compounds*, John Wiley & Sons, Ltd, Chichester; (b) Henderson, W., Nicholson, B.K., and McCaffrey, L.J. (1998) *Polyhedron*, **17**, 4291–4313; (c) Traeger, J.C. (2000) *Int. J. Mass Spectrom.*, **200**, 387–401; (d) Plattner, D.A. (2001) *Int. J. Mass Spectrom.*, **207**, 125–144.
7. Aliprantis, A.O. and Canary, J.W. (1994) *J. Am. Chem. Soc.*, **116**, 6985–6986.
8. (a) Whitehouse, C.M., Dreyer, R.N., Yamashita, M., and Fenn, J.B. (1985) *Anal. Chem.*, **57**, 675–679; (b) Fenn, J.B., Mann, M., Meng, C.K., Wong, S.F., and Whitehouse, C.M. (1989) *Science*, **246**, 64–71; (c) Fenn, J.B. (2003) *Angew. Chem. Int. Ed.*, **42**, 3871–3894.
9. (a) Miyaura, N. and Suzuki, A. (1995) *Chem. Rev.*, **95**, 2457–2483; (b) Suzuki, A. (1999) *J. Organomet. Chem.*, **576**, 147–168; (c) Kotha, S., Lahiri, K., and Kashinath, D. (2002) *Tetrahedron*, **58**, 9633–9695; (d) Miyaura, N. (2002) *Top. Curr. Chem.*, **219**, 11–59.
10. Aramendía, M.A., Lafont, F., Moreno-Mañas, M., Pleixats, R., and Roglans, A. (1999) *J. Org. Chem.*, **64**, 3592–3594.
11. Campi, E.M., Jackson, W.R., Marcuccio, S.M., and Naeslund, C.G.M. (1994) *J. Chem. Soc., Chem. Commun.*, 2395–2396; (b) Gillmann, T. and Weeber, T. (1994) *Synlett*, 649–651; (c) Song, Z.Z. and Wong, H.N.C. (1994) *J. Org. Chem.*, **59**, 33–41.
12. (a) Smith, C.A., Campi, E.M., Jackson, R., Marcuccio, S., Naeslund, C.G.M., and Deacon, G.B. (1997) *Synlett*, 131–132; (b) Wong, M.S. and Zhang, X.L. (2001) *Tetrahedron Lett.*, **42**, 4087–4089; (c) Kabalka, G.W. and Wang, L. (2002) *Tetrahedron Lett.*, **43**, 3067–3068; (d) Koza, D.J. and Carita, E. (2002) *Synthesis*, 2183–2186; (e) Parrish, J.P., Jung, Y.C., Floyd, R.J., and Jung, K.W. (2002) *Tetrahedron Lett.*, **43**, 7899–7902; (f) Yoshida, H., Yamaryo, Y., Ohshita, J., and Kunai, A. (2003) *Tetrahedron Lett.*, **44**, 1541–1544.
13. Moreno-Mañas, M., Pérez, M., and Pleixats, R. (1996) *J. Org. Chem.*, **61**, 2346–2351.
14. Adamo, C., Amatore, C., Ciofini, I., Jutand, A., and Lakmini, H. (2006) *J. Am. Chem. Soc.*, **128**, 6829–6836.
15. For a review on diazonium salts in palladium chemistry, see (a) Roglans, A., Pla-Quintana, A., and Moreno-Mañas, M., (2006) *Chem. Rev.*, **106**, 4622–4643; For reviews on organotrifluoroborates, see: (b) Molander, G.A. and Ellis, N., (2007) *Acc. Chem. Res.*, **40**, 275–286; (c) Stefani, H.A.,

Cella, R., and Vieira, A.S. (2007) *Tetrahedron*, **63**, 3623–3658; (d) Darses, S. and Gênet, J.-P. (2008) *Chem. Rev.*, **108**, 288–325.
16 Masllorens, J., González, I., and Roglans, A. (2007) *Eur. J. Org. Chem.*, 158–166.
17 Taccardi, N., Paolillo, R., Gallo, V., Mastrorilli, P., Nobile, C.F., Räisänen, M., and Repo, T. (2007) *Eur. J. Inorg. Chem.*, 4645–4652.
18 Chin, C.C.H., Yeo, J.S.L., Loh, Z.H., Vittal, J.J., Henderson, W., and Andy Hor, T.S. (1998) *J. Chem. Soc., Dalton Trans.*, 3777–3784.
19 Jutand, A. and Mosleh, A. (1995) *Organometallics*, **14**, 1810–1817.
20 Jutand, A. and Négri, S. (2003) *Organometallics*, **22**, 4229–4237.
21 (a) Heck, R.F. (1979) *Acc. Chem. Res.*, **12**, 146–151; (b) De Meijere, A. and Meyer, F.E. (1994) *Angew. Chem. Int. Ed. Engl.*, **33**, 2379–2411; (c) Cabri, W. and Candiani, I. (1995) *Acc. Chem. Res.*, **28**, 2–7; (d) Crisp, G.T. (1998) *Chem. Soc. Rev.*, **27**, 427–436; (e) Beletskaya, I.P. and Cheprakov, A.V. (2000) *Chem. Rev.*, **100**, 3009–3066; (f) Whitcombe, N.J., Kuok Hii, K., and Gibson, S.E. (2001) *Tetrahedron*, **57**, 7449–7476.
22 Brown, J.M. and Hii, K.K. (1996) *Angew. Chem. Int. Ed.*, **35**, 657–659.
23 Hii, K.K., Claridge, T.D.W., and Brown, J.M. (1997) *Angew. Chem. Int. Ed.*, **36**, 984–987.
24 Ripa, L. and Hallberg, A. (1996) *J. Org. Chem.*, **61**, 7147–7155.
25 Masllorens, J., Moreno-Mañas, M., Pla-Quintana, A., and Roglans, A. (2003) *Org. Lett.*, **5**, 1559–1561.
26 Pla-Quintana, A. and Roglans, A. (2005) *Arkivoc*, **IX**, 51–62; For interactions between Pd-complex 24 and phosphanes studied also by ESI-MS see: (b) Moreno-Mañas, M., Pleixats, R., Spengler, J., Chevrin, C., Estrine, B., Bouquillon, S., Hénin, F., Muzart, J., Pla-Quintana, A., and Roglans, A. (2003) *Eur. J. Org. Chem.*, 274–283.
27 Sabino, A.A., Machado, A.H.L., Correia, C.R.D., and Eberlin, M.N. (2004) *Angew. Chem. Int. Ed.*, **43**, 2514–2518.
28 Svennebring, A., Sjöberg, P.J.R., Larhed, M., and Nilsson, P. (2008) *Tetrahedron*, **64**, 1808–1812.
29 Enquist, P.A., Nilsson, P., Sjöberg, P., and Larhed, M. (2006) *J. Org. Chem.*, **71**, 8779–8786.
30 (a) Cho, C.S. and Uemura, S. (1994) *J. Organomet. Chem.*, **465**, 85–92;(b) Du, X., Suguro, M., Hirabayashi, K., Mori, A., Nishikata, T., Hagiwara, N., Kawata, K., Okeda, T., Wang, H.F., Fugami, K., and Kosugi, M. (2001) *Org. Lett.*, **3**, 3313–3316; (c) Jung, Y.C., Mishra, R.K., Yoon, C.H., and Jung, K.W. (2003) *Org. Lett.*, **5**, 2231–2234;(d) Andappan, M.M.S., Nilsson, P., and Larhed, M. (2003) *Mol. Div.*, **7**, 97–106.
31 Stahl, S.S. (2004) *Angew. Chem. Int. Ed.*, **43**, 3400–3420.
32 (a) Reetz, M.T. and Westermann, E. (2000) *Angew. Chem. Int. Ed.*, **39**, 165–168; (b) de Vries, A.H.M., Parlevliet, F.J., Schmieder-van de Vondervoort, L., Mommers, J.H.M., Henderickx, H.J.W., Walet, M.A.M., and de Vries, J.G. (2002) *Adv. Synth. Catal.*, **344**, 996–1002; (c) de Vries, A.H.M., Mulders, J.M.C.A., Mommers, J.H.M., Henderickx, H.J.W., and de Vries, J.G. (2003) *Org. Lett.*, **5**, 3285–3288; (d) Reetz, M.T. and de Vries, J.G. (2004) *Chem. Commun.*, 1559–1563; (e) de Vries, J.G. (2006) *Dalton Trans.*, 421–429.
33 Amatore, C. and Jutand, A. (2000) *Acc. Chem. Res.*, **33**, 314–321.
34 Milstein, D. and Stille, J.K. (1978) *J. Am. Chem. Soc.*, **100**, 3636–3638; (b) Stille, J.K. (1986) *Angew. Chem. Int. Ed.*, **25**, 508–524; (c) Mitchell, T.N. (1986) *J. Organomet. Chem.*, **304**, 1–16; (d) Mitchell, T.N. (1992) *Synthesis*, 803–815.
35 (a) Casado, A.L., Espinet, P., Gallego, A.M., and Martínez-Ilarduya, J.M. (2001) *Chem. Commun.*, 339–340; (b) Napolitano, E., Farina, V., and Persico, M. (2003) *Organometallics*, **22**, 4030–4037;

(c) Espinet, P. and Echavarren, A.M. (2004) *Angew. Chem. Int. Ed.*, **43**, 4704–4734.
36 Jutand, A. (2003) *Eur. J. Inorg. Chem.*, 2017–2040; (b) Amatore, C., Bahsoun, A.A., Jutand, A., Meyer, G., Ntepe, A.N., and Ricard, L. (2003) *J. Am. Chem. Soc.*, **125**, 4212–4222.
37 Santos, J.S., Rosso, G.B., Pilli, R.A., and Eberlin, M.N. (2007) *J. Org. Chem.*, **72**, 5809–5812.
38 (a) Zimmer, R., Dinesh, C.U., Nandanan, E., and Khan, F.A. (2000) *Chem. Rev.*, **100**, 3067–3126; (b) Hashmi, A.S.K. (2000) *Angew. Chem. Int. Ed.*, **39**, 3590–3593.
39 (a) Oh, C.H., Ahn, T.W., and Reddy, R. (2003) *Chem. Commun.*, 2622–2623; (b) Ma, S., Jiao, N., and Ye, L. (2003) *Chem. Eur. J.*, **9**, 6049–6056.
40 Ma, S. (2003) *Acc. Chem. Res.*, **36**, 701–712.
41 Qian, R., Guo, H., Liao, Y., Guo, Y., and Ma, S. (2005) *Angew. Chem. Int. Ed.*, **44**, 4771–4774.
42 Ma, S. and Jiao, N. (2002) *Angew. Chem. Int. Ed.*, **41**, 4737–4740.
43 Guo, H., Qian, R., Liao, Y., Ma, S., and Guo, Y. (2005) *J. Am. Chem. Soc.*, **127**, 13060–13064.
44 Zeni, G., Braga, A.L., and Stefani, H.A. (2003) *Acc. Chem. Res.*, **36**, 731–738.
45 Raminelli, C., Prechtl, M.H.G., Santos, L.S., Eberlin, M.N., and Comasseto, J.V. (2004) *Organometallics*, **23**, 3990–3996.
46 (a) Sonogashira, K. (2002) *J. Organomet. Chem.*, **653**, 46–49; (b) Chinchilla, R. and Nájera, C. (2007) *Chem. Rev.*, **107**, 874–922.
47 Fairlamb, I.J.S., Lee, A.F., Loe-Mie, F.E.M., Niemelä, E.H., O'Brien, C.T., and Whitwood, A.C. (2005) *Tetrahedron*, **61**, 9827–9838.
48 (a) Tsuji, J. (1996) *Palladium Reagents and Catalysts*, John Wiley & Sons, Ltd, Chichester; (b) Consiglio, G. and Waymouth, R. (1989) *Chem. Rev.*, **89**, 257–276; (c) Frost, C.G., Howard, J., and Williams, J.M.J. (1992) *Tetrahedron: Asymmetry*, **3**, 1089–1122; (d) Pfaltz, A. (1993) *Acc. Chem. Rev.*, **26**, 339–345; (e) Trost, B.M. (1996) *Acc. Chem. Rev.*, **29**, 355–364; (f) Trost, B.M. and Van Kanken, D.L. (1996) *Chem. Rev.*, **96**, 395–422.
49 Chevrin, C., Le Bras, J., Hénin, F., Muzart, J., Pla-Quintana, A., Roglans, A., and Pleixats, R. (2004) *Organometallics*, **23**, 4796–4799.
50 Chevrin, C., Le Bras, J., Roglans, A., Harakat, D., and Muzart, J. (2007) *New J. Chem.*, **31**, 121–126.
51 (a) Markert, C. and Pfaltz, A. (2004) *Angew. Chem. Int. Ed.*, **43**, 2498–2500; (b) Markert, C., Rösel, P., and Pfaltz, A. (2008) *J. Am. Chem. Soc.*, **130**, 3234–3235.
52 Müller, C.A. and Pfaltz, A. (2008) *Angew. Chem. Int. Ed.*, **47**, 3363–3366.
53 Chen, P. (2003) *Angew. Chem. Int. Ed.*, **42**, 2832–2847.
54 Thiery, E., Chevrin, C., Le Bras, J., Harakat, D., and Muzart, J. (2007) *J. Org. Chem.*, **72**, 1859–1862.
55 (a) Johnson, L.K., Killian, C.M., and Brookhart, M. (1995) *J. Am. Chem. Soc.*, **117**, 6414–6415; (b) Johnson, L.K., Mecking, S., and Brookhart, M. (1996) *J. Am. Chem. Soc.*, **118**, 267–268; (c) Ittel, S.D., Johnson, L.K., and Brookhart, M. (2000) *Chem. Rev.*, **100**, 1169–1203.
56 (a) Hinderling, C. and Chen, P. (1999) *Angew. Chem. Int. Ed.*, **38**, 2253–2256; (b) Hinderling, C. and Chen, P. (2000) *Int. J. Mass Spec.*, **195/196**, 377–383.
57 Wu, F., Foley, S.R., Burns, C.T., and Jordan, R.F. (2005) *J. Am. Chem. Soc.*, **127**, 1841–1853.

8
Practical Investigation of Molecular and Biomolecular Noncovalent Recognition Processes in Solution by ESI-MS

Kevin A. Schug

8.1
Introduction

One would be hard pressed to find an area of science in which potential contributions by mass spectrometry (MS) had not been examined [1]. Modern MS instrumentation and techniques (broadly defined) have few limits when they are used to obtain useful qualitative and quantitative information from almost any substance of interest regardless of size, state, or chemical nature. In the realm of materials and life sciences, such information is often sought to characterize and optimize interactions based on noncovalent forces. In the past 20 years, with the advent of soft ionization sources, MS has been investigated with increasing success as a physical/biophysical technique for elucidating associative/dissociative equilibria in the solid, liquid, and gaseous phases. The aim of this chapter is to critically evaluate state-of-the-art electrospray ionization – mass spectrometry (ESI-MS) [2, 3] as applied to the investigation of solution phase interactions. Although mention is made of a variety of related techniques to place the larger discussion in context, the main focus is on discussing the practical use of titration analysis for obtaining quantitative binding affinities in small-molecule and macromolecular interaction systems. Published work and personal experience suggest that this subset of MS techniques is still a maturing area: one which is worthy of advancement for its considerable advantages, but also one that is not without limitations.

Quantitative binding determinations can be carried out by a wide range of analytical and physical methods, including nuclear magnetic resonance (NMR) spectroscopic [4], UV/Vis and fluorescence spectroscopic [5], and calorimetric [6] titrations, as well as by other techniques including surface plasmon resonance (SPR) [7] and affinity capillary electrophoresis [8, 9]. More recently, ESI-MS has also been shown to be a practical choice. The soft nature of the ionization process [3] allows the preservation and transfer of solution-phase-based noncovalent complexes into the gas phase for mass analysis. In the past 15 years, a large number of review articles have been dedicated to discussing various applications of this technology in

Reactive Intermediates: MS Investigations in Solution. Edited by Leonardo S. Santos.
Copyright © 2010 WILEY-VCH Verlag GmbH & Co. KGaA, Weinheim
ISBN: 978-3-527-32351-7

both small-molecule and macromolecular interaction systems both in solution and in the gas phase [10–39]. In contrast to many common physical methods, ESI-MS is a label-free technique and allows direct observation of each component in equilibrium and interaction stoichiometries based on predictable mass intervals. ESI-MS has also been used to measure macromolecular and small-molecule binding in solution over a wide range of interaction affinities (dissociation constants (K_d) from pM to mM). The visualization of extremely large biomolecules and their associated complexes is facilitated by both multiple charging imparted during the ESI process [40] and the advancement of hybrid and high-resolution mass analyzers. Further, ESI-MS allows interrogation of binding under physiological and atypical solution conditions and simultaneous target discrimination among multiple ligands in noncompetitive and competitive binding experiments. Once an ionic complex is introduced into the gas phase, various gas-phase dissociation experiments (e.g., tandem mass spectrometry) can be performed to study noncovalent forces in the absence of solvation. The speed, sensitivity, specificity, and versatility of ESI-MS, in conjunction with its ideal marriage to liquid-phase sample introduction techniques and associated data handling systems provides a solid platform upon which various binding assays can be generated.

Realizing the potential advantages of using ESI-MS over other common methods for quantitative binding determinations paints a pretty picture. Indeed, it is fairly easy to mix two interactants together in solution, to flow them through an electrospray source, and to observe a noncovalent complex ion formed between them in a mass spectrum. The promise for rapid evaluation, of for example, large compound libraries by this method, requiring miniscule sample amounts and being easily automated, is intoxicating to contemplate. Such an approach already serves as the backbone for a substantial number of drug discovery efforts to date. However, the caution flag is raised when a discussion of the meaningfulness of these results turns from qualitative to quantitative aspects. The bulk of the remainder of this chapter will deal with these crucial aspects and will consider how quantitative binding information is obtained from ESI-MS experiments.

At this point, it is prudent to mention a few global aspects of the measurement technique, which can affect observed results regardless of whether qualitative or quantitative binding information is desired. Perhaps the most limiting factor is the nature of the interactions which can be observed. ESI is a dynamic phase transfer process, and noncovalent complexes need to be stable in both the solution and the gas phase in order to be studied. Assuming the equilibria of interest are stable within the time scale of the phase transfer process (μsec to msec) and that ion abundances are directly proportional to their solution-phase equilibrium concentrations, ESI-MS can provide an accurate snapshot of the distribution of the free and bound analytes present in the droplet. The first part of this assumption is reasonable according to experiments aimed at investigating the speed of droplet desolvation and its effect on various host-guest equilibria [41, 42]. The second part of the assumption is a matter of considerable debate and will be expanded upon later in this chapter. As a result of stripping of the solvent from each species, interactions based on electrostatic forces are strengthened as the dielectric constant of the medium is substantially

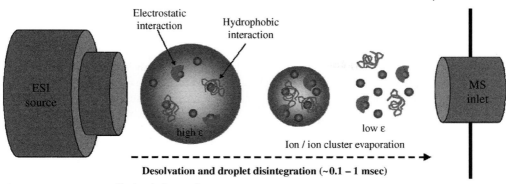

Figure 8.1 Preservation of hydrophobic vs electrostatic complexes during the electrospray process. Electrostatic forces are stabilized in the gas phase because of the low dielectric constant (ε) of the environment, whereas solvophobic forces are destabilized as solvent is removed.

reduced in the gas phase. In contrast, complexes held together strictly by solvophobic forces will not survive the phase transfer process as solvent is removed. Figure 8.1 illustrates this concept. In general, it is reasonable to expect that enthalpically-driven recognition processes are most amenable to analysis by ESI-MS, whereas those that are entropically-driven must be stabilized by some enthalpic component in order to survive the electrospray process. This leads to a system dependency in the application of ESI-MS for studying noncovalent interactions, which is perhaps the most significant limitation. That said, work continues to elucidate the possibility of ESI-MS for studying hydrophobic interactions [43, 44].

One must also realize that ESI is a competitive ionization process. This phenomenon is best described by the work of Enke [45], and later was extended to host-guest interactions by Brodbelt and coworkers [46, 47] on an equilibrium partitioning model (EPM). This model elegantly describes those factors which give rise to a limited linear dynamic range as a fixed number of charged sites on the surface of a droplet are taken up by surface-active species. This model also allows one to better understand matrix and ionization saturation effects, which can give rise to noncorrelation between gas-phase ion abundances and solution-phase concentrations as multiple analytes in the droplet compete for ionization. Species initially in the electrically neutral core of the droplet partition to the surface of the droplet according to their respective solvation energies and surface activities. The equilibrium partition model for host-guest complexation and competitive ionization is depicted in Figure 8.2. As a final general point, the preponderance of nonspecific interactions and adduct ion formation, often observed when relatively high concentrations of analytes and solution modifiers are employed, must be carefully considered, monitored, and controlled in order to avoid false positives and false negatives during both qualitative screening and quantitative binding experiments.

A basic understanding of the caveats involved in performing physical binding measurements with ESI-MS is a necessity. Good experimental design is also

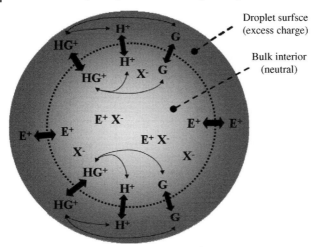

Figure 8.2 Competitive equilibrium host-guest partition and ionization model in an electrospray droplet. Adapted from reference [46].

important. The practical ease of use of modern ESI-MS instrumentation should not be taken for granted at the expense of understanding the underlying factors controlling what is observed in the mass spectra. It is my hope that this chapter provides tempered encouragement to the beginner as well as a gentle reminder to the seasoned user, helping both to successfully obtain useful and accurate quantitative binding information for whatever interaction systems they may be interested in studying. This chapter is not meant to be an exhaustive review of all of the contributions to the field, but rather is intended to highlight some prominent studies and examples of the application of various methods being discussed.

8.2
Methods and Applications

The practitioner has several levels of experimental set-ups from which to choose depending on the level of information desired. Figure 8.3 provides a general classification scheme for these choices. A recent review focusing on the use of mass spectrometry in drug discovery efforts breaks down choices related to 'affinity selection – mass spectrometry' (AS-MS) into direct and indirect methods [38]. This is a convenient place to start. Direct methods refer to those where both free and bound species in a particular interaction system are introduced together into the mass analyzer. The components are resolved by mass in the resulting spectra for analysis. This chapter is mainly dedicated to discussing direct methods. Indirect methods refer to experiments designed with the aim to first isolate compounds that have bound to a particular target and then to use the MS to identify and quantify the isolated compounds. Various applications using size-exclusion

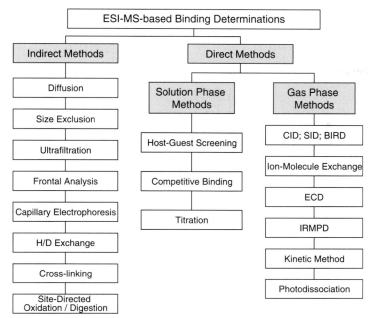

Figure 8.3 General classification of methods for the use of ESI-MS-related technologies for noncovalent binding determinations.

chromatography [48–50], affinity selection media (immobilized target) [51–53], immunoprecipitation [54], and ultrafiltration [55–57] have been demonstrated for studying a wide range of receptor-ligand interactions by indirect methods. In these set-ups, increased throughput is often a main focus, and some approaches, such as the SpeedScreen method pioneered at Novartis [50], can be used to screen as many as 400 000 compounds per week.

Direct methods can be classified into those designed to probe gas-phase and solution-phase interactions. Gas-phase methods refer to those approaches which mainly utilize tandem mass spectrometry for isolation and dissociation of ionic complexes by thermal, collisional, electronic, or photonic activation, and for investigating ion-molecule reactions. In these experiments, kinetic and thermodynamic properties of noncovalent associates can be studied in the absence of solvation [23, 25, 27, 34, 58–61]. Direct solution-phase methods include host-guest screening, competitive binding, and titration analysis. These experiments are expressly designed to investigate solution-phase phenomena, relying on the ability of the ESI-MS system to provide a snapshot of equilibria of interest.

Screening experiments can range from single host/single guest to multi-host/multi-guest (multiplexed; competitive) set-ups. Relative interaction affinities can be gauged by comparing the relative intensity of noncovalent complex ions observed. However, gleaning quantitative binding information from experiments run at a single concentration is less reliable, since the selectivity (relative affinity) values

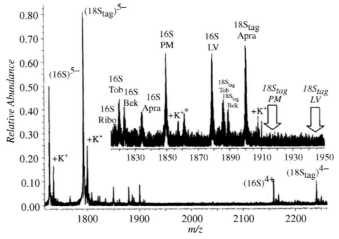

Figure 8.4 Mass spectrum for MASS evaluation of the binding of six aminoglycoside antibiotics (apramycin (Apra), ribostamycin (Ribo), tobramycin (Tob), bekanamycin (Bek), paromomycin (PM), and lividomycin (LV)) with two RNA constructs (16S and $18S_{tag}$). Incorporating a neutral mass tag on 18S allows the complexes formed with each RNA construct to be better visualized in this multiplex screening experiment. Reprinted with permission from reference [63] (Copyright 2000 Elsevier Science B.V.)

obtained may not be concentration independent [36, 62] and the relative ionization efficiencies for different systems can vary substantially. Additionally, if a large number of sequential experiments are to be compared, a contribution from run-to-run error due to instrumental instabilities can be expected to contribute to variance in the data. Even so, screening methods are very useful for optimizing conditions prior to performing more rigorous (e.g., titration) analyses or for use as a first-pass method to pin-point compounds of interest for further study. The multi-target affinity/specificity screening (MASS) approach pioneered by Hofstadler and coworkers is an excellent example of the latter method, where a high-resolution FT-ICR system is used to deconvolute and identify RNA binders based on a doubly multiplexed host-guest set-up [35, 63]. Figure 8.4 displays an example of a mass spectrum for such screening, demonstrating the binding between multiple aminoglycoside antibiotics and two RNA (untagged 16S and tagged 18S) constructs measured simultaneously [63]. In general, the speed, sensitivity, and specificity with which screening measurements can be made places ESI-MS as a versatile and powerful tool in these contexts.

In order to avoid run-to-run error and to more reliably judge the relative binding affinity of, for example, two hosts for a single guest or vice versa, a competitive binding experiment may be employed. Competitive binding analysis can be performed in the context of a basic screening format, as described above, to obtain qualitative and semi-quantitative information, or it can also be performed in the context of titration analysis, as described below, to obtain quantitative binding information. All competitive, uncompetitive, or noncompetitive (i.e., mixed inhibition) binding experiments can be considered for study. These terms can be defined on the basis of standard enzyme inhibition arguments as detailed in standard biophysical chemistry textbooks (i.e., the mode of competition is defined by the position to which the inhibitor binds to the free or bound enzyme). Alternatively, they may be defined in terms of the relative concentrations of the binding partners (e.g., two guests competing for one host) mixed in solution. Competitive binding can be probed with good sensitivity if there is a well-defined balance between the concentration of the host and the binding constants for each competing guest. Here, the binding by one competitor limits the availability of host for the other guest to bind. If the affinities of the guests for the host are similar, then the distribution of the limited amount of the host can be observed in the mass spectrum as ionic complexes. In contrast, if a large disparity in binding is present for the guests, only one complex might be observed. A competitive binding set-up is most suitable for quantitative analysis in terms of a competitive titration experiment. In a noncompetitive binding format, the host is present in large excess concentration relative to that of guests. Thus, association of multiple guests can all be observed, and a comparison of the relative abundance of the observed ionic complexes can be directly made. The noncompetitive binding experiment would be less influenced by large disparities in binding affinities of the guests for the host and would be a more suitable choice for qualitative screening analysis. In any case, the user is again cautioned that quantitative comparison of systems (both free and bound species) that have large disparities in terms of their physicochemical properties (and thus, ionization efficiencies) can be problematic. This is especially crucial when studying small-molecule binding by ESI-MS [64].

Competitive binding experiments have been used to study a wide range of small-molecule and macromolecular interaction systems. Small-molecule systems investigated include inclusion complexes [47, 65–69], self-assembling hosts [70], and chiral recognition systems [71], to name a few. In macromolecular systems, competitive binding experiments have been used to study oligonucleotide-ligand [72, 73], enzyme-ligand [74], and a variety of protein-ligand [75, 76] interaction systems. As stated previously, the line is often blurred if we wish to assign experiments to a single class of analysis techniques as described here. Researchers commonly employ qualitative screening and quantitative titration analyses in conjunction with competitive binding formats.

Titration analysis provides the most rigorous quantitative assessment of binding. Experiments are performed in a concentration-dependent fashion to sample the position of binding equilibria over a range of degrees of association in order to determine concentration-independent binding constants. Equilibrium binding constants (K) are among the most useful thermochemical data for evaluating noncovalent interactions because of their fundamental relationship to changes in state functions, such as Gibbs energy, enthalpy, and entropy. The binding constants are typically obtained by fitting a theoretical binding model, which can be derived from logical arguments according to the expected stoichiometry of interaction, to the experimental data over a carefully chosen range of concentrations. To reiterate, a notable advantage of ESI-MS binding determination experiments is that interaction stoichiometries can be directly elucidated based on signals in the mass spectra, alleviating the need to use methods, such as a Job's plot [77, 78], which would be commonly used in many spectroscopic titration analyses to obtain this information [45].

To briefly recount the different methods which are commonly employed for titration analysis, we can begin by considering a simple 1 : 1 binding equilibrium between a host species (H) and a guest species (G) to form a host-guest complex (HG), as shown in Eq. (8.1):

$$H + G \rightleftharpoons HG \tag{8.1}$$

The expression for the binding constant (or the dissociation constant, K_d) is then given by Eq. (8.2):

$$K = \frac{1}{K_d} = \frac{[HG]}{[H][G]} \tag{8.2}$$

where values in square brackets denote equilibrium concentrations of species. The equilibrium concentration of species can be related to initial concentration of reactants ($c_{0,X}$) through simple mass balance expressions:

$$c_{0,H} = [H] + [HG] \tag{8.3a}$$

$$c_{0,G} = [G] + [HG] \tag{8.3b}$$

It is also convenient to define a degree of association (α) or fraction-bound quantity, as shown in a variety of formats in Eq. (8.4):

$$\alpha = \frac{[HG]}{[H] + [HG]} = \frac{[HG]}{c_{0,H}} = \frac{[G]}{K_d + [G]} \tag{8.4}$$

Thus, a typical titration experiment may involve preparing a series of discrete concentrations with a range of $c_{0,G}$ and $c_{0,H}$ (or holding one constant and varying the other), followed by analysis of each solution in order to determine the position of equilibrium. Depending on the detection method, signals which are in some way proportional to the equilibrium solution concentration of each species are monitored and applied to a derived model. Well-established linearization models, such as the methods of Scatchard (Eq. (8.5); n indicates interaction stoichiometry) [79], Benesi and Hildebrand (Eq. (8.6)) [80], and Scott (not shown) [81], are still commonly employed in spectroscopic, as well as in mass spectrometric titration experiments. For example, Lim et al. compared stability constants determined by both UV and ESI-MS titration for a series of macrocyclic antibiotics binding depsipeptide analogs of Gram positive bacteria cell walls and found good agreement [82]. Other examples of combining Scatchard analysis with ESI-MS titration include the work of: Loo et al. who studied the binding of a series of phosphorylated and nonphosphorylated peptides to the 12.9 kDa Src SH2 domain protein [83], Greig et al. who studied the binding of oligonucleotides to bovine serum albumin [84], Sannes-Lowery et al. who studied the binding of aminoglycoside antibiotics to RNA [85], and Janis et al. who studied the binding of thioxylo-oligosaccharides to xylanase enzymes [86]. Even so, it has been well established that linearization methods, such as Scatchard analysis, are best suited for 1:1 binding systems, that data must be treated with great care in order to avoid systematic error from contributions by nonspecific binding, and that only a single host-guest system at a time can be evaluated by this model. Additionally, slight deviations from linearity can be indicative of cooperative and anti-cooperative binding effects. The work by Sannes-Lowery *et al.* has also demonstrated a case where Scatchard analysis fails if multiple nonequivalent binding modes are present in the analyzed system [85].

$$\frac{\alpha}{[G]} = -\frac{\alpha}{K_d} + \frac{n}{K_d} \tag{8.5}$$

$$\frac{1}{\alpha} = \frac{K_d}{[G]} + n \tag{8.6}$$

Many researchers have moved from the use of linearization techniques to the use of nonlinear models to allow for more diversity in terms of experimental observables, interaction stoichiometries, and data analysis. The availability of powerful computer platforms and the customizability of these models can facilitate automated and high-throughput schemes. Dotsikas and Loukas derived a sophisticated nonlinear model for the evaluation of binding between β-cyclodextrin and three small model ligands (phenol and hydrazine derivatives), comparing the result of this method to that using the Benesi-Hildebrand treatment by UV and ESI-MS titration [87]. They found better agreement between UV- and ESI-MS-determined values using the nonlinear model. Schug *et al.* derived a simple nonlinear 1:1 binding model to investigate the chiral recognition capacity of cinchonane-based chiral selector agents for discriminating between the enantiomers of dinitrobenzoyl-leucine by ESI-MS [62]. Results were compared with published data from microcalorimetry and chiral HPLC experiments. Although absolute binding constants determined by ESI-MS were shifted to increased affinity relative to solution-phase data for this small-molecule system, good empirical correlation was found between the relative binding affinities measured by the different methods. Wang *et al.* provide a rigorous assessment of instrumental parameters for their use of nanoelectrospray ionization (nESI) and a nonlinear titration model in studying the binding between a single chain fragment of a monoclonal antibody (scFv) and carbohydrate (trisaccharide) ligands [88]. Good correlation, evidenced by the data shown in Table 8.1, was shown between the nESI-FTICR-MS results and those obtained by isothermal titration calorimetry. The group of Zenobi and coworkers has been very active in their pursuit of new, reliable methods for noncovalent binding determinations by ESI-MS (see below). As a relevant example here, Daniel *et al.* have used a nonlinear model to successfully demonstrate the accurate binding affinity determination for two noncovalent inhibitors against adenylate kinase by ESI-MS [89].

The experimental set-up of titration experiments can be highly variable. In many cases, one of the components in the equilibrium of interest is held at a fixed concentration, while that of the other component is varied over one or more orders of magnitude. Conversely, it is also possible to perform titration experiments in

Table 8.1 Comparison of association constants (K) between the monoclonal antibody ScFv and four carbohydrate ligands determined by nESI-FTICR-MS and ITC [88].

Ligand	K (10^{-5} M^{-1}) (MS)	K (10^{-5} M^{-1}) (ITC)
Galα[Abe]Man	1.70 ± 0.05	1.6 ± 0.2
Abe(2-O—CH$_3$—Man)	1.50 ± 0.20	1.43 ± 0.05
Glcβ[Abe]Man	0.15 ± 0.04	0.30 ± 0.14
GlcGlcGalα[Abe]Man	5.30 ± 0.50	3.81 ± 0.13

which both the host and guest are present at equimolar concentration, which is then varied over a significant range. Jorgensen *et al.* have successfully applied this approach for studying binding between vancomycin group antibiotics and various bacterial cell wall analogs, including the investigation of pH and stereospecific binding effects [90]. Experiments were performed in a competitive titration format with three peptide ligands present (G1, G2, and G3) to bind to each antibiotic (both vancomycin and ristocetin). By virtue of using equimolar solutions, the model employed for extracting each binding constant was simplified to the form (for, e.g., one of the ligands, G1):

$$K_{HG1} = \frac{[HG1]}{[H]([H] + [HG2] + [HG3])} \tag{8.7}$$

Gabelica *et al.* have also used equimolar mixture-based titration analysis to study both DNA-ligand and cyclodextrin (CD) – dicarboxylic acid interaction systems [91]. Their approach included the simultaneous determination of the binding constant and relative response factor (R) by fitting relative ion abundance data to Eq. (8.8):

$$\frac{i_H}{i_{HG}} = \frac{1 + (1 + 4Kc_0)^{1/2}}{2RKc_0} \tag{8.8}$$

where $c_0 = c_{0,H} = c_{0,G}$ and R is the ratio of response factors of the complex to that of the free host. Interestingly, the DNA-ligand complexes were shown to provide an R factor close to unity, suggesting that the ionization efficiency of the DNA was similar to that of the DNA-ligand complex. A plot of the experimental data and fit of the data to Eq. (8.8) is shown in Figure 8.5. The R value obtained ($R = 0.5 \pm 0.2$) indicates that free DNA duplex is less than twice as responsive as the complex formed when the duplex binds with Hoechst 33 342. In contrast, values far from unity were obtained for the CD – dicarboxylic acid system ($R \approx 5 - 7$), indicating a more significant shift in ionization efficiency upon complexation by the CD. The authors rationalize this finding by considering the relatively poor ionization efficiency of the neutral CD molecule compared to the increased ionization efficiency provided by the inclusion of the dicarboxylic acid ligand. They show how such an approach can be useful for determining relative response factors, which can provide a critical correction factor when determining binding constants for small-molecule systems by ESI-MS. However, care must be taken to prepare strictly equimolar solutions,

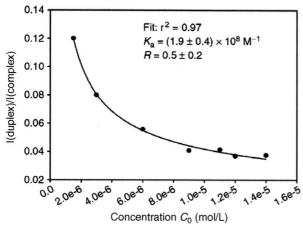

Figure 8.5 Equimolar ($c_{0,H} = c_{0,G} = c_0$) titration of binding affinity between d(CGCGAATTCGCG)$_2$ and Hoechst 33 342. Experimental data (ratio of ion intensities) were fitted to Eq. (8.8) for simultaneous determination of association constant and relative response factor (R). Reprinted with permission from Ref. [91] (Copyright 2003 John Wiley & Sons, Inc.)

as small discrepancies can provide biases to measured values. Furthermore, the simultaneous fitting of multiple variables to a set of data must be analyzed carefully in order to assess the potential presence of multiple minima in the fitting procedure.

In 2006, Krishnaswamy et al. reported an effective competitive binding 'laddering' approach to determine the relative free energies of binding for protein-protein interactions between variants of *Bacillus amyloliquefaciens* protein barstar and the RNAase barnase which bind in the nM to pM range [92]. Measurements of relative affinities for multiple combinations of two species binding to a common partner by ESI-MS are compared by referencing each to one or more intermediate species to construct a ladder of affinities which correlated very well with solution-phase binding constant measurements made using a fluorescence-based barnase activity assay. The method allows simultaneous comparison of binding for multiple species, making it suitable for high-throughput screening, and, furthermore, the method does not require highly purified samples or precisely known concentrations. The main limitation is the need to consider the presence of buffer salts in the medium which may give rise to adduct formation or deleterious matrix effects.

Approaches to removing the dependence of the measurement on the response factors of the interacting species of interest have also been described. Kempen and Brodbelt have described a method whereby a calibration is first set up for a well-characterized reference complex of known binding affinity (a crown ether – alkali metal binding system) [93]. Following this, different ligands and/or hosts can be evaluated in a competitive binding format to ascertain the change in the observed response of the reference complex. The obvious advantage of this method is that the equilibrium of interest need not be observed in the mass spectrum, since only a change in the reference system is monitored. The equilibrium solution concentration, and thus the response of the unknown system, is not directly considered. The disadvantage is the need for a suitable reference complex which should incorporate either a host or a guest in common with the system which is desired to be studied.

Very recently, Zenobi and coworkers described a variation on this type of internal standardization method using the relative response of two ligands (in equimolar concentration; 5 µM) in competition for binding a protein to determine dissociation constants in the low pM range [94]. This approach extends the applicability of the ESI-MS direct binding determination technique even further in terms of high-affinity systems, and also removes the need to consider response factors of noncovalent complexes. In this set-up, one ligand of known binding affinity is used as an internal standard. The ratio between the response of this ligand and that of a ligand of interest, competing for interaction with the same protein, is monitored relative to the concentration of protein added (1–10 µM). A mathematical model using standard equilibrium and mass balance expressions was derived to obtain the binding constant of interest. The method was validated studying biotin – avidin and related protein-ligand systems, and it was then applied for studying the binding of ligands to p38 mitogen-activated protein kinase. Figure 8.6 shows the results obtained from titrating SB202190 (reference ligand) and VX-745 with increasing

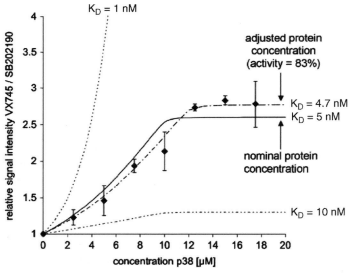

Figure 8.6 Titration of ligands SB202190 (reference ligand) and VX-745 with tyrosine p38α using an internal standardized titration model. Dissociation constants obtained for nominal and adjusted active protein concentrations are shown bracketed by model curves representing hypothetical stronger ($K_D = 1$ nM) and weaker ($K_D = 10$ nM) binding affinities. Reprinted with permission from Ref. [94] (Copyright 2008 John Wiley & Sons, Inc.)

concentrations of tyrosine p38α. A dissociation constant of 4.7 nM for VX-745 was obtained after correcting for protein activity (83%). Several advantages are noted, including the removal of response factor consideration for the protein-ligand complex, the ability to make measurements in the low mass range considering free ligand ion responses only, and the demonstrated capability of addressing very high affinity systems. Noted disadvantages include the need for a suitable reference system, the need to work under nondissociative ESI-MS conditions (limited sensitivity with ionization under physiological conditions), and the need for excessive equilibration times to avoid a kinetic bias in the measurements. Zenobi and coworkers have also previously described a method whereby an internal reference compound is used to normalize the response of all ions incorporated in a study of zinc-binding beta-peptides [95]. An initial calibration is performed for the free peptide, and then the response of the free peptide is monitored and related back to the calibration curve to determine free peptide concentration as zinc is titrated into the system. In this manner, a model was derived whereby the free peptide concentration is plotted against the total metal ion concentration to determine the association constant for the interaction.

Besides the derivation of novel models for extraction of binding constant information, novel experimental set-ups have also recently been described to increase the throughput of ESI-MS-based titration approaches. Fryčák and Schug recently reported two variations of a ligand band-dispersion titration approach, termed 'dynamic titration', which allows for the determination of binding constants from single solutions of host and guest [96, 97]. The basis of the method is the injection of a guest into a flowing stream (constant concentration) of host. The mixture is passed through a band-broadening element (length of tubing comprising substantial dead volume) so that the guest component disperses into a characterizable distribution. When the mixture enters the ESI source, a natural titration is presented with an increase in concentration of guest on the up-slope of the peak and a decrease in concentration on the down-slope. The method was demonstrated in an initial manner by fitting a Gaussian function to the band-broadened profile in order to discern the initial concentration of guest at each point on the distribution. Simultaneously, the initial concentrations were related to a ratio of ion abundances following a standard 1 : 1 binding model. The method was demonstrated for single host – single guest equilibria comprising (a) a previously studied chiral recognition system and (b) CDs binding sulfonated azo dyes [96]. Good agreement was shown with results obtained by the traditional titration approach using discrete solutions. In the second-generation approach, a more general method whereby the time- and concentration-dependent binding distribution was integrated to obtain dissociation constants was presented [97]. This approach incorporated the ability to use HPLC for separation of guest mixtures (with post-column addition of host) to multiplex binding determinations in a competitive titration format with no theoretical limitation on the number of simultaneous interactions which can be processed (e.g., coeluting peaks) and to remove the need for a particularly well-characterized distribution. The method was applied to study binding between various CD host molecules and a series of nonsteroidal anti-inflammatory drugs and showed good agreement with data from both traditional titration approaches using ESI-MS and other standard spectroscopic techniques. In-house software programs were generated for both the Gaussian and the general dynamic titration procedures to analyze the data.

Other approaches, such as those based on solution-phase H/D exchange (PLIMSTEX [98], SUPREX [99, 100], and others [101]) and diffusion-based measurements [102, 103] represent promising experimental set-ups for studying a wide range of molecular and macromolecular interaction systems. These are indirect methods based on the definition given previously and rely on detecting the dissociated receptor-ligand interaction partners through the use of either solution (e.g., for quenching of H/D exchange) or ionization conditions not amenable for detecting noncovalent complexes. H/D exchange measurements, in particular, can also be useful for determining interaction sites in protein-ligand systems [104, 105]. Furthermore, they can be performed in either the solution or the gas phase [106, 107], providing significant versatility and complementarity in binding information. It is also worth noting that the Fitzgerald group very recently reported the analysis of protein stability and ligand binding using a technique termed SPROX (stability of proteins from rates of oxidation) [108]. Conceptually similar to SUPREX, SPROX utilizes the irreversible oxidation rates of globally protected sites, instead of reversible H/D exchange, to obtain thermochemical and site-localized information in protein-ligand binding systems. Prior to this, radical oxidation of complexes to study protein inter- and intramolecular interactions has been reported [109, 110]. Related site-localization and binding determination experimental variants include the use of enzyme hydrolysis of complexes [111, 112] and photoaffinity labeling [113, 114]. Recently, the use of chemical cross-linking, especially for studying protein-protein interactions in combination with the some of the above-described methods, has become quite popular [39, 115, 116].

It is apparent from the range of direct and indirect experimental set-ups and data analysis procedures employed that ESI-MS provides a versatile platform for studying noncovalent interactions in both small- and large-molecule interaction systems. The vast majority of reported literature demonstrates suitable correlation between ESI-MS-based affinity measurements and those performed by more established solution-phase binding determination methods. The fact that validation is often presented (and necessary) in the case of these complementary methods indicates dependence of success on the system. Indeed, it is certain that many studies have been performed whereby insufficient correlation has been observed, but these studies may not be disseminated or highlighted in the literature. Some studies have reported poor correlation [117], offering a variety of explanations for such results. Therefore it is prudent to discuss some of these aspects in more detail.

8.3
Practical Aspects of Titration Analysis

The practical aspects of titration analysis are perhaps best considered and addressed in a stepwise fashion starting from the application to a particular interaction system of interest, moving to sample preparation and experimental design, then to instrumental analysis, and finally to data analysis. Mention has already been made of the versatility of interaction systems which can be potentially studied, but also of the limitations in terms of the noncovalent forces which are necessary to preserve the interaction during the dynamic ESI phase transfer process. In the end, for direct relative or absolute affinity determinations by titration and competitive binding methods, quantitative correlation between gas-phase ion abundances and solution-phase equilibrium concentrations is sought. This requires careful optimization of sample concentrations, solvent conditions, instrumental parameters, and data handling procedures.

Once the specific interaction system has been identified, a simple screening procedure may be run to ensure that binding between the host(s) and guest(s) of interest can be observed. This will undoubtedly involve some instrumental optimization, which is discussed below. Assuming that the complex of interest can be observed, one would then proceed to considering how best to set up the titration experiment. The goal is to suitably sample the equilibria of interest while remaining in a stable nondissociative regime within the linear dynamic range of the instrument. For reliable determination of binding constants, the equilibria of interest should be sampled such that the degree of association (α) is within the range 20–80% (see Eq. (8.4)). An excellent and detailed discussion of this consideration is given by Hirose in his practical guide to binding constant determination (by UV and NMR titrations) [5]. The main point is that binding constants based on sampling solely where $\alpha < 20\%$ or $\alpha > 80\%$ are subject to a much higher degree of variance, which can contribute significant error to the measured binding constant. In practice, the relative starting concentrations of host(s) and guest(s) can be varied across a range of concentrations in diverse ways, as illustrated in Figure 8.7, in order to sample the equilibria most effectively.

From the instrumental side, the linear dynamic range of most ESI-MS instruments can be a limiting factor. The sensitivity (limit of detection) of the instrument

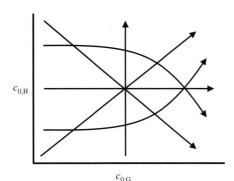

Figure 8.7 Potential variations in relative starting host and guest concentrations ($c_{0,H}$ and $c_{0,G}$, respectively) for performing titration analysis. Adapted from Ref. [5].

provides a limitation at lower concentrations. This lower limit, of course, depends on the analyte under investigation, the solution conditions from which the analyte is sprayed, and the instrumental settings (e.g., in terms of efficiency of desolvation), and can be highly variable. The upper limit of the linear dynamic range is controlled by the onset of droplet saturation or ion suppression. This process is, as mentioned earlier, best described by considering the competitive nature of the ionization process in terms of the equilibrium partitioning model [45–47]. The consequences are twofold: (i) the limited linear dynamic range can introduce significant imprecision in relative ion abundance ratios recorded where the components being compared have more than, for example, a 100-fold difference in absolute abundance; and (ii) in multiplexed, competitive, or internal reference-based binding experiments, the limited dynamic range substantially reduces the range of affinities which can be reliably elucidated in a single experiment. Conventional ESI sources can be expected to produce linear dynamic ranges on the order of 10^3–10^4. In contrast, nanoelectrospray ionization (nESI) [118] and electrosonicspray ionization (ESSI) [119] sources can exhibit a wider linear dynamic range (up to 10^6), making them very attractive for diverse binding determination set-ups, among other advantages (see below).

Figure 8.8 provides a means to visualize the limits placed by the linear dynamic range of an ESI-MS system on the titration of a noncovalent interaction system. These data are for a small-molecule interaction system comprising the binding of *tert*-butylcarbamoylquinidine (tBuCQD, host, held at a constant concentration of 10 μM), a chiral selector employed as a bonded stationary phase in enantioselective HPLC, to (*R*)-dinitrobenzoyl-leucine (DNB-Leu; guest). The experimental data, comprising the ratio of the ion abundance of the host (i_H) to that of the complex (i_C), where $I = i_H/i_C$, plotted against the initial concentration of guest, is fitted using

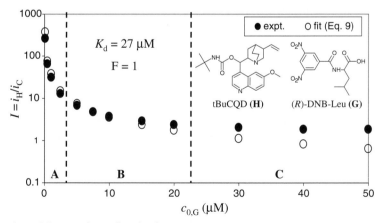

Figure 8.8 Example small-molecule ESI-MS titration experiment demonstrating instrumental limitations in the fit of experimental data to a standard titration model (Eq. (8.9)) at low (A) and high (C) concentrations.

a least squares minimization procedure to the derived 1:1 binding model given in Eq. (8.9) to obtain a dissociation constant ($K_d = 27\,\mu M$ in this example) [62].

$$I^2 F^2 c_{0,G} + IF(c_{0,G} - c_{0,H} - K_d) - K_d = 0 \tag{8.9}$$

Here, F is the ratio of the response factors for the complex and the free host ($F = f_C/f_H$, which is identical to R in Eq. (8.8)). The value of F is assumed to be unity in this example. The measurement was performed in 50/50 MeOH/water with 10 µM NaOAc on a Thermo LCQ Deca XP instrument with a conventional ESI ion source. At low concentrations (region A), the ratio of ion abundances is very high. As I approaches 1000, the relative standard deviation in this measured point increases dramatically. It is possible to observe up to 100% RSD in replicate data points when the relative abundance of ions differs by more than three orders of magnitude. At this level, the response of the complex is very small and is hardly distinguishable from the noise. In the middle concentrations (region B), the data fit very well to the derived model. In region C, the experimental curve begins to flatten relative to the values predicted by the theoretical model. This is attributed to the onset of saturation of ionization sites at the droplet surface, which happens to correspond closely with the limit of linearity. It is necessary to carry the titration into this region to some extent in order to access a significant degree of association for evaluating the equilibrium ($\alpha \approx 37\%$ at 20 µM; $\alpha \approx 61\%$ at 50 µM). Once a K_d is obtained, the degree of association can be determined by numerically solving Eq. (8.10) to find the value of α at which the right and left hand sides of the equation are equal.

$$K_d + c_{0,G} + c_{0,H} = \frac{c_{0,G}}{\alpha} + \alpha c_{0,H} \tag{8.10}$$

The above example demonstrates important considerations for choosing appropriate concentrations to perform the titration experiment. Another consideration is the need for suitable replication (triplicate determination is the recommended minimum). Additional replicates mean additional analysis time, which can impose a bottleneck in such experiments if a large number of discrete solutions need to be prepared to be analyzed where high throughput is desired. A potential solution to this problem is to employ the ligand band-dispersion titration technique described previously [96, 97]. Additionally, if high-affinity systems (especially protein-protein and protein-ligand) are to be investigated, an extended equilibration time (from minutes to days depending on the kinetics of the system) may need to be considered once the components are mixed [92, 94].

The choice of solvent system is an important consideration. The choice should most closely reflect the native solution in which the interaction of interest takes place. Nearly all inter- and intramolecular noncovalent forces are affected in some manner (either strengthened or weakened) by a change in solvent composition [23]. For physiologically relevant interactions, traditional nonvolatile phosphate buffers and detergents are poorly compatible with ESI-MS analysis. A physiologically relevant ionic strength may be achieved by using volatile buffers such as ammonium acetate or ammonium bicarbonate. In 100% aqueous systems, sensitivity is sacrificed because of the relatively larger droplets produced during ESI, which are less efficiently desolvated in the source region. In the above example, a 50/50 MeOH/water mixture was chosen to mimic conditions employed in chiral separations using such immobilized chiral selector compounds. Different instrument configurations have different tolerances for solvents and additives. Most systems can handle a wide range of solvents, from purely aqueous to purely organic, sacrificing sensitivity to some extent at each of the extremes. The use of nonvolatile buffers is more amenable to instruments with special inlet configurations (e.g., orthogonal or Z-spray interfaces). Additionally, the use of an nESI source may be more tolerable to such conditions. In any case, matrix effects (the presence of co-analytes which are difficult to define and which affect reproducibility and sensitivity of ion response as a result of competitive ionization effects) are always a potential problem if unpurified samples are analyzed.

Following sample preparation, the samples are taken to the instrument for analysis. By this time, hopefully the source conditions have been optimized utilizing some basic screening procedures; for discussion purposes, these considerations are given here. Figure 8.9 shows an idealized view of a conventional electrospray source,

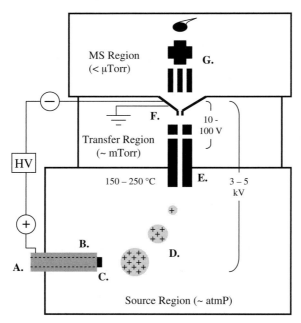

Figure 8.9 Important instrumental parameters for source optimization to detect noncovalent complex ions. The letters in the diagram correspond to specific components (and variables to be optimized): A. Inlet (solution composition, flow rate); B. N$_2$ sheath gas (T, flow rate); C. spray capillary (dimensions, position relative to E., V); D. shrinking droplet process; E. transfer capillary (T, V); F. 'skimmer' (or other similar) region (V, P); G. ion transfer and mass analyzer region. Figure adapted from Ref. [36].

highlighting the various components in the source which require optimization, mainly in terms of voltages and temperatures. Many instruments have factory-built software which can be used to optimize voltages, but frequently these automatic optimization functions do not control gas flow rates and temperatures in the source, parameters which are also very important for observing noncovalent complex ions.

The samples are introduced at the inlet (A). This introduction is typically performed via manual injection, direct infusion, or autosampler. The solution composition (discussed previously) and the flow rate are the key considerations. Conventional ESI typically offers better sensitivity at low flow rates (low $\mu L\, min^{-1}$). The ionization process in conventional ESI is also commonly assisted by introducing a nitrogen sheath gas (B) in a coaxial arrangement with the sprayer. The flow rate (and in some cases the temperature) of the nitrogen should be optimized to match the flow rate of the solution to aid optimal desolvation without destroying the ionic complexes of interest. The spray capillary (C) is perhaps one of the most influential parameters to be considered. Jecklin et al. recently reported a detailed comparison of conventional ESI, nESI, and ESSI sources for studying protein-ligand interactions [120]. Whereas quantitative binding determinations can still be performed with a conventional source, the advantages of nESI and ESSI are many. These include lower flow rate, lower sample consumption, lower spray voltage, wider dynamic range, higher tolerance to nonvolatile buffers, and better sensitivity. Additionally, it was shown that nESI and ESSI may be considered 'softer' ionization sources in terms of preserving noncovalent interactions during their transfer into the gas phase and providing consistent affinity values across a range of multiple charge state complexes. Table 8.2 demonstrates this comparison for dissociation constants obtained for the binding of ADP with adenylate kinase, including analysis of results obtained from individual charge states and two separate binding sites, using ESI, nESI, and ESSI. Overall, all values are in good agreement with solution-phase binding data obtained by traditional methods, but those obtained using ESSI were judged to be the most accurate.

Table 8.2 Charge state-dependent and mean dissociation constants obtained by titration for ADP-binding adenylate kinase evaluated using ESI, nESI, and ESSI ionization sources [120].

Charge state	K_d Binding Site 1 (μM)			K_d Binding Site 2 (μM)		
	ESSI	nESI	ESI	ESSI	nESI	ESI
+11			3.8 ± 0.2			36.6 ± 1.8
+10	1.9 ± 0.1	2.0 ± 0.2	4.7 ± 0.1	15.0 ± 0.9		40.8 ± 1.4
+9	1.8 ± 0.1	2.8 ± 0.1	5.7 ± 0.2	17.7 ± 0.9	23.4 ± 0.8	43.4 ± 1.1
+8	2.1 ± 0.1	3.6 ± 0.2		19.7 ± 0.8	23.0 ± 0.8	
+7	2.3 ± 0.1			23.1 ± 0.6		
+6	2.7 ± 0.1			22.1 ± 0.8		
Mean K_d	2.2 ± 0.8	2.8 ± 1.0	4.7 ± 1.0	19.5 ± 8.0	23.2 ± 3.2	40.3 ± 5.0

Letter D in Figure 8.9 refers to the droplet desolvation process. Besides the competitive ionization process inherent to ESI and the effect of different solvent compositions on ionization efficiency, it is also feasible to conjecture that the shrinking droplet may impart a concentration gradient which could cause a shift in the equilibria of interest. However, prior hypotheses and recent evidence suggest that if the host-guest association is kinetically stable on the time scale of the ESI process (μsec – msec), then a reliable snapshot of the solution phase equilibrium may be obtained [10, 42]. Additional studies in this area may shed more light on the system dependence of this potentially deleterious effect.

Another consequence of the shrinking droplet process is the potential for observing nonspecific interactions. These may occur in the form of simple adduct ions or as seemingly feasible molecular interactions, depending on the nature of the system and the experimental set-up. Nonspecific interactions are defined as interactions which are observed as a by-product of the analysis but which are not actually present in bulk solution (or are low affinity interactions which are amplified during the ionization process). If these observations are considered as 'real' interactions, this can introduce a systematic error in the binding constant calculation or lead to false positives. The appearance of nonspecific interactions can be minimized by working with low concentrations of interactants or by increasing the ionic strength of the solution to limit 'chance' electrostatic binding. Stringent guidelines for assessing the presence of nonspecific interactions have been given by Cuniff and Vouros in their study of CD binding systems [121]. They specifically recommend the investigation of a range of analytes which resemble the host and guest in every manner possible with the exception of exhibiting binding. Klassen and coworkers have also studied the phenomenon of nonspecific interactions extensively in their investigation of protein-carbohydrate binding systems [88, 122, 123]. They have devised a method for assessing the propensity for nonspecific interactions in a given experimental set-up by incorporating a nonbinding protein in their assay system. With this method they demonstrated the ability to correct for nonspecific interactions to determine protein-ligand association constants under conditions where nonspecific interactions were prevalent. It is also worthy of note that they have investigated the use of small organic compounds, specifically imidazole, to stabilize protein-ligand interactions which are particularly susceptible to dissociation during the ionization process [124].

Ions and ionic complexes which are mostly desolvated and released from the electrosprayed droplet (either by ion evaporation or as a charged residue) encounter a potential gradient, which influences their passage into the transfer capillary (Figure 8.9, letter E.) connecting the source region with the mass spectrometer. In many instrument configurations, the transfer capillary is heated. The temperature reached must be optimized in order to achieve removal of residual solvent clusters attached to the ion of interest while not imparting so much energy that dissociation of noncovalent complexes is caused. Typically a range of 150–250 °C is employed. After passing through the transfer line, the ions encounter the 'skimmer' region (E). The lenses in this region have a wide array of configurations and names depending on the manufacturer, but the key point here is to optimize the potential gradient experienced by the ion. Because this region is still characterized by a substantially high pressure relative to the high vacuum in the mass analyzer, if a noncovalent complex ion is pulled through this region too vigorously, dissociation of the complex can take place. Increasing the potential gradient in this region to induce fragmentation or dissociation is often denoted as 'in-source' or 'poor man's' CID. For noncovalent complexes, the potential difference in this region should be minimized, but it should also be set at a value which allows for robust ion transmission and avoids discrimination due to collisional dissociation [83, 125]. After passing through this region, the ions are subjected to mass analysis and detection in the high-vacuum mass analyzer region (G). A wide variety of capabilities are available from a wide variety of mass analyzer types on the market, a full discussion of which is beyond the scope of this chapter.

In the previous section we have described a range of different data analysis options for extracting quantitative binding information in the form of binding constants from titration and competitive binding experiments. Obviously, the experimental set-up and optimization method is dictated to a large degree by the binding model which will be used. Furthermore, some of the caveats in terms of expected ion intensity and reproducibility related to the dynamic response range of the instrument have already been discussed. What remains to be discussed, however, is perhaps one of the most important considerations in titration analysis: the response factor of the ionic complex.

For determinations of binding constant in unknown systems, when the host(s) and guest(s) are mixed in solution, the equilibrium solution concentration of the complex is not known. Because ion response in ESI-MS is intimately tied to the physicochemical attributes of the analytes, if the ionic complex exhibits a character that is markedly different from that of the free host from which it is formed, then the responses will not be equivalent. It is common to assume that the response of the free host and the host-guest complex exhibit a similar ionization efficiency in order to simplify data analysis (i.e., F or $R \approx 1$). For a large protein host binding a small ligand, the change in mass and physicochemical attributes will likely be small, validating this approximation. Another aspect to consider is whether or not the free host and ionic complex(es) are observed in equivalent charge states. If this is the case, then this also lends some weight to the assumption that the relative response factor ratio is approximately unity. In fact, for protein-ligand binding, it is prudent to check the dependence of measured binding constants on charge states (a consistent affinity value should be observed across all charge states) to ensure that the system is well-behaved [88, 92, 94]. In small-molecule interaction systems, the assumption of an equivalent response is less valid. In many cases the host-guest complex can be as much as twice the size of the free host. Several methods whereby internal references or multi-parameter fitting procedures have been employed to address this problem or elucidate relative response factors are described above. Doubtless, other data treatment methods will also be forthcoming, but this still remains a problem of considerable magnitude when performing traditional titration experiments.

8.4
Summary and Outlook

In most well-conceived binding determination studies using ESI-MS, validation of results requires comparison with data obtained by established solution-phase binding determination methods, such as spectroscopic titration or calorimetry. This fact points to the system dependence of success and validates the scrutiny of the scientific community for the use of ESI-MS to study various interaction systems. However, coupled with strict consideration of the topics presented in this chapter, the literature base will continue to expand. The advantages of mass spectrometry-based technology over more traditional but cumbersome and time-consuming techniques make it a powerful approach worthy of serious consideration and further development. Already, the ESI-MS instrument is an essential component of most academic, industrial, and governmental research programs. Its maturing use for routine physical and biophysical binding measurements will be aided by instrumental breakthroughs, new methods, and new applications, as the fundamental understanding of the electrospray process continues to grow. With increased emphasis being placed on drug discovery efforts, the '-omics' areas (specifically, the emerging field of interactomics), environmental science, and food science, to name a few, ESI-MS sits poised to provide rapid qualitative and quantitative information to aid these research efforts in a more than substantial manner.

Acknowledgments

The author would like to thank Dr. Petr Fryčák and Dr. Misjudeen Raji for their critical proof-reading and suggestions during the preparation of this manuscript.

References

1 Grayson, M.A. (ed.) (2002) *Measuring Mass: From Positive Rays to Proteins*, Chemical Heritage Press, Philadelphia.
2 Yamashita, M. and Fenn, J.B. (1984) Electrospray ion source: another variation on a free-jet theme. *J. Phys. Chem.*, **88**, 4451–4459.
3 Cole, R.B. (ed.) (1997) *Electrospray Ionization Mass Spectrometry: Fundamentals, Instrumentation, and Applications*, John Wiley & Sons, Inc.., New York.
4 Fielding, L. (2000) Determination of association constants (K_a) from solution NMR Data. *Tetrahedron*, **56**, 6151–6170.
5 Hirose, K. (2001) A practical guide for the determination of binding constants. *J. Incl. Phenom. Macrocyc. Chem.*, **39**, 193–209.
6 Jelesarov, I. and Bosshard, H.R. (1999) Isothermal titration calorimetry and differential scanning calorimetry as complementary tools to investigate the energetics of biomolecular recognition. *J. Molec. Recog.*, **12**, 3–18.
7 Huber, W. and Mueller, F. (2006) Biomolecular interaction analysis in drug discovery using surface plasmon resonance technology. *Curr. Pharm. Des.*, **12**, 3999–4021.

8 Rundlett, K.L. and Armstrong, D.W. (2001) Methods for the determination of binding constants by capillary electrophoresis. *Electrophoresis*, **22**, 1419–1427.

9 Tanaka, Y. and Terabe, S. (2002) Estimation of binding constants by capillary electrophoresis. *J. Chromatogr. B*, **768**, 81–92.

10 Smith, R.D. and Light-Wahl, K.J. (1993) The observation of non-covalent interactions in solution by electrospray ionization mass spectrometry: promise, pitfalls, and prognosis. *Biol. Mass Spectrom.*, **22**, 493–501.

11 Smith, D.L. and Zhang, Z. (1994) Probing noncovalent structural features of proteins by mass spectrometry. *Mass Spectrom. Rev.*, **13**, 411–429.

12 Colton, R., D'Agostino, A., and Traeger, J.C. (1995) Electrospray mass spectrometry applied to inorganic and organometallic chemistry. *Mass Spectrom. Rev.*, **14**, 79–106.

13 Przybylski, M. and Glocker, M.O. (1996) Electrospray mass spectrometry of biomacromolecular complexes with noncovalent interactions - new analytical perspectives for supramolecular chemistry and molecular recognition processes. *Angew. Chem. Int. Ed.*, **35**, 806–826.

14 Loo, J.A. (1997) Studying noncovalent protein complexes by electrospray ionization mass spectrometry. *Mass Spectrom. Rev.*, **16**, 1–23.

15 Winston, R.L. and Fitzgerald, M.C. (1997) Mass spectrometry as a readout of protein structure and function. *Mass Spectrom. Rev.*, **16**, 165–179.

16 Smith, R.D., Bruce, J.E., Wu, Q., and Lei, Q.P. (1997) New mass spectrometric methods for the study of noncovalent associations of biopolymers. *Chem. Soc. Rev.*, **26**, 191–202.

17 Pramanik, B.N., Bartner, P.L., Mirza, U.A., Liu, Y.-H., and Ganguly, A.K. (1998) Electrospray ionization mass spectrometry for the study of non-covalent complexes: an emerging technology. *J. Mass Spectrom.*, **33**, 911–920.

18 Veenstra, T.D. (1999) Electrospray ionization mass spectrometry in the study of biomolecular non-covalent interactions. *Biophys. Chem.*, **79**, 63–79.

19 Brodbelt, J.S. (2000) Probing molecular recognition by mass spectrometry. *Int. J. Mass Spectrom.*, **200**, 57–69.

20 Loo, J.A. (2000) Electrospray ionization mass spectrometry: a technology for studying noncovalent macromolecular complexes. *Int. J. Mass Spectrom.*, **200**, 175–186.

21 Hofstadler, S.A. and Griffey, R.H. (2001) Analysis of noncovalent complexes of DNA and RNA by mass spectrometry. *Chem. Rev.*, **101**, 377–390.

22 Schalley, C.A. (2001) Molecular recognition and supramolecular chemistry in the gas phase. *Mass Spectrom. Rev.*, **20**, 253–309.

23 Daniel, J.M., Friess, S.D., Rajagopalan, S., Wendt, S., and Zenobi, R. (2002) Quantitative determination of noncovalent binding interactions using soft ionization mass spectrometry. *Int. J. Mass Spectrom.*, **216**, 1–27.

24 Nesatyy, V.J. (2002) Mass spectrometry evaluation of the solution and gas-phase properties of noncovalent protein complexes. *Int. J. Mass Spectrom.*, **221**, 147–161.

25 Finn, M.G. (2002) Emerging methods for the rapid determination of enantiomeric excess. *Chirality*, **14**, 534–540.

26 Ganem, B. and Henion, J.D. (2003) Going gently into flight. *Bioorg. Med. Chem.*, **11**, 311–314.

27 Tao, W.A. and Cooks, R.G. (2003) Chiral analysis by MS. *Anal. Chem.*, **75**, 25A–31.

28 Heck, A.J.R. and van den Heuvel, R.H.H. (2004) Investigation of intact protein complexes by mass spectrometry. *Mass Spectrom. Rev.*, **23**, 368–389.

29 Peschke, M., Verkerk, U.H., and Kebarle, P. (2004) Features of the ESI mechanism that affect the observation of multiply

charged noncovalent protein complexes and the determination of the association constant by the titration method. *J. Am. Soc. Mass Spectrom.*, **15**, 1424–1434.
30 Heck, A.J.R. and Jorgensen, T.J.D. (2004) Vancomycin in vacuo. *Int. J. Mass Spectrom.*, **236**, 11–23.
31 Breuker, K. (2004) The study of protein-ligand interactions by mass spectrometry – a personal view. *Int. J. Mass Spectrom.*, **239**, 33–41.
32 Breuker, K. (2004) New mass spectrometric methods for the quantification of protein-ligand binding in solution. *Angew. Chem. Int. Ed.*, **43**, 22–25.
33 Di Tullio, A., Reale, S., and De Angelis, F. (2005) Molecular recognition by mass spectrometry. *J. Mass Spectrom.*, **40**, 845–865.
34 Schug, K.A. and Lindner, W. (2005) Chiral molecular recognition for the detection and analysis of enantiomers by mass spectrometric methods. *J. Sep. Sci.*, **28**, 1932–1955.
35 Hofstadler, S.A. and Sannes-Lowery, K.A. (2006) Applications of ESI-MS in drug discovery: interrogation of noncovalent complexes. *Nature Rev.*, **5**, 585–595.
36 Schug, K.A. (2007) Solution phase enantioselective recognition and discrimination by electrospray ionization – mass spectrometry: state-of-the-art, methods, and an eye towards increased throughput measurements. *Combin. Chem. High Throughput Screen.*, **10**, 301–316.
37 Zehender, H. and Mayr, M. (2007) Application of mass spectrometry technologies for the discovery of low-molecular weight modulators of enzymes and protein-protein interactions. *Curr. Opin. Chem. Biol.*, **11**, 511–517.
38 Annis, D.A., Nickbarg, E., Yang, X., Ziebell, M.R., and Whitehurst, C.E. (2007) Affinity selection – mass spectrometry screening techniques for small molecule drug discovery. *Curr. Opin. Chem. Biol.*, **11**, 518–526.
39 Gingras, A.-C., Gstaiger, M., Raught, B., and Aebersold, R. (2007) Analysis of protein complexes using mass spectrometry. *Nature Rev.*, **8**, 645–654.
40 Fenn, J.B., Mann, M., Meng, C.K., Wong, S.F., and Whitehouse, C.M. (1989) Electrospray ionization for mass spectrometry of large biomolecules. *Science*, **246**, 64–71.
41 Wang, H. and Agnes, G.R. (1999) Kinetically labile equilibrium shifts induced by the electrospray process. *Anal. Chem.*, **71**, 4166–4172.
42 Wortmann, A., Kistler-Momotova, A., Zenobi, R., Heine, M.C., Wilhelm, O., and Pratsinis, S.E. (2007) Shrinking droplets in electrospray ionization and their influence on chemical equilibria. *J. Am. Soc. Mass Spectrom.*, **18**, 385–393.
43 Harron, A., Bentzley, C., Moore, P., and Davis, D. (2008) The analysis of the interactions and complexation of polycyclic aromatic hydrocarbons and cyclodextrin using electrospray ionization mass spectrometry. Proc. 56th ASMS Conference on Mass Spectrometry and Allied Topics, June 1–5, Denver, CO.
44 Liu, L. and Klassen, J.S. (2008) Quantifying protein-hydrophobic ligand interactions by ES-MS. Proc. 56th ASMS Conference on Mass Spectrometry and Allied Topics, June 1–5, Denver, CO.
45 Enke, C.G. (1997) A predictive model for matrix and analyte effects in electrospray ionization of singly-charged ionic analytes. *Anal. Chem.*, **69**, 4885–4893.
46 Sherman, C.L. and Brodbelt, J.S. (2003) An equilibrium partitioning model for predicting response to host-guest complexation in electrospray ionization mass spectrometry. *Anal. Chem.*, **75**, 1828–1836.
47 Sherman, C.L. and Brodbelt, J.S. (2005) Partitioning model for competitive host-guest complexation in ESI-MS. *Anal. Chem.*, **77**, 2512–2523.

48 Annis, D.A., Nazef, N., Chuang, C.C., Scott, M.P., and Nash, H.M. (2004) A general technique to rank protein-ligand binding affinities and determine allosteric versus direct binding site competition in compound mixtures. *J. Am. Chem. Soc.*, **126**, 15495–15503.

49 Annis, D.A., Chuang, C.-C., and Nazef, N. (2007) ALIS: an affinity selection – mass spectrometry system for the discovery and characterization of protein-ligand interactions, in *Mass Spectrometry in Medicinal Chemistry* (eds K. Wanner and G. Höfner), Wiley-VCH Verlag GmbH, Weinheim, pp. 121–184.

50 Brown, N., Zehender, H., Azzaoui, K., Schuffenhauer, A., Mayr, L.M., and Jacoby, E. (2006) A chemoinformatics analysis of hit lists obtained from high-throughput affinity-selection screening. *J. Biomol. Screen.*, **11**, 123–130.

51 Ng, E.S., Yang, F., Kameyama, A., Palcic, M.M., Hindsgaul, O., and Schriemer, D.C. (2005) High-throughput screening for enzyme inhibitors using frontal analysis chromatography with liquid chromatography and mass spectrometry. *Anal. Chem.*, **77**, 6125–6133.

52 Slon-Usakiewicz, J.J., Ng, W., Dai, J.R., Pasternak, A., and Redden, P.R. (2005) Frontal affinity chromatography with MS detection (FAC-MS) in drug discovery. *Drug Discov. Today*, **10**, 409–416.

53 Gullo, V.P., McAlpine, J., Lam, K.S., Baker, D., and Petersen, F. (2006) Drug discovery from natural products. *J. Ind. Microbiol. Biotechnol.*, **33**, 523–531.

54 Sydor, J.R., Scalf, M., Sideris, S., Mao, G.D., Pandey, Y., Tan, M., Mariano, M., Moran, M.F., Nock, S., and Wagner, P. (2003) Chip-based analysis of protein-protein interactions by fluorescence detection and on-chip immunoprecipitation combined with μLC-MS/MS analysis. *Anal. Chem.*, **75**, 6163–6170.

55 Comess, K.M. and Schurdak, M.E. (2004) Affinity-based screening techniques for enhancing lead discovery. *Curr. Opin. Drug Discov Devel.*, **7**, 411–416.

56 Comess, K.M., Schurdak, M.E., Voorbach, M.J., Coen, M., Trumbull, J.D., Yang, H., Gao, L., Tang, H., Cheng, X., Lerner, C.G., Mccall, J.O., Burns, D.J., and Beutel, B.A. (2006) An ultra-efficient affinity-based high throughput screening process: application to bacterial cell wall biosynthesis enzyme MurF. *J. Biomol. Screen.*, **11**, 743–754.

57 Cheng, X. and van Breemen, R.B. (2005) Mass spectrometry-based screening for inhibitors of beta-amyloid protein aggregation. *Anal. Chem.*, **77**, 7012–7015.

58 Armentrout, P.B. (1992) Thermochemical measurements by guided ion beam mass spectrometry, in *Advances in Gas Phase Ion Chemistry*, vol. 1 (eds N.G. Adams and L.M. Babcock), JAI Press, Greenwich, pp. 83–119.

59 Liang, Y., Bradshaw, J.S., Izatt, R.M., Pope, R.M., and Dearden, D.V. (1999) *Int. J. Mass Spectrom.*, **185/186/187**, 977–988.

60 Grigorean, G., Gronert, S., and Lebrilla, C.B. (2002) Enantioselective gas-phase ion-molecule reactions in a quadrupole ion trap. *Int. J. Mass Spectrom.*, **219**, 79–87.

61 Xie, Y., Zhang, J., Yin, S., and Loo, J.A. (2006) Top-down ESI-ECD-FT-ICR mass spectrometry localizes noncovalent protein-ligand binding sites. *J. Am. Chem. Soc.*, **128**, 14432–14433.

62 Schug, K.A., Frycák, P., Maier, N.M., and Lindner, W. (2005) Measurement of solution phase chiral molecular recognition in the gas phase using electrospray ionization – mass spectrometry. *Anal. Chem.*, **77**, 3660–3670.

63 Sannes-Lowery, K.A., Drader, J.J., Griffey, R.H., and Hofstadler, S.A. (2000) Fourier transform ion cyclotron resonance mass spectrometry as a high throughput affinity screen to identify

RNA binding ligands. *Trends Anal. Chem.*, **19**, 481–491.

64 Schug, K. and Lindner, W. (2005) Using electrospray ionization – mass spectrometry/tandem mass spectrometry and small molecules to study guanidinium – anion interactions. *Int. J. Mass Spectrom.*, **241**, 11–23.

65 Wang, K. and Gokel, G.W. (1996) Correlation of solution and gas phase complexation assessed by electrospray ionization mass spectrometry: application to one-, two-, and three-ring macrocycles. *J. Org. Chem.*, **61**, 4693–4697.

66 Leize, E., Jaffrezic, A., and Van Dorsselaer, A. (1996) Correlation between solvation energies and electrospray mass spectrometric response factors. study by electrospray mass spectrometry of supramolecular complexes in thermodynamic equilibrium in solution. *J. Mass Spectrom.*, **31**, 537–544.

67 Blair, S.M., Kempen, E.C., and Brodbelt, J.S. (1998) Determination of binding selectivities in host-guest complexation by electrospray/quadrupole ion trap mass spectrometry. *J. Am. Soc. Mass Spectrom.*, **9**, 1049–1059.

68 Reyzer, M.L., Brodbelt, J.S., Marchand, A.P., Chen, Z., Huang, Z., and Namboothiri, I.N.N. (2001) *Int. J. Mass Spectrom.*, **204**, 133–142.

69 Ventola, E., Hyyrylainen, A., and Vainiotalo, P. (2006) Complex formation between a tetramesityl sulfonated resorcarene and alkylammonium ions: a mass spectrometric study of noncovalent interactions. *Rapid Commun. Mass Spectrom.*, **20**, 1218–1224.

70 Schalley, C.A., Martin, T., Obst, U., and Rebek, J. Jr (1999) Characterization of self-assembling encapsulation complexes in the gas phase and solution phase. *J. Am. Chem. Soc.*, **121**, 2133–2138.

71 Schug, K.A., Maier, N.M., and Lindner, W. (2006) Chiral recognition mass spectrometry: remarkable effects observed from the relative ion abundances of ternary diastereomeric complexes using electrospray ionization. *Chem. Commun.*, 414–416.

72 Sannes-Lowery, K.A., Mei, H.-Y., and Loo, J.A. (1999) Studying aminoglycoside binding to HIV-1 TAR RNA by electrospray ionization mass spectrometry. *Int. J. Mass Spectrom.*, **193**, 115–122.

73 Wan, K.X., Shibue, T., and Gross, M.L. (2000) Non-covalent complexes between DNA-binding drugs and double-stranded oligodeoxynucleotides: a study by ESI ion-trap mass spectrometry. *J. Am. Chem. Soc.*, **122**, 300–307.

74 Gao, J., Cheng, X., Chen, R., Sigal, G.B., Bruce, J.E., Schwartz, B.L., Hofstadler, S.A., Anderson, G.A., Smith, R.D., and Whitesides, G.M. (1996) Screening derivatized peptide libraries for tight binding inhibitors to carbonic anhydrase II by electrospray ionization-mass spectrometry. *J. Med. Chem.*, **39**, 1949–1955.

75 Bruce, J.E., Anderson, G.A., Chen, R., Cheng, X., Gale, D.C., Hofstadler, S.A., Schwartz, B.L., and Smith, R.D. (1995) Bio-affinity characterization mass spectrometry. *Rapid Commun. Mass Spectrom.*, **9**, 644–650.

76 Robinson, C.V., Chung, E.W., Kragelund, B.B., Knudsen, J., Aplin, R.T., Poulsen, F.M., and Dobson, C.M. (1996) Probing the nature of noncovalent interactions by mass spectrometry. A study of protein-CoA ligand binding and assembly. *J. Am. Chem. Soc.*, **118**, 8646–8653.

77 Job, P. (1925) Spectrographic study of the formation of complexes in solution and of their stability. *Compt. rend.*, **180**, 928–930.

78 Job, P. (1928) Formation and stability of inorganic complexes in solution. *Ann. Chim. Appl.*, **9**, 113–203.

79 Scatchard, G. (1949) The attractions of proteins for small molecules and ions. *Ann. NY Acad. Sci.*, **51**, 660–672.

80 Benesi, H.A. and Hildebrand, J.A. (1949) A spectrophotometric investigation of the interaction of iodine with aromatic hydrocarbons. *J. Am. Chem. Soc.*, **71**, 2703–2707.

81 Scott, R.L. (1956) Some comments on the Benesi-Hildebrand equation. *Rec. Trav. Chim.*, **75**, 787–789.

82 Lim, H.-K., Hsieh, Y.L., Ganem, B., and Henion, J. (1995) Recognition of cell-wall peptide ligands by vancomycin group antibiotics: studies using ion spray mass spectrometry. *J. Mass Spectrom.*, **30**, 708–714.

83 Loo, J.A., Hu, P., McConnell, P., Mueller, W.T., Sawyer, T.K., and Thanabal, V. (1997) A study of the SRC SH2 domain protein-phosphopeptide binding interactions by electrospray ionization mass spectrometry. *J. Am. Soc. Mass Spectrom.*, **8**, 234–243.

84 Greig, M.J., Gaus, H., Cummins, L.L., Sasmor, H., and Griffey, R.H. (1995) Measurement of macromolecular binding using electrospray mass spectrometry: determination of dissociation complexes for oligonucleotide – serum albumin complexes. *J. Am. Chem. Soc.*, **117**, 10765–10766.

85 Sannes-Lowery, K.A., Griffey, R.H., and Hofstadler, S.A. (2000) Measuring dissociation constants of RNA and aminoglycoside antibiotics by electrospray ionization mass spectrometry. *Anal. Biochem.*, **280**, 264–271.

86 Janis, J., Hakanpaa, J., Hakulinen, N., Ibatullin, F.M., Hoxha, A., Derrick, P.J., Rouvinen, J., and Vainiotalo, P. (2005) Determination of thioxylo-oligosaccharide binding to family 11 xylanases using electrospray ionization Fourier transform ion cyclotron resonance mass spectrometry and X-ray crystallography. *FEBS J.*, **272**, 2317–2333.

87 Dotsikas, Y. and Loukas, Y.L. (2003) Efficient determination and evaluation of model cyclodextrin complex binding constants by electrospray mass spectromtery. *J. Am. Soc. Mass Spectrom.*, **14**, 1123–1129.

88 Wang, W., Kitova, E.N., and Klassen, J.S. (2003) Influence of solution and gas phase processes on protein-carbohydrate binding affinities determined by nanoelectrospray Fourier transform ion cyclotron resonance mass spectrometry. *Anal. Chem.*, **75**, 4945–4955.

89 Daniel, J.M., McCombie, G., Wendt, S., and Zenobi, R. (2003) Mass spectrometric determination of association constants of adenylate kinase with two noncovalent inhibitors. *J. Am. Soc. Mass Spectrom.*, **14**, 442–448.

90 Jorgensen, T.J.D., Roepstorff, P., and Heck, A.J.R. (1998) Direct determination of solution binding constants for noncovalent complexes between bacterial cell wall analogues and vancomycin group antibiotics. *Anal. Chem.*, **70**, 4427–4432.

91 Gabelica, V., Galic, N., Rosu, F., Houssier, C., and De Pauw, E. (2003) Influence of response factors on determining the equilibrium association constants of non-covalent complexes by electrospray ionization mass spectrometry. *J. Mass Spectrom.*, **38**, 491–501.

92 Krishnaswamy, S.R., Williams, E.R., and Kirsch, J.F. (2006) Free energies of protein-protein association determined by electrospray ionization mass spectrometry correlate accurately with values obtained by solution methods. *Protein Sci.*, **15**, 1465–1475.

93 Kempen, E.C. and Brodbelt, J.S. (2000) A method for the determination of binding constants by electrospray ionization mass spectrometry. *Anal. Chem.*, **72**, 5411–5416.

94 Wortmann, A., Jecklin, M.C., Touboul, D., Badertscher, M., and Zenobi, R. (2008) Binding constant determination of high-affinity protein-ligand complexes by electrospray ionization mass spectrometry and ligand competition. *J. Mass Spectrom.*, **43**, 600–608.

95 Wortmann, A., Rossi, F., Lelais, G., and Zenobi, R. (2005) Determination of zinc to beta-peptide binding constants with electrospray ionization mass spectrometry. *J. Mass Spectrom.*, **40**, 777–784.

96 Fryčák, P. and Schug, K.A. (2007) On-line dynamic titration: determination of dissociation constants for noncovalent complexes using Gaussian concentration profiles by electrospray ionization mass spectrometry. *Anal. Chem.*, **79**, 5407–5413.

97 Fryčák, P. and Schug, K.A. (2008) Dynamic titration: determination of dissociation constants for noncovalent complexes in multiplexed format using HPLC-ESI-MS. *Anal. Chem.*, **80**, 1385–1393.

98 Zhu, M.M., Rempel, D.L., Du, Z., and Gross, M.L. (2003) Quantification of protein-ligand interactions by mass spectrometry, titration, and H/D exchange: PLIMSTEX. *J. Am. Chem. Soc.*, **125**, 5252–5253.

99 Powell, K.D., Ghaemmaghami, S., Wang, M.Z., Ma, L., Oas, T.G., and Fitzgerald, M.C. (2002) A general mass spectrometry-based assay for the quantitation of protein-ligand binding interactions in solution. *J. Am. Chem. Soc.*, **124**, 10256–10257.

100 Tang, L., Roulhac, P.L., and Fitzgerald, M.C. (2007) H/D exchange and mass spectrometry-based method for biophysical analysis of multidomain proteins at the domain level. *Anal. Chem.*, **79**, 8728–8739.

101 Lorenz, S.A., Maziarz, E.P. III, and Wood, T.D. (2001) Using solution phase hydrogen/deuterium exchange to determine the origin of non-covalent complexes observed by electrospray ionization mass spectrometry: in solution or in vacuo. *J. Am. Soc. Mass Spectrom.*, **12**, 795–804.

102 Clark, S.M. and Konermann, L. (2003) Diffusion measurements by electrospray mass spectrometry for studying solution-phase noncovalent interactions. *J. Am. Soc. Mass Spectrom.*, **14**, 430–441.

103 Clark, S.M. and Konermann, L. (2004) Screening for ligand-protein dissociation constants by electrospray mass spectrometry-based diffusion measurements. *Anal. Chem.*, **76**, 1257–1263.

104 Akashi, S. and Takio, K. (2000) Characterization of the interface structure of enzyme-inhibitor complex by using hydrogen-deuterium exchange and electrospray ionization Fourier transform ion cyclotron resonance mass spectrometry. *Protein Sci.*, **9**, 2497–2505.

105 Lanman, J. and Prevelige, P.E. Jr (2004) High sensitivity mass spectrometry for imaging subunit interactions: hydrogen/deuterium exchange. *Curr. Opin. Struct. Biol.*, **14**, 181–188.

106 Lifschitz, C. (2004) A review of gas-phase H/D exchange experiments: the protonated arginine dimer and bradykinin nonapeptide systems. *Int. J. Mass Spectrom.*, **234**, 63–70.

107 Kellersberger, K.A., Dejsupa, C., Liang, Y., Pope, R.M., and Dearden, D.V. (1999) Gas phase studies of ammonium – cyclodextrin compounds using fourier transform ion cyclotron resonance. *Int. J. Mass Spectrom.*, **193**, 181–195.

108 West, G.M., Tang, L., and Fitzgerald, M.C. (2008) Thermodynamic analysis of protein stability and ligand binding using a chemical modification- and mass spectrometry-based strategy. *Anal. Chem.*, **80**, 4175–4185.

109 Maleknia, S.D. and Downard, K. (2001) Radical approaches to probe protein structure, folding, and interactions by mass spectrometry. *Mass Spectrom. Rev.*, **20**, 388–401.

110 Wong, J.W.H., Maleknia, S.D., and Downard, K.M. (2003) Study of the ribonuclease-S-protein-peptide complex using a radical probe and electrospray ionization mass spectrometry. *Anal. Chem.*, **75**, 1557–1563.

111 Kriwacki, R.W., Wu, J., Siuzdak, G., and Wright, P.E. (1996) Probing protein/protein interactions with mass spectrometry and isotopic labeling: analysis of the p21/Cdk2 complex. *J. Am. Chem. Soc.*, **118**, 5320–5321.

112 Shields, S.J., Oyeyemi, O., Lightstone, F.C., and Balhorn, R. (2003) Mass spectrometry and non-covalent protein-ligand complexes: confirmation of binding sites and changes in tertiary structure. *J. Am. Soc. Mass Spectrom.*, **14**, 460–470.

113 Jahn, O., Eckart, K., Tezval, H., and Spiess, J. (2004) Characterization of peptide-protein interactions using photoaffinity labeling and LC/MS. *Anal. Bioanal. Chem.*, **378**, 1031–1036.

114 Robinette, D., Neamati, N., Tomer, K.B., and Borchers, C.H. (2006) Photoaffinity labeling combined with mass spectrometric approaches as a tool for structural proteomics. *Expert Rev. Proteomics*, **3**, 399–408.

115 Rhode, B.M., Hartmuth, K., Urlaub, H., and Lührmann, R. (2003) Analysis of site-specific protein-RNA crosslinks in isolated RNP complexes. Combining affinity selection and mass spectrometry. *RNA*, **9**, 1542–1551.

116 Hurst, G.B., Lankford, T.K., and Kennel, S.J. (2004) Mass spectrometric detection of affinity purified crosslinked peptides. *J. Am. Soc. Mass Spectrom.*, **15**, 832–839.

117 Raji, M.A., Frycak, P., Beall, M., Sakrout, M., Ahn, J.-M., Bao, Y., Armstrong, D.W., and Schug, K.A. (2007) Development of an ESI-MS screening method for evaluating binding affinity between integrin fragments and RGD-based peptides. *Int. J. Mass Spectrom.*, **262**, 232–240.

118 Wilm, M. and Mann, M. (1996) Analytical properties of the nanoelectrospray ion source. *Anal. Chem.*, **68**, 1–8.

119 Takats, Z., Wiseman, J.M., Gologan, B., and Cooks, R.G. (2004) Electrosonic spray ionization. A gentler technique for generating folded proteins and protein complexes in the gas phase and for studying ion-molecule reactions at atmospheric pressure. *Anal. Chem.*, **76**, 4050–4058.

120 Jecklin, M.C., Touboul, D., Bovet, C., Wortmann, A., and Zenobi, R. (2008) Which electrospray-based ionization method best reflects protein-ligand interactions found in solution? a comparison of ESI, nanoESI, and ESSI for the determination of dissociation constants with mass spectrometry. *J. Am. Soc. Mass Spectrom.*, **19**, 332–343.

121 Cuniff, J.B. and Vouros, P. (1995) False positives and the detection of cyclodextrin inclusion complexes by electrospray mass spectrometry. *J. Am. Soc. Mass Spectrom.*, **6**, 437–447.

122 Wang, W., Kitova, E.N., and Klassen, J.S. (2005) Nonspecific protein-carbohydrate complexes produced by nanoelectrospray ionization. Factors influencing their formation and stability. *Anal. Chem.*, **77**, 3060–3071.

123 Sun, J., Kitova, E.N., Wang, W., and Klassen, J.S. (2006) Method for distinguishing specific from nonspecific protein-ligand complexes in nanoelectrospray ionization mass spectrometry. *Anal. Chem.*, **78**, 3010–3018.

124 Sun, J., Kitova, E.N., and Klassen, J.S. (2007) Method for stabilizing protein-ligand complexes in nanoelectrospray ionization mass spectrometry. *Anal. Chem.*, **79**, 416–425.

125 Loo, J.A., Holsworth, D.D., and Root-Bernstein, R.S. (1994) Use of electrospray ionization mass spectrometry to probe antisense peptide interactions. *Biol. Mass Spectrom.*, **23**, 6–12.

Index

a

activation
– barrier 43
– C–H 162f.
– peroxide 164
addition
– cycloaddition of *trans*-anethole 145ff.
– /elimination 200
– gradual 84
– Lewis acid-catalyzed 162
– ligand 213
– Michael 162
– nucleophilic 97
– olefin 76, 78
– oxidative 76, 81ff.
additive 21, 26
– ionic 5
– NH$_4$Ac (ammonium acetate) 21, 29
– quenching 230
aerosol
– atmospheric 93f., 173f.
– science 2
alkynilation of tellurides mediated by Pd (II) 84, 86f., 158ff.
analyte
– aggregation 44
– concentration 19f.
– large ions 22ff.
– organic 21f.
– protonated 21f.
– quantitative analysis 18
– surface-active 17
APCI, *see* atmospheric-pressure chemical ionization
APCI-MS, *see* atmospheric-pressure chemical ionization mass spectroscopy
APTDI, *see* atmospheric pressure thermal desorption ionization
ASAP, *see* atmospheric solids analysis probe
association
– cation 88
– constant 286f.
atmospheric oxidation 10, 93
– of isoprene 93f., 173ff.
atmospheric-pressure chemical ionization (APCI) 40, 51, 133, 143, 199
– desorption (DAPCI) 41
– off-line monitoring, *see* ESI/MS
– on-line monitoring, *see* ESI/MS
atmospheric-pressure chemical ionization mass spectroscopy (APCI-MS) 113, 117, 176
– microreactor system 117
– radical cation chain reaction 145ff.
– /MS 117, 147
atmospheric pressure thermal desorption ionization (APTDI) 51
atmospheric solids analysis probe (ASAP) 41
Avogadro's number 24

b

Baylis–Hillman reaction 45, 47, 121, 166ff.
– co-catalyzed by ionic liquids 167f.
Benesi–Hildebrand treatment 285
bifunctional mimicking 104, 175
binding
– affinity 287
– competitive 281ff.
– constants 283f., 286, 289, 291
– equilibrium 284
– macromolecule 279
– noncompetitive 282f.
– noncovalent 281
– nonspecific 285
– quantitative 279ff.
– small-molecule 278
– solution-phase 287, 290, 295

Reactive Intermediates: MS Investigations in Solution. Edited by Leonardo S. Santos.
Copyright © 2010 WILEY-VCH Verlag GmbH & Co. KGaA, Weinheim
ISBN: 978-3-527-32351-7

- stereospecific 286
- uncompetitive 282
biomimetic systems 96
biomolecules, see molecules
black body radiation 41
bond
- activation/bond formation reactions 212f., 219
- C–C, see carbon–carbon
- C–H 213, 219, 221f.
- C–I 208
- C–N 213
- C–OH 213
- Cu–C 207f.
- heterolysis 215f.
- homolysis of metal nitrides 216ff.
- ion-neutral 31
- metal–H 211
- N–H 221f.
- N–O 217
- nonspecific 31
bromocyclization 164f.
Brookhart polymerization 181f., 269

c

C_5 isoprene hypothesis 174
Cannizzaro reaction 44
capillary
- borosilicate 26
- electrolytic half-cell 134
- entrance 26
- exit 5
- glass 9, 14
- inner 137f.
- micro- 26
- mixer 137f.
- outer 137f.
- outlet 137f.
- puller 26
- silica 27, 137, 140
- tubing 4
capillary tip 3ff.
- nanospray 140, 153
- radius 5, 9, 26, 140
- stainless steel 8, 137, 139
- Zn 8
carbopalladation 84, 115, 257
carbon–carbon
- bond-coupling 199, 207f., 212f., 231, 240
- bond cross coupling 208
- chain cleavage 39
- oxidative cleavage 105ff.
catalysis
- co-catalyzed 67f., 124, 167f.

- homogeneous 1
- Lewis base 65
- silver surfaces 219
catalyst
- acid/base 103f.
- amino acids 103
- cationic carbene 126
- chiral 265ff.
- homogeneous 179
- libraries 136
- multicatalyst screening 266
- neutral 123
- organo- 70f.
- precursors 91
- ruthenium 123, 126
- screening 49
- (thio)ureas 70f.
catalytic cycles
- bifurcation step 165
- MBH reaction 65f., 71
- metal-mediated decarboxylation of acetic acid 204
- metal-mediated decarboxylation of formic acid 210
- α-methylenation 73
- Pd-catalyzed addition 116
- two-step 204
charge
- balance 7, 21
- charge-charge repulsion 152
- dispersal 22
- exchange 42f.
- molecular concentration 21
- preservation 134
- separation 134
- stabilization 22, 135
charged residual model (CRM) 15, 16, 23, 26f., 30
Chauvin mechanism 126
chemical ionization (CI) 40, 42
chemistry
- biophysical 283
- bulk solution 50
- condensed-phase organic 38
- electro- 45
- gas-phase ion 28, 37ff.
- metal-mediated 51, 199
- organometallic 142
- – substrate species 162
- synthetic organic 105, 114
chiral
- centers 74
- selector 292, 294
chromatography

- gas chromatography mass spectroscopy (GC-MS) 39, 95, 237
- high performance liquid (HPLC) 1, 74, 266, 285, 289
- size-exclusion 281
CID (collision-induced dissociation) 37, 41, 67, 76, 91, 119, 148, 170f., 181, 199ff.
- -in-source 201, 210, 297
- low-energy spectrum 222
- multiple stages 212
Claisen rearrangement 44
Claisen–Schmidt reaction 44
cluster 42f.
- cleavage 214
- fragmentation 214
- gold hydride 214
- ions 15f., 50
- isotopologous ions 177
- nickel hydride 210
- organocopper 88, 162
- silver-amino acid 212
- silver hydride 213
- silver iodide 214
co-catalyzed, see catalysis
collision
- cell 49, 133, 143, 186
- inelastic 41
- octopole 127
collisional
- activation 4f., 203
- stabilization 42
complexes
- anionic 251, 254
- binuclear 216, 238
- cationic 161f., 240
- chiral ruthenium 162
- dialkyl glutarate–Lewis acid 149
- dimeric ion 149, 151
- Fischer carbene 209f.
- fully ring-deligated 153
- heterodimeric 151
- host-guest 40
- Lewis acid–ester 149
- Meisenheimer 169f., 172
- metal 88, 114, 162
- metathesis-active 123
- mixed-ring sandwich 153
- Mn^{IV} 166
- monomeric ion 149, 151
- mononuclear 216
- noncovalent 31, 278, 281, 288, 294, 296
- nonspecific 31
- palladium 237ff.
- palladium enolate 262
- precursor metal 210
- protein 24
- protein–carbohydrate 31
- protein–ligand 288f.
- protein–substrate 30
- radical complex ion 150f.
- (Schiff base) vanadium(V) 163f.
- solution-phase-based noncovalent 277
- specific 31
- transition metal 44, 49, 215
condensation 42f., 103
conductivity
- solution 10, 17
- techniques 239
continuous flow apparatus, see ESI-MS microreactor
coordination
- ligand 238
- mode 238, 263
- number 135
- Pd 237f., 263
Cope rearrangement 44
coulomb fission 6, 11ff.
CRM, see charged residue model
cumulene synthesis 44
current
- analyte 21
- detection technologies 7
- ion 7, 9, 18, 21
- oscillations 7
- time-ion 152
- total droplet (TDC) 21
- total electrospray 18f.
cyclodextrin mimics of RNase A 104

d

DABCO, see 1,4-diazabicyclo[2.2.2]octane
DAPPI, see desorption atmospheric pressure photoionization
DART, see direct analysis in real time
DBDI, see dielectric barrier discharge ionization
decarboxylation 201ff.
- competing 206
- double 205f.
dendrimers 22f.
density functional theory 70f., 101, 203f.
- exothermicities of decarboxylation 205ff.
- Knoevenagel pathways 103
DESI, see desorption electrospray ionization
desorption atmospheric pressure photoionization 41
desorption electrospray ionization 41, 51f.
desorption sonic spray ionization (DeSSI) 41
DFT, see density functional theory

Index

1,4-diazabicyclo[2.2.2]octane   65f., 68, 121
dielectric barrier discharge ionization   41
dielectric constant   10, 278
Diels–Adler reaction   44f., 47f., 148f.
diffusion limit rates   30
direct analysis in real time   41
dissociation constants   288f., 293, 295
downfield   5f.
droplets
– charge   2f., 10f., 13f.
– desolvation   278, 295f.
– discharge   13
– disintegration   3, 279
– emission   7
– evaporation   1f., 11, 13f.
– evolution   11, 13ff.
– fission   11f., 13f.
– nano-   4
– polarization   15
– production   5ff.
– progeny   3, 6, 11, 14, 20f., 27, 30
– radius   6, 10ff.
– saturation   292
– shrinking   296
– size   6, 10

e

Eberlin reaction   52
ECD, see electron capture dissociation
EESI, see extractive electrospray ionization
electric double layer   3
electric potential
– difference   5
– long-range attractive   43
electrical
– breakdown   10
– gas discharges   8ff.
electrochemical cell, see ESI-MS microreactor
electrode
– counter   5
– disk   141
– metal   8f.
– separation   5
electrolyte
– aqueous   155
– background   27
– concentration   17f.
– dissociation   10
– Na^+ 2, 8
– positive   6
electron
– abstraction   176
– acceleration   10
– azapalladation   115

– capture   10, 43
– charges   21
– collision frequency   10
– deprotonation   176
– emission   9
– free   10
– protonation   176
– thermal   43
electron capture dissociation (ECD)   41, 43
electron transfer   200, 216
– dissociation   41
– photo-induced   152
– -single   134
electrophile   65, 81, 231, 242, 245, 252
– $TeCl_4$ 181, 185
electrophilic fluorination   45
electrophoresis
– capillary   277
– mechanism   6, 8
electrosonic spray ionization (ESSI)   52, 292, 295
electrospray (ES)
– atmospheric pressure region   3, 9f.
– cone jet mode   6f., 10, 17
– mass spectrometry (ESMS)   45
– micro-   26
– multijet modes   6
– nano-   4, 9, 14, 26f., 285, 292, 295
– negative ion mode (ES(−))   9, 66
– positive ion mode (ES(+))   3f., 7, 18, 22, 66
– self-flow   26
– stable   9
electrospray ionization (ESI)   1, 9ff.
electrospray ionization Fourier Transform mass spectroscopy (ESI-FTMS)   84
electrospray ionization Fourier Transform ion cyclotron mass spectroscopy (ESI-FTICRMS)   113ff.
electrospray ionization mass spectroscopy (ESI-MS)
– clean up stages   29, 31
– dissociation   28, 31
– ESI(−)-MS   93, 144, 158, 166, 169, 171, 230, 236, 250, 254, 259, 260
– ESI(+)-MS   71ff.
– HR (high resolution)   40, 237, 278
– hybrid   278
– ion-fishing technique   49, 63f., 66f., 77, 94, 133, 144, 162f., 230
– microreactor   115, 117, 121, 125, 136ff.
– off-line reaction screening   64, 135f.
– on-line reaction screening   64, 66f., 125, 136ff.
– qualitative screening   279, 283

- rapid screening 45
- real-time 136, 140
- sequential monitoring 144
- time-resolved 137f.
electrospray ionization time-of-flight mass spectroscopy (ESI/TOF-MS) 118f.
electrostatic
- dispersion of liquids 2, 27
- spraying 2
elimination
- concomitant 204f.
- β-hydrogen 87f., 179, 240, 248, 251
- MBH reaction 67ff.
- reductive 82f., 158, 252f.
- three-component tandem double addition-cyclization reaction 84
elucidating reaction mechanism 43f., 123
enantiodiscrimination 265
enantioselectivity 70, 266
- allylation 265f.
- Pd-catalyzed allylation reaction 91f.
- synthesis of optically active epoxides 164
endothermicities 203
- alkaline earth acetates 203
- ligand fragmentation 220, 222
- organometalates 203
epoxidation 164f.
equilibrium
- constant 28ff.
- partitioning model 292
- shift 30
ESI/MS(/MS) 63f., 74ff.
- direct infusion 97, 158
- ESI(−)-MS(/MS) 89f., 94, 107, 169f., 174f.
- ESI(+)-MS(/MS) 99, 101ff.
- in situ probing 139, 174
- off-line screening 136, 143
- on-line reaction screening 95, 97, 174
- real-time 95, 174
evaporation/ionization step 29
extended x-ray absorption fine structure (EXAFS) 250
extractive electrospray ionization 41

f

FAB, *see* fast-atom bombardment
Faraday's constant 21
fast time-lapse imaging 7
fast-atom bombardment 40, 44
FD, *see* field desorption
Fenton systems 96, 171
- heterogeneous reaction 93f., 96, 144f.
FI, *see* field ionization
field desorption (FD) 40, 44

field-induced droplet ionization (FIDI) 41, 51
field ionization (FI) 40
Fischer carbene complex 209f.
Fischer indole synthesis 44f., 48
flow
- continuous 137f.
- laminar 138
- radial- 141
flow rate
- solution 2, 6f., 10, 21, 26, 294f.
- syringe 137
- volumetric 140
formation
- alkaline earth organometalates 203ff.
- metal carbenes 208ff.
- metal hydrides 210ff.
- metal nitrides 220ff.
- metal oxide 215ff.
- organoargenates 205ff.
- organocuprates 205ff.
- organolithium 202f.
- organometallics 201ff.
forces
- attractive 15
- cohesive 11
- electrostatic 278
- noncovalent 278, 293
- repelled 15
- repulsion 3, 11, 15, 134, 152
- solvophobic 279
Fourier Transform (FT)
- ion cyclotron resonance (FT-ICR) 50
- method 39
fragmentation
- ion 5, 39ff.
- organic compound 39
Friedel–Craft acylation 44, 96, 186
functional group
- protonated 31
- unprotonated basic 21

g

GABA (γ-amino butyric acid) 73f.
gas chromatography mass spectroscopy (GC-MS) 39, 95, 237
- offline 95
gas-phase basicity 22, 26
Gilman reaction 45, 47
Grignard reagents 204
Grubbs metathesis 45f.

h

Haber–Weiss reaction 44
Heck reaction 45, 47, 75f., 78, 154, 156, 249ff.

heterolytic 201
high performance liquid chromatography (HPLC) 1, 74, 266, 285, 289
high-throughput screening 49, 136, 265, 285, 287
Hofmann rearrangement 44
homocoupling 234, 237
homolytic 201, 206, 216ff.
host-guest
– complexation 279, 284
– concentrations 291
– equilibria 278, 280, 289
– screening 281f.
hygroscopic polar groups 94, 173
β-hydrogen elimination 87f., 179, 240, 248, 251
hydrogenation
– asymmetric 162f.
– de- 200
hydrolysis 74, 104f.
hydropalladation 255

i

ICR, see ion cycloton resonance
iminium mechanism 102
impurity
– Cl⁻ 28
– constant 18
– electrolyte 18
– Na⁺ 28f.
insertion
– n-butyl vinyl ether 247
– intermolecular 242
– migratory 271
– olefin 240, 248f., 251
– oxidative 200
– polymerization 271
interaction
– disruption 134
– host-guest 279
– hydrophobic 279
– ion–pairing 43
– metal–ligand 30
– neutral molecule–electron 43
– noncovalent 40, 134, 283, 295
– nonspecific 279, 296
– protein–protein 287, 290
– receptor–ligand 281, 290
– small-molecule–macromolecular 277f., 283, 290
– stoichiomeries 278, 283ff.
interface
– electrospray/mass spectrometer 4
– liquid/metal 7

– solution/solid 52
intermediates
– anionic 107, 144
– Biginelli 101
– cationic 76, 79, 175, 246
– cleavage 234
– degradation 95f.
– neutral 114, 144
– organocatalytic 121
– organocopper cluster 88, 162
– oxidative addition 237ff.
– palladium 229ff.
– reactive metallic 8, 199, 201ff.
– synthetic 71
– transient ionic 45, 48, 63, 97, 120
– unmasking 199, 201ff.
– zwitterions 66
ion
– aggregate 15
– artefact 120
– association 1
– beam 40
– clean-up 4
– counter- 63, 135, 179
– depletion 19
– desolvation 5, 134
– discharge-generated 9
– dissociation methods 41f.
– -electron multipliers 7
– evaporation model (IEM) 11, 15ff.
– exchange 27
– -fishing technique, see ESI-MS
– gas phase 2, 4, 7f., 11, 15, 21, 27f.
– intensity 19ff.
– mass-selected 143
– metastable 38f.
– mobility 2, 40, 51
– mode, see electrospray (ES)
ion cycloton resonance (ICR) 39
ion/molecule reaction (IMR) 37f., 40ff.
– equilibria 50
– high-pressure 51f.
– in-source 52
– low-pressure 50f.
ion
– multiply charged 134f.
– precursor 184
– selection 45, 49
– transient 44, 49, 123
– transmission 19f., 50, 297
– -trap analyzers 39
– -trap cut-off 203
ionization
– cold-spray 199

– competitive 296
– efficiency 286, 296
– electron-addition 39
– energy (IE) 42, 44
– intramolecular 152
– laser desorption 40
– saturation 279, 293
– selective 42
– soft 40, 42, 123, 231
IR (infrared) 239
– multiphoton excitation 41
isomerization of a double bond 268
isothermal titration calorimetry 285
isotope
– Cu 162, 259
– β-deuterium 208
– inverse effect 208
– patterns 45, 63, 89, 124, 162
– Pd 76, 230f.

j

JeDI (jet desorption ionization) 41

k

ketoesters 71ff.
kinetic
– barrier 204
– corroborated 90, 171
– energy analysis 2
– hydrolysis of phosphate diester 104f.
– intensity time profiles 137
– ion 50
kinetic isotopic effect (KIE) 67, 219
kinetic model 128
Knoevenagel mechanism 102f.
Kolbe reaction 44

l

labels
– double mass-labeling strategy 49, 266
– equilibrium label-free technique 278
– isotope 40, 221f.
– mass-labeled enantiomers 91
laser desorption/ionization 199f.
– matrix-assisted 199
– spray-assisted (SALDI) 41
lifetime
– few minutes 138, 140
– long 229
– millisecond range 141, 153
– short 136
ligand
– auxiliary 202, 204, 215, 222
– bridging 239

– chelating 245f.
– chiral 49
– connectivity 201
– enzyme- 283
– exchange 87, 178, 238, 246, 266
– exchange equilibrium 162
– fragmentation 199ff.
– -free systems 249ff.
– hydrophilic 260, 263, 268
– loss 214, 216
– migration 214
– oligonucleotide- 283
– protonated 217, 220
– steric hindrance 271
– – substrate ratio 29
– switching reactions 219
– water-soluble 263, 268
linearization methods 284f.
liquid microjunction surface sampling probe method 41
LMJ-SSP, see liquid microjunction surface sampling probe method
long-range potential capture models 43
Lossen rearrangement 44

m

macromolecules 15, 277ff.
MALDESI, see matrix-assisted laser desorption electrospray ionization
MALDI, see matrix-assisted laser desorption/ionization
Mannich-type reaction 71f., 162, 175ff.
mass-analyzed ion kinetic energy spectrometer 39
mass-to-charge ratio (m/z) 4, 30, 42, 45, 91, 101, 105ff.
mass spectrometry (MS)
– affinity selection (AS-MS) 280f.
– 3D quadrupole 51
– direct screening 136
– field-portable 40
– Fourier Transform (FT) 200
– high-pressure 50
– history 37ff.
– low-pressure 43
– low-resolution 45
– MASS (multi-target affinity/specificity screening) 282
– miniature 40
– /MS dissociations 167
– MS^3 (triple-stage MS) 50f.
– multiplex screening 282
– pentaquadrupoles 50f.
– pulsed-electron high-pressure 50f.

- quadrupole 19f., 50ff.
- quadrupole ion trap (QIT) 50, 143
- run-to-run error 282
- sequential 142
- tandem 50, 63, 113, 143
- tandem-in-space mode 50
- time-resolved atmospheric pressure ionization 51
- triple quadrupole 50, 143
matrix-assisted laser desorption electrospray ionization 41
matrix-assisted laser desorption/ionization 40, 44
- -TOF-MS 123
mechanistic cycle 255f., 264
Meerwein–Ponndorf–Verley reduction/Oppenauer oxidation 44
Meisenheimer complex 169f., 172
- TNT 52, 89, 91
meniscus
- destabilization 3
- distortion 5
- polarized 4
metal
- alkali 202ff.
- alkaline earth 202
- carbenes 208ff.
- center 153f., 201, 249, 251
- cluster 200
- coordination 204
- hydride cations 200
- – ligand system 30
- -mediated reaction 199
Michael addition 44, 162
MIKES, see mass-analyzed ion kinetic energy spectrometer
Mitsunobu reaction 199
Mizoroki-Heck catalytic cycle 240f.
molecular mass
- dendrimer 23
- protein 24f.
molecule
- anionized 63
- bicyclic 96
- bio- 44, 51, 201
- cationized 63, 66
- deprotonated 63, 134
- neutral 42f., 52, 63, 134, 142
- protonated 63, 66, 134, 173, 175
- small molecule 277f., 283, 290
Morita–Baylis–Hillman (MBH) reaction 65ff.
multinuclear metal hydrides 212ff.

n
nanoparticles 249, 251
NMR, see nuclear magnetic resonance
nuclear magnetic resonance 45, 89, 161, 179, 234, 237, 239f., 252
nuclear fission 40
nucleophilic 266
- attack 67, 69

o
oligonucleotide ladder sequencing 40
organometallics 44, 81, 114, 134, 176, 199
- coordination compounds 230
- fragmentation 201ff.
- ionization 44
- nucleophilic 247
- reactivity 201ff.
oxidation 7f., 45, 47, 144f., 163f., 173ff.
- photo- 93f., 173ff.
- potential 8
- OH radical-mediated 94f., 174f.
- -sensitve analytes 8
oxidative
- addition 113f.
- caffeine degradation 95, 171ff.
- homocoupling 234
ozonation 164, 166

p
PADI (plasma-assisted desorption ionization) 41
Pd-catalyzed
- allylic substitution 260ff.
- arylation 76f., 255f.
- C–C bond formation 47f., 51, 75f.
- coupling of electrophiles 81, 161f.
- coupling of olefins 240
- cross-coupling 81
- enantioselective allylation reaction 156f.
- enantioselective Mannich-type 162
- intramolecular arylation 241f.
- oxidation of 2-allylphenols 268f.
- polymerization 269ff.
- reaction of allenes 254ff.
- self-coupling 81, 156
- three-component tandem double addition-cyclization 83f., 86, 114, 117
PDI, see phase doppler interferometry
permittivity 9ff.
Petasis olefination 45, 47, 176, 178
pH 8, 144
phase transfer 278f.
phase doppler interferometry 11

photoconversion 152
photoirradiation 138f.
pinacol rearrangement 44, 48
plasma desorption (PD) 40
polymerization 123
– Brookhart 181f.
– catalysts 136
– epoxides 153
– homogeneous 265
– insertion 270f.
– Pd-catalyzed 269ff.
– photo-induced 153f.
– Ziegler–Natta 124f., 178ff.
positive-ray parabola apparatus 42
product-ion spectrum 125, 127
– catalytic ion 127
– ethene 184
propargyl alcohols 181, 185
protein
– affinity 42
– charge 25f.
– conformation change probing 40
– gas-phase 23
– globular 23ff.
– -like lysozyme 22
– mass mapping 40
– nondenaturated 22ff.
– radius 24
proteomics 40f.
proton abstraction 67
proton transfer 26, 42f., 52, 200, 216
– equilibrium 175
– intra-cluster 134
– intramolecular 29
– promotor 70
– rate-limiting 71
prototropic shift 66

q
QIT (quadrupole ion trap), see mass spectrometry

r
radical
– cation 252f., 255
– cation chain mechanism 115, 117f., 145ff.
– chain reactions 149ff.
– complex ion 150f.
– Fenton reaction 143f.
– free 95, 149
– hydroxyl 95f., 144f., 173f.
– ions of biomolecules 201
– methyl 203

– OH 94f.
– resonance-stabilized allyl 219
– transient 143
– traps 118
radiolytic technique 51
Raney Nickel-catalyzed coupling reaction 45f.
rate-determining step (RDS) 67f., 70f., 128, 234, 251, 293
Rayleigh
– curve 16
– equation 11, 24
– fissions 12
– limit 11, 13, 15f., 23f.
– instability condition 30
reaction
– acid–base 203, 210
– aldol 121f.
– alkyne-nitrile metathesis 223
– allylic substitution 49, 91f.
– association 43
– Au-catalyzed cyclopropanation 48
– autocatalytic 67, 69
– back-reaction screening 266
– bimolecular 28
– C–H activation 48
– concomitant reduction 202
– condensed phase 42, 44
– consecutive 141
– cyclopropanation 45
– displacement 45f., 51
– endoergic 43
– endothermic 43
– enzyme 40
– exothermic 42f.
– ion/ion 28f., 37
– kinetics 15
– low-energy ion/molecule 43
– metal-catalyzed cross-coupling 229
– α-methylenation of ketoesters 45, 48, 71f., 175ff.
– microscale 45
– non-metal catalyted 121
– nucleophilic substitution 44, 47f., 89, 109ff.
– organocatalysis 121
– organometallic 44f.
– oxa-Heck arylation 77, 79f.
– oxidation 7f., 45, 47, 144f., 163ff.
– oxidative cleavage of terminal C=C bonds 105ff.
– parallel 141
– photoallylation 152
– photochemical 138f., 151f.

– photochemical switching   151f.
– photo-induced polymerization   153f.
– radical reactions   115ff.
– redox   134
– ring-contruction   73f., 169
– second order   67
– S_N2 89ff.
– stereoselective   49, 74
– sulfoxidation   48
– tandem double addition-cyclization   256ff.
– three-component Biginelli   98ff.
– transition metal-catalyzed   49, 123ff.
reactivity
– alkaline earth organometalates   203ff.
– gas-phase   199
– intrinsic   63
– ionic   43f., 51
– metal hydrides   210ff.
– metal nitrides   220ff.
– metal oxide   215ff.
– organolithium ions   202f.
– organometallics   201ff.
– probing   142f.
– – selectivity studies   163
– solution-phase   128
– two-state reactivity model   219
recognition
– enthalpy-driven   279
– molecular   277
– noncovalent   277ff.
recombining energy (RE)   42
redox-induced loss of N_2 221
Reformatsky reaction   44
regioselective   114, 223
– dimerization   115
– hydroarylation   114
– Pd-catalyzed addition   254
residence time   128
ribonuclease A (RNase A) mechanism   103f.
ring-opening metathesis polymerization (ROMP)   49, 123f., 126ff.
ruthenium(II)-catalyzed asymmetric reduction   46, 48

S

secondary organic addition (SOA)   94, 174
selectivity   142, 202
sensitivity   142
– coefficients   18f., 29
– electrospray ionization mass spectroscopy (ESI-MS)   45, 74
SID (surface-induced dissociation)   41
SIFT (selected-ion flow tube)   50
signal-to-noise ratio   7

SIMS (secondary-ion mass spectrometry)   40
size-exclusion chromatography   281
skimmer   5, 7
SOA, see secondary organic addition
solution
– aqueous   8, 22
– basicity   22
– concentrations   18
– dilute   1
– in-solution method   30
– nonviscous   26
– -phase equilibrium concentrations   278
– -phase reactivity   128, 135, 277ff.
– -phase-based noncovalent   277
– polastyrene   2
– self-flow   26
solvation
– energy   20, 279
– shell   17
SORI-CID (sustained off-resonance collosion-induced dissociation)   115
Sonogashira cross-coupling reaction   259
spontaneous rearrangement reaction   89, 170
SPR, see surface plasmon resonance
SPROX (stability of proteins from rates of oxidation)   290
Staudinger reaction   199
stereoselective   74, 149
– dia-   165
– dimerization   115
– hydroarylation   114
– Pd-catalyzed addition   254
stereospecificity   88, 114
Stille reaction   48, 81f., 157f., 254
– cross-coupling   251ff.
structural information   45, 63
substrate   29f.
– neutral   142
– quasi-enantiomeric   49
– structure   213
supramolecular
– coordination   67, 168
– system   26, 67, 135, 167
surface
– glass   27
– ion-exchange problem   27
– potential energy   205, 207
surface activity   279
– analytes   27
– ions   19f.
surface plasmon resonance (SPR)   277
surface tension
– droplet   5f.
– solvent   9ff.

– water   24
surface-to-volume ratio   21
Suzuki elimination reaction   46, 48, 81,113f., 156
Suzuki-Miyaura cross-coupling   231ff.

t
Taylor cone   7f., 26
Tebbe olefination   176ff.
TeCl4 addition to propargyl alcohols   88f.
TEM (transmission electron microscopy)   250
thermal
– declustering   4
– desomposition   206
– disproportionation   153
– dissociation   41
– stability   142
thermospray ionization source (TSP)   40
titration method   30, 281f., 284, 290ff.
– calorimetric   277
– competitive   283
– dynamic   289
– equimolar mixture-based   286
– isothermal titration calorimetry   285
– ligand band-dispersion approach   289
– nonlinear method   285
– quantitative   283
– sensitivity   291
– small-molecule   292
– UV   284
TOF (time-of-flight) instrument   39, 118
transition metal   134, 200
– -catalyzed polymerization   123
– complexes   44, 49, 135
– oxide coordination complexes   215
transition state
– centered   203
– energy   206ff.
– theory   15
– T-shaped   208
transmetalation   237, 252f.
– alkynilation   162
– Stille reaction   81f.
– Suzuki reaction   113f., 231
Tröger's bases   96f., 99, 186f.
TSP, *see* thermospray ionization source
Tsuji-Trost reaction   260

u
UV
– cut-off filter   138f.
– H_2O_2/UV   96, 171, 173
– irradiation   95, 152, 172, 277
– laser excitation   41
– TiO_2/UV   96, 171, 173

v
virus analyzes   40
voltammetry   161
volatility   142

w
water
– background   216
– density   24
– elimination   103
– -soluble   263, 268
Wittig reaction   44, 48, 199
Wolff rearrangement   44, 46, 48, 209

x
XPS (X-ray photoelectron spectroscopy)   179
X-ray crystallography   237

z
Ziegler–Natta olefin oligomerization   46, 48, 124f., 178ff.
zwitterionic species   63, 66f., 168